Advances in Intelligent and Soft Computing 83

Editor-in-Chief: J. Kacprzyk

Advances in Intelligent and Soft Computing

Editor-in-Chief

Prof. Janusz Kacprzyk
Systems Research Institute
Polish Academy of Sciences
ul. Newelska 6
01-447 Warsaw
Poland
E-mail: kacprzyk@ibspan.waw.pl

Further volumes of this series can be found on our homepage: springer.com

Vol. 69. E. Piętka and J. Kawa (Eds.)
Information Technologies in Biomedicine, 2010
ISBN 978-3-642-13104-2

Vol. 70. Y. Demazeau, F. Dignum, J.M. Corchado, J. Bajo Pérez (Eds.)
Advances in Practical Applications of Agents and Multiagent Systems, 2010
ISBN 978-3-642-12383-2

Vol. 71. Y. Demazeau, F. Dignum, J.M. Corchado, J. Bajo, R. Corchuelo, E. Corchado, F. Fernández-Riverola, V.J. Julián, P. Pawlewski, A. Campbell (Eds.)
Trends in Practical Applications of Agents and Multiagent Systems, 2010
ISBN 978-3-642-12432-7

Vol. 72. J.C. Augusto, J.M. Corchado, P. Novais, C. Analide (Eds.)
Ambient Intelligence and Future Trends, 2010
ISBN 978-3-642-13267-4

Vol. 73. J.M. Corchado, P. Novais, C. Analide, J. Sedano (Eds.)
Soft Computing Models in Industrial and Environmental Applications, 5th International Workshop (SOCO 2010), 2010
ISBN 978-3-642-13160-8

Vol. 74. M.P. Rocha, F.F. Riverola, H. Shatkay, J.M. Corchado (Eds.)
Advances in Bioinformatics, 2010
ISBN 978-3-642-13213-1

Vol. 75. X.Z. Gao, A. Gaspar-Cunha, M. Köppen, G. Schaefer, and J. Wang (Eds.)
Soft Computing in Industrial Applications, 2010
ISBN 978-3-642-11281-2

Vol. 76. T. Bastiaens, U. Baumöl, and B.J. Krämer (Eds.)
On Collective Intelligence, 2010
ISBN 978-3-642-14480-6

Vol. 77. C. Borgelt, G. González-Rodríguez, W. Trutschnig, M.A. Lubiano, M.Á. Gil, P. Grzegorzewski, and O. Hryniewicz (Eds.)
Combining Soft Computing and Statistical Methods in Data Analysis, 2010
ISBN 978-3-642-14745-6

Vol. 78. B.-Y. Cao, G.-J. Wang, S.-Z. Guo, and S.-L. Chen (Eds.)
Fuzzy Information and Engineering 2010
ISBN 978-3-642-14879-8

Vol. 79. A.P. de Leon F. de Carvalho, S. Rodríguez-González, J.F. De Paz Santana, and J.M. Corchado Rodríguez (Eds.)
Distributed Computing and Artificial Intelligence, 2010
ISBN 978-3-642-14882-8

Vol. 80. N.T. Nguyen, A. Zgrzywa, and A. Czyzewski (Eds.)
Advances in Multimedia and Network Information System Technologies, 2010
ISBN 978-3-642-14988-7

Vol. 81. J. Düh, H. Hufnagl, E. Juritsch, R. Pfliegl, H.-K. Schimany, and Hans Schönegger (Eds.)
Data and Mobility, 2010
ISBN 978-3-642-15502-4

Vol. 82. B.-Y. Cao, G.-J. Wang, S.-L. Chen, and S.-Z. Guo (Eds.)
Quantitative Logic and Soft Computing 2010
ISBN 978-3-642-15659-5

Vol. 83. J. Angeles, B. Boulet, J.J. Clark, J. Kövecses, and K. Siddiqi (Eds.)
Brain, Body and Machine, 2010
ISBN 978-3-642-16258-9

Jorge Angeles, Benoit Boulet, James J. Clark,
József Kövecses, and Kaleem Siddiqi (Eds.)

Brain, Body and Machine

Proceedings of an International Symposium
on the Occasion of the 25th Anniversary
of the McGill University Centre
for Intelligent Machines

Springer

Editors

Prof. Jorge Angeles
Department of Mechanical Engineering &
Centre for Intelligent Machines
McGill University
817 Sherbrooke St. W.
Montreal, Quebec
Canada H3A 2K6
E-mail: angeles@cim.mcgill.ca

Prof. Benoit Boulet
Department of Electrical and
Computer Engineering &
Centre for Intelligent Machines
McGill University
3480 University Street
Montreal, Quebec
Canada H3A 2A7
E-mail: benoit.boulet@mcgill.ca

Prof. James Clark
Department of Electrical and
Computer Engineering &
Centre for Intelligent Machines
McGill University
3480 University Street
Montreal, Quebec
Canada H3A 2A7
E-mail: clark@cim.mcgill.ca

Prof. József Kövecses
Department of Mechanical Engineering &
Centre for Intelligent Machines
McGill University
817 Sherbrooke St. West
Montreal, Quebec
Canada H3A 2K6
E-mail: jozsef.kovecses@mcgill.ca

Prof. Kaleem Siddiqi
School of Computer Science &
Centre for Intelligent Machines
McGill University
3480 University Street
Montreal, QC
Canada H3A 2A7
E-mail: siddiqi@cim.mcgill.ca

ISBN 978-3-642-16258-9 e-ISBN 978-3-642-16259-6

DOI 10.1007/978-3-642-16259-6

Advances in Intelligent and Soft Computing ISSN 1867-5662

Library of Congress Control Number: 2010937343

Typeset & Cover Design: Scientific Publishing Services Pvt. Ltd., Chennai, India.

Printed on acid-free paper
5 4 3 2 1 0
springer.com

Preface

The *Centre for Intelligent Machines* (CIM), McGill University, was officially created on November 13th, 1985, upon approval by the University Senate. The mission of the centre is, since its inception, "to excel in the field of intelligent systems, stressing basic research, technology development and education." *Intelligent machines* are understood, along these lines, as systems "capable of adapting their goal-oriented behavior by sensing and interpreting their environment, making decisions and plans, and then carrying out those plans using physical actions." Research activities of interest to CIM include robot design, mechanical system dynamics, robot control, computer vision, visual perception, medical imaging, haptics, system theory and applications, and virtual environments. These fields can be succinctly described by the title of the international symposium recorded in this book, *Brain, Body and Machine* (BBM).

CIM's object of research is thus reflected in the book, where readers will find that most papers cover at least two of the three main thrusts. Papers reflect the state-of-the-art of the multidiscipline of intelligent machines, nowadays highly diversified. A broad spectrum of applications is covered, from humanoid robots and autonomous planetary rovers, to innovative procedures and systems for shape-reproduction and on to systems or devices for medical diagnosis. Applications are all geared to one common goal, a better quality of life for highly challenged patients or simply for the healthy individual. Methods to face the challenges include not only sophisticated deterministic algorithms, but also their heuristic counterparts, at the roots of what is known as *soft computing*.

The reader will find here papers on human-robot interaction as well as human-safety algorithms; haptic interfaces; innovative instruments and algorithms for the sensing of motion and the identification of brain neoplasms; even a paper on a saxophone-playing robot.

In its first 25 years, CIM has produced hundreds of graduates at the Master's, Ph.D. and postdoctoral levels. These alumnae and alumni are now successful professors, researchers, developers and managers who are at the forefront of many industries, and entrepreneurs and business developers. CIM's graduates, moreover, are distributed all over the world. This rich distribution made the *International Symposium on Brain, Body and Machine* possible.

Montreal,
August 2010

The Editors
Jorge Angeles
Benoit Boulet
James J. Clark
József Kövecses
Kaleem Siddiqi

Acknowledgements

McGill University's Centre for Intelligent Machines (CIM) is the product of a visionary, Professor Pierre Belanger, who started promoting the idea as Chairman of Electrical Engineering in the early eighthies. He gathered a group of colleagues in his department, in the School of Computer Science and in the Department of Mechanical Engineering, in an effort that culminated with the creation of "the centre" on November 13, 1985. Since 2007, CIM has been stregthened by the financial and strategic support provided by Quebec's *Fonds de recherche sur la nature et les technologies* via a grant to the *Regroupement stratégique pour l'étude des environnements partagés intelligents répartis* (RÉPARTI), a network of five Quebec universities sharing the same research interests as McGill in the domain of intelligent systems. In fact, RÉPARTI has enhanced CIM's networking activities beyond Quebec and Canada.

For the successful celebration of the first 25 years of CIM, we are indebted to the Deans of Engineering, Prof. Christophe Pierre, and of Science, Prof. Martin Grant, to the Chairmen of Electrical and Computer Engineering, Prof. David Plant, and of Mechanical Engineering, George Haller, as well as to the Director of the School of Computer Science, Prof. Gregory Dudek, who supported and encouraged our effort. The logistic support provided by Jan Binder, Systems Manager, not only during the celebration, but also during most of the past 25 years, has played a decisive role in making of CIM a success story. Marlene Gray, Manager, Cynthia Davidson, Secretary, and Patrick McLean, Systems Administrator, have provided excellent support in keeping the centre running throughout the years, and certainly during this celebratory year. The celebration activities have run through the year, with six public seminars and one Beatty Lecture, given by prominent researchers in the areas of brain, body and machine.

Last, but not least, the Editors want to acknowledge those who participated in the production of this book: all CIM members, who composed the Technical Committee of the *International Symposium on Brain, Body and Machine*; the anonymous reviewers who provided not only their expertise, but also their precious time; and the key role played by Dr. Seyedhossein Hajzargarbashi, the manager of the symposium website. All these individuals contributed to securing a high quality of our final product: this book.

Acknowledgments

Contents

List of Contributors

Karim Abou-Moustafa
McGill University, Montreal,
Canada
karimt@cim.mcgill.ca

Xiangjing An
National University of
Defense Technology, P.R. China
anxiangjing@yahoo.com

Jorge Angeles
McGill University, Montreal, Canada
angeles@cim.mcgill.ca

Amin Atrash
McGill University, Montreal, Canada
aatras@cs.mcgill.ca

Carmen E. Au
McGill University, Montreal, Canada
au@cim.mcgill.ca

Timothy D. Barfoot
University of Toronto, Toronto, Canada
tim.barfoot@utoronto.ca

Eric Barnett
McGill University, Montreal, Canada
ebarnett@cim.mcgill.ca

Mathieu Beauchemin-Turcotte
École de Technologie Suprieure,
Montreal, Canada
mathieu.beauchemin-turcotte.1@
ens.etsmtl.ca

Philippe Bidaud
ISIR-UPMC, Paris, France
bidaud@isir.upmc.fr

Benoit Boulet
McGill University, Montreal,
Canada
benoit.boulet@mcgill.ca

Peter E. Caines
McGill University, Montreal,
Canada
peterc@cim.mcgill.ca

Philippe Cardou
Laval University, Quebec,
Canada
pcardou@gmc.ulaval.ca

Patrick J.F. Carle
University of Toronto, Toronto, Canada
pat.carle@
robotics.utias.utoronto.ca

Stéphane Caro
Institut de Recherche en Communications et Cybernétique de Nantes, France
stephane.caro@irccyn.ec-nantes.fr

Damien Chablat
Institut de Recherche en Communications et Cybernetique de Nantes, France
damien.chablat@irccyn.ec-nantes.fr

Sha Chang
University of North Carolina at Chapel Hill, Chapel Hill, NC, USA
sxchang@med.unc.edu

Chao Chen
Monash University, Australia
chao.chen@eng.monash.edu.au

Chen-Rui Chou
University of North Carolina at Chapel Hill, Chapel Hill, NC, USA
cchou@email.unc.edu

James J. Clark
McGill University, Montreal, Canada
clark@cim.mcgill.ca

D. Louis Collins
McGill University, Montreal, Canada
louis.collins@mcgill.ca

Sylvain Comtois
Laval University, Quebec, Canada
sylvain.comtois@gel.ulaval.ca

Benjamin Corn
Tel-Aviv Sourasky Medical Center, Tel-Aviv, Israel
bencorn@tasmc.health.gov.il

Fernando De La Torre
Carnegie Mellon University, Pittsburgh, PA, USA
ftorre@cs.cmu.edu

Agostino De Santis
Università degli Studi di Napoli Federico II, Naples, Italy
agodesa@unina.it

Rolando Del Maestro
McGill University, Montreal, Canada
rolando.delmaestro@mcgill.ca

Gregory Dudek
McGill University, Montreal, Canada
dudek@cim.mcgill.ca

John P. Enright
Ryerson University, Toronto, Canada
jenright@ryerson.ca

Frank Ferrie
McGill University, Montreal, Canada
ferrie@cim.mcgill.ca

Vladimir Fonov
McGill University, Montreal, Canada
vfonov@bic.mni.mcgill.ca

C. Brandon Frederick
University of North Carolina at Chapel Hill, Chapel Hill, NC, USA
cbfreder@email.unc.edu

Paul T. Furgale
University of Toronto, Toronto, Canada
paul.furgale@utoronto.ca

Yi Gao
Georgia Institute of Technology, Atlanta, GA, USA
yi.gao@gatech.edu

Guy Gauthier
École de Technologie Suprieure,
Montreal, Canada
guy.gauthier@etsmtl.ca

Thibault Gayral
Institut de Recherche en
Communications et
Cybernétique de Nantes, France
thibault_gayral@
hotmail.fr

Clément Gosselin
Laval University, Quebec, Canada
gosselin@gmc.ulaval.ca

Christophe Grand
ISIR-UPMC, Paris, France
christophe.grand@upmc.fr

Ahmad Haidar
McGill University, Montreal, Canada
ahhai@cim.mcgill.ca

Kunimatsu Hashimoto
Toyota Motor Corporation, Aichi, Japan
kunimatsu_hashimoto@
mail.toyota.co.jp

Vincent Hayward
Universite de Paris, Paris, France
hayward@cim.mcgill.ca

Hangen He
National University of
Defense Technology, P.R. China
hehangen@yahoo.com

Minyi Huang
Carleton University, Ottawa, Canada
mhuang@math.carleton.ca

Sungkyu Jung
University of North Carolina at
Chapel Hill, Chapel Hill, NC, USA
sungkyu@email.unc.edu

Robert Kaplow
McGill University, Montreal, Canada
rkaplo@cs.mcgill.ca

Paul Kry
McGill University, Montreal, Canada
kry@cs.mcgill.ca

Micheal Langer
McGill University, Montreal, Canada
langer@cim.mcgill.ca

Denis Laurendeau
Laval University, Quebec, Canada
laurend@gel.ulaval.ca

Pascal Lee
Mars Institute, Vancouver, Canada
pascal.lee@marsinstitute.net

Simon Lefrançois
Laval University, Quebec, Canada
simon.lefrancois.1@ulaval.ca

Martin Levine
McGill University, Montreal, Canada
levine@cim.mcgill.ca

Jian Li
National University of
Defense Technology, P.R. China
jian.li6@mail.mcgill.ca

Vincenzo Lippiello
Università degli Studi di Napoli
Federico II, Naples, Italy
vincenzo.lippiello@unina.it

Xiaoxiao Liu
University of North Carolina at
Chapel Hill, Chapel Hill, NC, USA
sharonxx@cs.unc.edu

Eric Lucet
Robosoft, Technopole d'Izarbel, Bidart,
France
lucet@isir.upmc.fr

Roland P. Malhamé
École Polytechnique
de Montréal, Canada
roland.malhame@polymtl.ca

Fahim Mannan
McGill University, Montreal, Canada
fmannan@cim.mcgill.ca

J.S. Marron
University of North Carolina at
Chapel Hill, Chapel Hill, NC, USA
marron@email.unc.edu

Laurence Mercier
McGill University, Montreal, Canada
laurence@bic.mni.mcgill.ca

Hannah Michalska
McGill University, Montreal, Canada
hannah.michalska@mcgill.ca

Guillaume Moroz
Institut de Recherche en
Communications et Cybernetique de
Nantes, France
guillaume.moroz@
irccyn.ec-nantes.fr

Mojtaba Nourian
McGill University, Montreal, Canada
mnourian@cim.mcgill.ca

Lasse Øestergaard
Aalborg University, Denmark
lasse@hst.aau.dk

Martin J.-D. Otis
McGill University, Montreal, Canada
martin.j.d.otis@ieee.org

Damiano Pasini
McGill University, Montreal, Canada
damiano.pasini@mcgill.ca

Kevin Petrecca
McGill University, Montreal, Canada
kevin.petrecca@mcgill.ca

Joelle Pineau
McGill University, Montreal, Canada
jpineau@cs.mcgill.ca

Stephen M. Pizer
University of North Carolina at
Chapel Hill, Chapel Hill, NC, USA
pizer@cs.unc.edu

Ioannis Rekleitis
McGill University, Montreal, Canada
yiannis@cim.mcgill.ca

Dan Schifter
Tel-Aviv Sourasky
Medical Center, Tel-Aviv, Israel
dansch@tasmc.health.gov.il

Inna Sharf
McGill University, Montreal, Canada
inna.sharf@mcgill.ca

Bruno Siciliano
Università degli Studi di Napoli
Federico II, Naples, Italy
bruno.siciliano@unina.it

Kaleem Siddiqi
McGill University, Montreal, Canada
siddiqi@cim.mcgill.ca

Pieter Sijpkes
McGill University, Montreal,
Canada
pieter.sijpkes@mcgill.ca

Robert Sim
Microsoft Corporation
rsim@microsoft.com

Jorge Solis
Waseda University, Tokyo,
Japan
j.solis@aoni.waseda.jp

Braden E. Stenning
University of Toronto, Toronto, Canada
braden.stenning@utoronto.ca

Svetlana Stolpner
McGill University, Montreal, Canada
sveta@cim.mcgill.ca

Zhenping Sun
National University of
Defense Technology,
P.R. China
sunzp1976@163.com

Atsuo Takanishi
Waseda University, Tokyo, Japan
contact@
takanishi.mech.waseda.ac.jp

Allen Tannenbaum
Georgia Institute of
Technology, Atlanta, GA, USA
tannenba@ece.gatech.edu

Luigi Villani
Università degli Studi di Napoli
Federico II, Naples, Italy
luigi.villani@unina.it

Julien Villemure
McGill University, Montreal, Canada
julien@cs.mcgill.ca

Xinjilefu
McGill University, Montreal, Canada
xxinji@cim.mcgill.ca

Kamel Zaarouri
McGill University, Montreal, Canada
kzaaro@cim.mcgill.ca

Nathaniel Zoso
Laval University, Quebec, Canada
nathaniel.zoso.1@ulaval.ca

Force and Visual Control for Safe Human-Robot Interaction

Bruno Siciliano, Luigi Villani, Vincenzo Lippiello, and Agostino De Santis

Abstract. Unlike the industrial robotics domain where the workspace of machines and humans can be segmented, applications of intelligent machines that work in contact with humans are increasing, which involve e.g. haptic interfaces and tele-operators, cooperative material-handling, power extenders and such high-volume markets as rehabilitation, physical training and entertainment. Force and vision play a fundamental role to increase the autonomy of a robotic system, especially in the presence of humans. Vision provides global information on the surrounding environment to be used for motion planning and obstacle avoidance, while force allows adjusting the robot motion so that the local constraints imposed by the environment are satisfied. In order to avoid dangerous collisions and ensure a safe interaction, suitable control strategies based on force and visual feedback can be used while tracking human motion. This paper surveys such strategies and presents some experimental results in a number of significant case studies.

1 Introduction

The extension of application domains of robotics from factories to human environments leads to implementing proper strategies for close interaction between people and robots. On the one hand, small-scale industrial robots have to learn to get along with human co-workers in factories, while, on the other hand, service robots are a solution for automating common daily tasks in domestic environments, due to lack or high cost of human workforce.

The size of an industrial robot, or the necessary autonomous behaviour of a service robot, can result in dangerous situations for humans co-habiting the robot operational domain. Therefore, physical issues must be carefully considered, since "natural" or unexpected behaviours of people during interaction with robots can

Bruno Siciliano · Luigi Villani · Vincenzo Lippiello · Agostino De Santis
PRISMA Lab, Dipartimento di Informatica e Sistemistica, Università degli Studi di Napoli Federico II, Naples, Italy
e-mail: {bruno.siciliano, luigi.villani}@unina.it, {vincenzo.lippiello, agodesa}@unina.it

J. Angeles et al. (Eds.): Brain, Body and Machine, AISC 83, pp. 1–16.
springerlink.com

result in injuries, which may be severe, when considering the current mechanical structure of robots available on the market [1].

One possible issue to consider, in order to increase safety, is the proper use of the two main "senses": vision and touch. Vision-and-force-based control for physical interaction may include collision avoidance, control of close interaction and fusion with other sensory modes, which all may lead to improving robots performance, without necessarily considering a novel mechanical design.

However, the need for safety suggests complementing the control system with the adoption of compliant components in the structure. Compliance can be introduced at the contact point by a soft covering of the whole arm with visco-elastic materials or by adopting compliant transmissions at the robot joints. Increasing in this way the robot mechanical compliance while reducing its overall apparent inertia can be realized through different elastic actuation/transmission arrangements which include: relocation of actuators close to the robot base and transmission of motion through steel cables and pulleys, combination of harmonic drives and lightweight link design, and use of parallel and distributed macro-mini [2] or variable-impedance [3] actuation. Other improvements for anticipating and reacting to collisions can be achieved through the use of combinations of external/internal robot sensing, electronic hardware and software safety procedures, which intelligently monitor, supervise, and control manipulator operation.

This paper focuses on techniques for augmenting safety by means of control systems. Human-like capabilities in close interaction can be considered as mimicking human sensing and actuation. This leads to considering fully integrated vision-and-force-based control. Thanks to visual perception, the robotic system may achieve global information on the surrounding environment that can be used for task planning and obstacle avoidance. On the other hand, the perception of the force applied to the robot allows adjusting the motion so that the local constraints imposed by the environment during the interaction are satisfied. Moreover, a vision system may substitute the complex infrastructure needed for "intelligent environments" [4] to detect and track people in the operational domain.

In the last years, several papers on this subject have been presented. Some of them combine force or vision in the same feedback control loop, such as hybrid visual/force control [5], shared and traded control [6, 7] or visual impedance control [8, 9]. These algorithms improve classical interaction control schemes [10], e.g., impedance control, hybrid force/position control, parallel force/position control, where only force and joint position measurements are used.

The approach presented here, for the part concerning interaction control based on force and vision, was developed in previous publications [11, 12]. The collision avoidance technique is based on the "skeleton algorithm" reported in [13].

2 Modeling

For physical human-robot interaction (pHRI) it is necessary to model, or track human motion, to formulate a model of robot motion and of the objects to interact with. Consider a robot in contact with an object, a wrist force sensor and a camera mounted on the end-effector (eye-in-hand configuration) or fixed in the workspace (eye-to-hand configuration). Some modeling assumption concerning the human user, the environment, the robot and the camera are illustrated below.

2.1 Human User

Positioning of critical parts of a human body may be addressed, like for robots, considering the kinematics of the structure. However, joint measures are not available on the human body; therefore, exteroceptive sensing by means of cameras is used, obtaining the position in the space of some relevant features (hands, head etc.). This leads to finding a simplified kinematic model, to be updated in real time, with the novel "skeleton algorithm" [13]. This algorithm considers a skeleton, composed of segments, as a simplified model of a human (or a robot or even an object), exploiting the simple geometric structures in order to compute the distances between the segments, which can be used for collision avoidance, considering all the points of the articulated structure of humans and robots which may collide. For every link of the skeleton of a human figure, the closest point to the robot or the object to be avoided is computed. The distance information between the two closest points of human and obstacle can be used to avoid a collision, using "spheres" located in the selected closest points as protective hulls: these spheres can have a finite or infinite radius and can be the source of repelling forces shaped as effects of virtual springs or potential fields. Summarizing, the steps of the algorithm are:

- Create a skeleton of the human body, by using vision, and of the robot, by using direct kinematics in order to find the extremal point of the segments.
- Compute the distances between the different segments, finding also the two closest points for each pair of links.
- Define intensity and shape of repelling forces between these two points and use them as reference values in the position/force control system.

Almost all structures can be encapsulated by a finite skeleton with spheres, as sketched in Fig. 1 for the DLR arm [14]. The position of the closest point on each link (continuous curves) varies continuously, preserving continuity of reference values for any kind of control scheme. The key point of the proposed approach is that only the two closest points (on each link) of the structure are considered each time, leading to a simple generation of the Cartesian desired velocity (or force) for only one of these points, which eventually is transformed in the corresponding joint trajectory via proper inverse kinematics (or kinetics). Any point on the structure can be considered as a control point. To simplify the problem, there is also the possibility to choose only a subset of control points, e.g., the articulation of the robot [15]. Moreover, it is possible to use an inverse kinematics, an impedance control or whatever is desired, since the algorithm just adds with continuity repelling forces or velocity, preserving stability of the control loops used for the system.

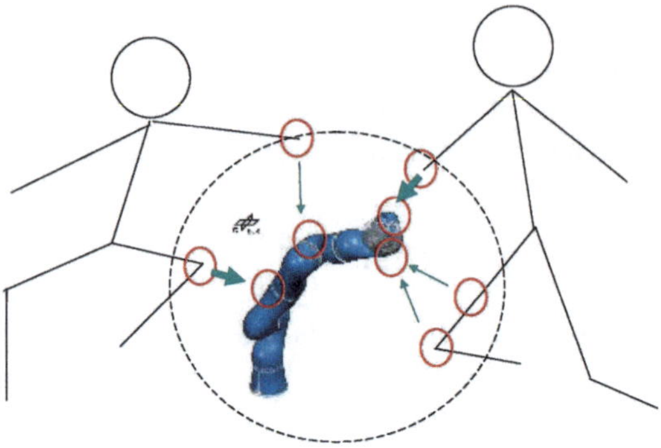

Fig. 1. Exemplification of the skeleton algorithm for the DLR lightweight arm

2.2 Environment

The position and orientation of a frame O_o–$x_o y_o z_o$ attached to the environment (assumed as a rigid object) with respect to a base coordinate frame O–xyz can be expressed in terms of the coordinate vector of the origin $\boldsymbol{o}_o$ and of the rotation matrix $\boldsymbol{R}_o(\boldsymbol{\varphi}_o)$, where $\boldsymbol{\varphi}_o$ is a $(p \times 1)$ vector corresponding to a suitable parametrization of the orientation. In the case that a minimal representation of the orientation is adopted, e.g., Euler angles, $p = 3$, while $p = 4$ if unit quaternions are used. Hence, the $(m \times 1)$ vector $\boldsymbol{x}_o = \left[\boldsymbol{o}_o^{\mathrm{T}}\ \boldsymbol{\varphi}_o^{\mathrm{T}}\right]^{\mathrm{T}}$ defines a representation of the object pose with respect to the base frame in terms of $m = 3 + p$ parameters.

The homogeneous coordinate vector $\tilde{\boldsymbol{p}} = \left[\boldsymbol{p}^T\ 1\right]^{\mathrm{T}}$ of a point P of the object with respect to the base frame can be computed as $\tilde{\boldsymbol{p}} = \boldsymbol{H}_o(\boldsymbol{x}_o)^o\tilde{\boldsymbol{p}}$, where $^o\tilde{\boldsymbol{p}}$ is the homogeneous coordinate vector of P with respect to the object frame and $\boldsymbol{H}_o$ is the homogeneous transformation matrix representing the pose of the object frame referred to the base frame.

It is assumed that the geometry of the object is known and that the interaction involves a portion of the external surface which satisfies a twice differentiable scalar equation $\varphi(^o\mathbf{p}) = 0$. Hence, the unit vector normal to the surface at the point $^o\boldsymbol{p}$ and pointing outwards can be computed as:

$$^o\boldsymbol{n}(^o\mathbf{p}) = \frac{(\partial\varphi(^o\mathbf{p})/\partial\, ^o\mathbf{p})^T}{\|(\partial\varphi(^o\mathbf{p})/\partial\, ^o\mathbf{p}\|}, \tag{1}$$

where $^o\boldsymbol{n}$ is expressed in the object frame.

Notice that the object pose $\boldsymbol{x}_o$ is assumed to be unknown and may change during the task execution. As an example, a compliant contact can be modelled assuming that $\boldsymbol{x}_o$ changes during the interaction according to an elastic law.

A further assumption is that the contact between the robot and the object is of point type and frictionless. Therefore, when in contact, the tip point P_q of the robot instantaneously coincides with a point P of the object, so that the tip position $^o\mathbf{p}_q$ satisfies the surface equation $\varphi(^o\mathbf{p}_q) = 0$. Moreover, the (3×1) contact force $^o\boldsymbol{h}$ is aligned to the normal unit vector $^o\boldsymbol{n}$.

2.3 Robot

The case of an n-joint robot manipulator is considered, with $n \geq 3$. The tip position $\boldsymbol{p}_q$ can be computed via the direct kinematics equation $\boldsymbol{p}_q = \boldsymbol{k}(\boldsymbol{q})$, where $\boldsymbol{q}$ is the $(n \times 1)$ vector of the joint variables. Also, the velocity of the robot tip $\boldsymbol{v}_{P_q}$ can be expressed as $\boldsymbol{v}_{P_q} = \boldsymbol{J}(\boldsymbol{q})\dot{\boldsymbol{q}}$ where $\boldsymbol{J}(\boldsymbol{q})$ is the robot Jacobian matrix. The vector $\boldsymbol{v}_{P_q}$ can be decomposed as

$$ {}^{o}\boldsymbol{v}_{P_q} = {}^{o}\dot{\boldsymbol{p}}_q + \boldsymbol{\Lambda}({}^{o}\boldsymbol{p}_q)\,{}^{o}\boldsymbol{\nu}_o, \tag{2} $$

with $\boldsymbol{\Lambda}(\cdot) = \begin{bmatrix}\boldsymbol{I}_3 & -\boldsymbol{S}(\cdot)\end{bmatrix}$, where $\boldsymbol{I}_3$ is the (3×3) identity matrix and $\boldsymbol{S}(\cdot)$ denotes the (3×3) skew-symmetric matrix operator, also known as cross-product matrix. In Eq. (2), ${}^{o}\dot{\boldsymbol{p}}_q$ is the relative velocity of the tip point P_q with respect to the object frame while ${}^{o}\boldsymbol{\nu}_o = \begin{bmatrix}{}^{o}\boldsymbol{v}_{O_o}^T & {}^{o}\boldsymbol{\omega}_o^T\end{bmatrix}^T$ is the velocity screw characterizing the motion of the object frame with respect to the base frame in terms of the translational velocity of the origin $\boldsymbol{v}_{O_o}$ and of the angular velocity $\boldsymbol{\omega}_o$. When the robot is in contact with the object, the normal component of the relative velocity ${}^{o}\dot{\boldsymbol{p}}_q$ is null, i.e., ${}^{o}\boldsymbol{n}^T({}^{o}\boldsymbol{p}_q)\,{}^{o}\dot{\boldsymbol{p}}_q = 0$.

2.4 Camera

A frame O_c–$x_c y_c z_c$ attached to the camera is considered. By using the classical pinhole model, a point P of the object with coordinates ${}^{c}\boldsymbol{p} = \begin{bmatrix}x\ y\ z\end{bmatrix}^{\mathrm{T}}$ with respect to the camera frame is projected onto the point of the image plane with coordinates $\begin{bmatrix}X\ Y\end{bmatrix}^{\mathrm{T}} = \lambda_c \begin{bmatrix}x/z\ y/z\end{bmatrix}^{\mathrm{T}}$ where λ_c is the lens focal length.

The homogeneous coordinate vector of P with respect to the camera frame can be expressed as ${}^{c}\tilde{\boldsymbol{p}} = {}^{c}\boldsymbol{H}_o(\boldsymbol{x}_o, \boldsymbol{x}_c)\,{}^{o}\tilde{\boldsymbol{p}}$. Notice that $\boldsymbol{x}_c$ is constant for eye-to-hand cameras; moreover, the matrix ${}^{c}\boldsymbol{H}_o$ does not depend on $\boldsymbol{x}_c$ and $\boldsymbol{x}_o$ separately but on the relative pose of the object frame with respect to the camera frame.

The velocity of the camera frame with respect to the base frame can be characterized in terms of the translational velocity of the origin $\boldsymbol{v}_{O_c}$ and of angular velocity $\boldsymbol{\omega}_c$. These vectors, expressed in camera frame, define the velocity screw ${}^{c}\boldsymbol{\nu}_c = \begin{bmatrix}{}^{c}\boldsymbol{v}_{O_c}^T & {}^{c}\boldsymbol{\omega}_c^T\end{bmatrix}^T$. It can be shown that the velocity screw ${}^{c}\boldsymbol{\nu}_o = \begin{bmatrix}{}^{c}\boldsymbol{v}_{O_o}^T & {}^{c}\boldsymbol{\omega}_o^T\end{bmatrix}^T$ corresponding to the absolute motion of the object frame can be expressed as

$$ {}^{c}\boldsymbol{\nu}_o = {}^{c}\boldsymbol{\nu}_{o,c} + \boldsymbol{\Gamma}({}^{c}\boldsymbol{o}_o)\,{}^{c}\boldsymbol{\nu}_c \tag{3} $$

where ${}^{c}\boldsymbol{\nu}_{o,c} = \begin{bmatrix}{}^{c}\dot{\boldsymbol{o}}_o^T & {}^{c}\boldsymbol{\omega}_{o,c}^T\end{bmatrix}^T$ is the velocity screw corresponding to the relative motion of the object frame with respect to camera frame, and the matrix $\boldsymbol{\Gamma}(\cdot)$ is

$$ \boldsymbol{\Gamma}(\cdot) = \begin{bmatrix}\boldsymbol{I}_3 & -\boldsymbol{S}(\cdot) \\ \boldsymbol{O}_3 & \boldsymbol{I}_3\end{bmatrix}, $$

being $\boldsymbol{O}_3$ the (3×3) null matrix.

The velocity screw ${}^{r}\boldsymbol{\nu}_s$ of a frame s with respect to a frame r can be expressed in terms of the time derivative of the vector $\boldsymbol{x}_s$ representing the pose of frame s through the equation

$$ {}^{r}\boldsymbol{\nu}_{s} = {}^{r}\boldsymbol{L}(\boldsymbol{x}_{s})\dot{\boldsymbol{x}}_{s} \tag{4} $$

where ${}^{r}\boldsymbol{L}(\cdot)$ is a Jacobian matrix depending on the particular choice of coordinates for the orientation.

3 Use of Vision, Force and Joint Positions Measurements

When the robot moves in free space, the unknown object pose and the position of the head of a human user can be estimated online by using the data provided by the camera; when the robot is in contact with the object, also the force measurements and the joint position measurements are used. Joint values are used for evaluating the position of the links for collision avoidance. The equations mapping the measurements to the unknown position and orientation of the object are derived below.

3.1 Vision

Vision is used to measure the image features, characterized by a set of scalar parameters f_j grouped in a vector $\boldsymbol{f} = \begin{bmatrix} f_1 & \cdots & f_k \end{bmatrix}^T$. The mapping from the position and orientation of the object to the corresponding image feature vector can be computed using the projective geometry of the camera and can be written in the form

$$ \boldsymbol{f} = \boldsymbol{g}_f({}^{c}\boldsymbol{H}_o(\boldsymbol{x}_o, \boldsymbol{x}_c)), \tag{5} $$

where only the dependence from the relative pose of the object frame with respect to camera frame has been explicitly evidenced. For the estimation of the object pose, the computation of the Jacobian matrix $\boldsymbol{J}_f = \partial \boldsymbol{g}_f / \partial \boldsymbol{x}_o$ is requred. To this end, the time derivative of Eq. (5) can be computed in the form

$$ \dot{\boldsymbol{f}} = \frac{\partial \boldsymbol{g}_f}{\partial \boldsymbol{x}_o}\dot{\boldsymbol{x}}_o + \frac{\partial \boldsymbol{g}_f}{\partial \boldsymbol{x}_c}\dot{\boldsymbol{x}}_c, \tag{6} $$

where the second term on the right-hand side is null for eye-to-hand cameras. On the other hand, the time derivative of Eq. (5) can be expressed also in the form $\dot{\boldsymbol{f}} = \boldsymbol{J}_{o,c}{}^{c}\boldsymbol{\nu}_{o,c}$ where the matrix $\boldsymbol{J}_{o,c}$ is the Jacobian mapping the relative velocity screw of the object frame with respect to the camera frame into the variation of the image feature parameters. The expression of $\boldsymbol{J}_{o,c}$ depends on the choice of the image features; examples of computation can be found in [16]. By taking into account the velocity composition in Eq. (3), vector $\dot{\boldsymbol{f}}$ can be rewritten in the form

$$ \dot{\boldsymbol{f}} = \boldsymbol{J}_{o,c}{}^{c}\boldsymbol{\nu}_o - \boldsymbol{J}_c{}^{c}\boldsymbol{\nu}_c \tag{7} $$

where $\boldsymbol{J}_c = \boldsymbol{J}_{o,c}\boldsymbol{\Gamma}({}^{c}\boldsymbol{o}_o)$ is the Jacobian corresponding to the contribution of the absolute velocity screw of the camera frame, known in the literature as interaction matrix [16]. In view of Eq. (4), comparing Eq. (7) with Eq. (6) yields

$$ \boldsymbol{J}_f = \boldsymbol{J}_{o,c}{}^{c}\boldsymbol{L}(\boldsymbol{x}_o). \tag{8} $$

3.2 Force

In the case of frictionless point contact, the measure of the force $\boldsymbol{h}$ at the robot tip during the interaction can be used to compute the unit vector normal to the object surface at the contact point ${}^o\boldsymbol{p}_q$, i.e.,

$$\boldsymbol{n}_h = \boldsymbol{h}/\|\boldsymbol{h}\|. \tag{9}$$

On the other hand, vector $\boldsymbol{n}_h$ can be expressed as a function of the object pose $\boldsymbol{x}_o$ and of the robot position $\boldsymbol{p}_q$ in the form

$$\boldsymbol{n}_h = \boldsymbol{R}_o {}^o\boldsymbol{n}({}^o\boldsymbol{p}_q) = \boldsymbol{g}_h(\boldsymbol{x}_o, \boldsymbol{p}_q), \tag{10}$$

with ${}^o\boldsymbol{p}_q = \boldsymbol{R}_o^T(\boldsymbol{p}_q - \boldsymbol{o}_o)$. For the estimation of the object pose, the computation of the Jacobian matrix $\boldsymbol{J}_h = \partial \boldsymbol{g}_h / \partial \boldsymbol{x}_o$ is required. To this end, the time derivative of Eq. (10) can be expressed as

$$\dot{\boldsymbol{n}}_h = \frac{\partial \boldsymbol{g}_h}{\partial \boldsymbol{x}_o}\dot{\boldsymbol{x}}_o + \frac{\partial \boldsymbol{g}_h}{\partial \boldsymbol{p}_q}\dot{\boldsymbol{p}}_q, \tag{11}$$

but also as

$$\dot{\boldsymbol{n}}_h = \dot{\boldsymbol{R}}_o {}^o\boldsymbol{n}({}^o\boldsymbol{p}_q) + \boldsymbol{R}_o {}^o\boldsymbol{N}({}^o\boldsymbol{p}_q){}^o\dot{\boldsymbol{p}}_q, \tag{12}$$

where ${}^o\boldsymbol{N}({}^o\boldsymbol{p}_q) = \partial {}^o\boldsymbol{n} / \partial {}^o\boldsymbol{p}_q$ depends on the surface curvature and ${}^o\dot{\boldsymbol{p}}_q$ can be computed as

$${}^o\dot{\boldsymbol{p}}_q = \boldsymbol{R}_o^{\mathrm{T}}(\dot{\boldsymbol{p}}_q - \dot{\boldsymbol{o}}_o + \boldsymbol{S}(\boldsymbol{p}_q - \boldsymbol{o}_o)\boldsymbol{\omega}_o).$$

Hence, by replacing the above expression in Eq. (12) and taking into account the equality $\dot{\boldsymbol{R}}_o {}^o\boldsymbol{n}({}^o\boldsymbol{p}_q) = -\boldsymbol{S}(\boldsymbol{n}_h)\boldsymbol{\omega}_o$, the following equality holds:

$$\dot{\boldsymbol{n}}_h = \boldsymbol{N}\dot{\boldsymbol{p}}_q - \boldsymbol{N}\dot{\boldsymbol{o}}_o - (\boldsymbol{S}(\boldsymbol{n}_h) - \boldsymbol{N}\boldsymbol{S}(\boldsymbol{p}_q - \boldsymbol{o}_o))\boldsymbol{\omega}_o \tag{13}$$

where $\boldsymbol{N} = \boldsymbol{R}_o {}^o\boldsymbol{N}({}^o\boldsymbol{p}_q)\boldsymbol{R}_o^T$. In view of Eq. (4), comparing Eq. (11) with Eq. (13) yields

$$\boldsymbol{J}_h = -\left[\boldsymbol{N} \;\; \boldsymbol{S}(\boldsymbol{n}_h) - \boldsymbol{N}\boldsymbol{S}(\boldsymbol{p}_q - \boldsymbol{o}_o)\right]\boldsymbol{L}(\boldsymbol{x}_o). \tag{14}$$

3.3 Joint Positions

The measurement of the joint position vector $\boldsymbol{q}$ can be used not only to evaluate the configuration of the robot, which can possibly collide with a user, but also to evaluate the position of the point P of the object when in contact with the robot tip point P_q, using the direct kinematics equation. In particular, it is significant to compute the scalar

$$\delta_{hq} = \boldsymbol{n}_h^T \boldsymbol{p}_q = g_{hq}(\boldsymbol{x}_o, \boldsymbol{p}_q), \tag{15}$$

using also the force measurements via Eq. (9). This quantity represents the component of the position vector of the robot tip point along the constrained direction $\boldsymbol{n}_h$. The corresponding infinitesimal displacement is the same for the robot and for the object (assuming that contact is preserved). This is not true for the other components of the tip position, which do not provide any useful information on the object motion because the robot tip may slide on the object surface; therefore, these components are not useful to estimate the pose of the object.

For the estimation of the object pose, the computation of the Jacobian matrix $\boldsymbol{J}_{hq} = \partial g_{hq}/\partial \boldsymbol{x}_o$ is required. The time derivative of δ_{hq} can be expressed as

$$\dot{\delta}_{hq} = \frac{\partial g_{hq}}{\partial \boldsymbol{x}_o}\dot{\boldsymbol{x}}_o + \frac{\partial g_{hq}}{\partial \boldsymbol{p}_q}\dot{\boldsymbol{p}}_q, \tag{16}$$

but also as $\dot{\delta}_{hq} = \dot{\boldsymbol{n}}_h^T \boldsymbol{p}_q + \boldsymbol{n}_h^T \boldsymbol{R}_o({}^o\dot{\boldsymbol{p}}_q + \boldsymbol{\Lambda}({}^o\boldsymbol{p}_q)^o\boldsymbol{\nu}_o)$, where the expression of the absolute velocity of the point P_q in Eq. (2) has been used. Using the identity ${}^o\boldsymbol{n}^T({}^o\boldsymbol{p}_q)^o\dot{\boldsymbol{p}}_q = 0$, this equation can be rewritten as

$$\dot{\delta}_{hq} = \boldsymbol{p}_q^T \dot{\boldsymbol{n}}_h + \boldsymbol{n}_h^T \boldsymbol{\Lambda}(\boldsymbol{p}_q - \boldsymbol{o}_o)\boldsymbol{\nu}_o. \tag{17}$$

Hence, by comparing Eq. (16) with Eq. (17) and taking into account Eqs. (12), (14) and (4), the expression below can be found

$$\boldsymbol{J}_{hq} = \boldsymbol{p}_q^T \boldsymbol{J}_h + \boldsymbol{n}_h^T \boldsymbol{\Lambda}(\boldsymbol{p}_q - \boldsymbol{o}_o)\boldsymbol{L}(\boldsymbol{x}_o). \tag{18}$$

4 Vision-Based Pose Estimation

4.1 Human Operator's Pose Estimation

In order to use the skeleton algorithm [13] for collision avoidance, simple fixed cameras are employed to detect the positions of face and hands of an operator in the operational space of the robot. In assembly tasks in cooperation with the robot, the operator does not move fast, thus simplifying the tracking by means of cameras. In preliminary experiments, markers are used to help the detection and tracking. The detected positions of the human operator are to be tracked in order to keep a safety volume around him/her, repelling the robot when it approaches too much. Cameras mounted on the robot can be used as well. Potential fields or optimization techniques are then to be designed, in order to create modifications to the robot trajectory aimed at avoiding dangerous approaches. Simple virtual springs or more complex modifications to trajectories, using null-space motion if possible, can be adopted also while using an interaction control with an object, which is considered below. The shape of the computed repelling force or velocity must preserve continuity of reference values for the robot controllers.

4.2 Object Pose Estimation

The pose vector $\boldsymbol{x}_o$ of the object with respect to the base frame can be estimated using an Extended Kalman Filter (EKF).

To this end, a discrete-time state space dynamic model has to be considered, describing the object motion. The state vector of the dynamic model is chosen as $\boldsymbol{w} = \left[\boldsymbol{x}_o^{\mathrm{T}}\ \dot{\boldsymbol{x}}_o^{\mathrm{T}}\right]^{\mathrm{T}}$. For simplicity, the object velocity is assumed to be constant over one sample period T_s. This approximation is reasonable in the hypothesis that T_s is sufficiently small. Hence, the discrete-time dynamic model can be written as $\boldsymbol{w}_k = \boldsymbol{A}\boldsymbol{w}_{k-1} + \boldsymbol{\gamma}_k$, where $\boldsymbol{\gamma}$ is the dynamic modeling error described by zero mean Gaussian noise with covariance $\boldsymbol{Q}$ and $\boldsymbol{A}$ is the $(2m \times 2m)$ block matrix

$$\boldsymbol{A} = \begin{bmatrix} \boldsymbol{I}_m & T_s\boldsymbol{I}_m \\ \boldsymbol{O}_m & \boldsymbol{I}_m \end{bmatrix}.$$

The output of the EKF, in the case that all the available data can be used, is the vector $\boldsymbol{\zeta}_k = \left[\boldsymbol{\zeta}_{f,k}^T\ \boldsymbol{\zeta}_{h,k}^T\ \zeta_{hq,k}^{\ \ T}\right]^T$ of the measurements at time kT_s, where $\boldsymbol{\zeta}_{f,k} = \boldsymbol{f}_k + \boldsymbol{\mu}_{f,k}$, $\boldsymbol{\zeta}_{h,k} = \boldsymbol{h}_k + \boldsymbol{\mu}_{h,k}$, and $\zeta_{hq,k} = \delta_k + \mu_{hq,k}$, and $\boldsymbol{\mu}$ is the measurement noise, which is assumed to be zero-mean Gaussian noise with covariance $\boldsymbol{\Pi}$.

By taking into account Eqs. (5), (10) and (15), the output equation of the EKF can be written as

$$\boldsymbol{\zeta}_k = \boldsymbol{g}(\boldsymbol{w}_k) + \boldsymbol{\mu}_k,$$

with $\left[\boldsymbol{\mu}_{f,k}^T\ \boldsymbol{\mu}_{h,k}^T\ \mu_{hq,k}^T\right]^T$ and $\boldsymbol{g}(\boldsymbol{w}_k) = \left[\boldsymbol{g}_f^T(\boldsymbol{w}_k)\ \boldsymbol{g}_h^T(\boldsymbol{w}_k)\ g_{hq}^T(\boldsymbol{w}_k)\right]^T$, where only the explicit dependence on the state vector $\boldsymbol{w}_k$ has been evidenced. The EKF requires the computation of the Jacobian matrix of the output equation

$$\boldsymbol{C}_k = \left.\frac{\partial \boldsymbol{g}(\boldsymbol{w})}{\partial \boldsymbol{w}}\right|_{w=\hat{w}_{k,k-1}} = \left[\frac{\partial \boldsymbol{g}(\boldsymbol{w})}{\partial \boldsymbol{x}_o}\ \ \boldsymbol{O}\right]_{w=\hat{w}_{k,k-1}},$$

where $\boldsymbol{O}$ is the null matrix of proper dimensions corresponding to the partial derivative of $\boldsymbol{g}$ with respect to the velocity variables. In view of Eqs. (8), (14) and (18), the Jacobian matrix $\partial \boldsymbol{g}(\boldsymbol{w})/\partial \boldsymbol{x}_o$ has the expression $\partial \boldsymbol{g}(\boldsymbol{w})/\partial \boldsymbol{x}_o = \left[\boldsymbol{J}_f^T\ \boldsymbol{J}_h^T\ \boldsymbol{J}_{hq}^T\right]^T$. The equations of the recursive form of the EKF, being standard, are omitted here for brevity, but can be found, e.g., in [17].

5 Interaction Control

The proposed algorithm can be used to estimate online the pose of an object in the workspace; this allows computing the surface equation with respect to the base frame in the form $\varphi(\boldsymbol{R}_o^{\mathrm{T}}(\boldsymbol{p}_q - \boldsymbol{o}_o)) = \varphi(\boldsymbol{q}, t) = 0$, where the last equality holds in view of the direct kinematic equation of the robot. In the following, it is assumed that the object does not move; the general case of a moving object is more complex but can be analyzed in a similar way. Hence, the constraint equation is $\varphi(\boldsymbol{q}) = 0$; moreover $\boldsymbol{J}_\varphi(\boldsymbol{q})\dot{\boldsymbol{q}} = 0$, where $\boldsymbol{J}_\varphi = \partial\varphi/\partial\boldsymbol{q}$ is a $(1 \times n)$ vector.

The dynamic model of the manipulator in contact with the environment is

$$\boldsymbol{B}(\boldsymbol{q})\ddot{\boldsymbol{q}} + \boldsymbol{n}(\boldsymbol{q}, \dot{\boldsymbol{q}}) = \boldsymbol{\tau} - \boldsymbol{J}_\varphi^{\mathrm{T}}(\boldsymbol{q})\lambda,$$

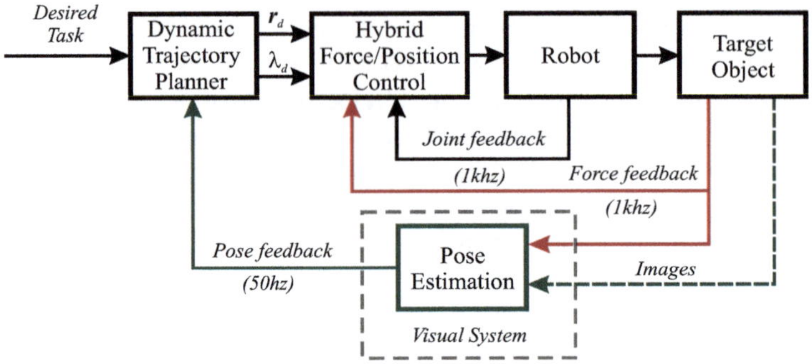

Fig. 2. Hybrid force/position control

where $\boldsymbol{B}$ is the $(n \times n)$ symmetric and positive definite inertia matrix, $\boldsymbol{n}(\boldsymbol{q}, \dot{\boldsymbol{q}})$ is the $(n \times 1)$ vector taking into account Coriolis, centrifugal, friction and gravity torques, $\boldsymbol{\tau}$ is the $(n \times 1)$ vector of the joint torques, and $\boldsymbol{\lambda}$ is the lagrangian multiplier associated to the constraint.

The online computation of the constraint equations can be suitably exploited for interaction control. The cases of hybrid force/position control and of impedance control are considered below.

5.1 *Hybrid Force/Position Control*

Following the formulation introduced in [18] the configuration of the robot subject to the constraint $\varphi(\boldsymbol{q}) = 0$ can be described in terms of a suitable vector $\boldsymbol{r}$ of $(n-1)$ independent variables. From the implicit function theorem, this vector can be defined as $\boldsymbol{r} = \boldsymbol{\psi}(\boldsymbol{q})$, where $\boldsymbol{\psi}(\boldsymbol{q})$ is any $((n-1) \times 1)$ twice differentiable vector function such that the $(n-1)$ components of $\boldsymbol{\psi}(\boldsymbol{q})$ and function $\varphi(\boldsymbol{q})$ are linearly independent at least locally in a neighborhood of the operating point. This implies the existence of an $(n \times 1)$ vector function $\boldsymbol{\rho}(\boldsymbol{r})$ such that $\boldsymbol{q} = \boldsymbol{\rho}(\boldsymbol{r})$, at least locally. Moreover, the Jacobian $\boldsymbol{J}_\rho = \partial \boldsymbol{\rho} / \partial \boldsymbol{r}$ of the differential mapping $\dot{\boldsymbol{q}} = \boldsymbol{J}_\rho(\boldsymbol{r})\dot{\boldsymbol{r}}$ is a full-rank matrix. The matrices

$$\boldsymbol{S}_f = \boldsymbol{J}^{-\mathrm{T}}(\boldsymbol{q})\boldsymbol{J}_\varphi^{\mathrm{T}}(\boldsymbol{q}), \qquad \boldsymbol{S}_v = \boldsymbol{J}(\boldsymbol{q})\boldsymbol{J}_\rho(\boldsymbol{r}),$$

by virtue of the equalities $\boldsymbol{h} = \boldsymbol{S}_f \boldsymbol{\lambda}$ and $\boldsymbol{v}_{P_q} = \boldsymbol{S}_v \dot{\boldsymbol{r}}$, play the role of selection matrices.

A hybrid force/position control task can be assigned by specifying the desired force $\lambda_d(t)$ and the $n-1$ components of the vector $\boldsymbol{r}_d(t)$. An inverse dynamics control law can be adopted, by choosing the control torque $\boldsymbol{\tau}$ as

$$\boldsymbol{\tau} = \boldsymbol{B}(\boldsymbol{q})\boldsymbol{\alpha} + \boldsymbol{n}(\boldsymbol{q}, \dot{\boldsymbol{q}})\dot{\boldsymbol{q}} + \boldsymbol{J}_\varphi^{\mathrm{T}}(\boldsymbol{q})\boldsymbol{S}_f h_\lambda,$$

with $\boldsymbol{\alpha} = \boldsymbol{J}^{-1}(\boldsymbol{q})(\boldsymbol{S}_v \boldsymbol{\alpha}_v + \dot{\boldsymbol{S}}_v \dot{\boldsymbol{r}} - \dot{\boldsymbol{J}}\dot{\boldsymbol{q}})$,

$$\boldsymbol{\alpha}_v = \ddot{\boldsymbol{r}}_d + \boldsymbol{K}_{Dr}(\dot{\boldsymbol{r}}_d - \dot{\boldsymbol{r}}) + \boldsymbol{K}_{Pr}(\boldsymbol{r}_d - \boldsymbol{r}) \tag{19}$$

and

$$h_\lambda = \lambda_d + k_{I\lambda} \int_0^t (\lambda_d(\tau) - \lambda(\tau))\mathrm{d}\tau, \tag{20}$$

where $\boldsymbol{K}_{Dr}$, $\boldsymbol{K}_{Pr}$ and $k_{I\lambda}$ are suitable feedback gains.

From the block scheme of Figure 2 it can be observed that the algorithm has an inner/outer structure, where the inner loop implements hybrid control whereas the outer loop computes the estimation of the object pose as well as the desired force and motion trajectories, on the basis of force, joint and visual measurements. Usually, the outer loop runs at a frequency lower than the inner loop, due to the limitation in the maximum camera frame rate (between 25 Hz and 60 Hz for low-cost cameras used in standard applications).

5.2 Impedance Control

The above scheme can be easily modified by replacing the inner hybrid force/ position control with a different interaction control scheme. For instance, a position-based impedance control algorithm, based on the concept of compliant frame [10], can be adopted (see Figure 3).

In detail, on the basis of the current estimate of the constraint equation, the Dynamic Trajectory Planner generates a pose trajectory for a desired end-effector frame specified in terms of the position of the origin $\boldsymbol{p}_d$ and orientation matrix $\boldsymbol{R}_d$. Moreover, a compliant frame r is introduced, specified in terms of $\boldsymbol{p}_r$ and $\boldsymbol{R}_r$. Then, a mechanical impedance between the desired and the compliant frame is considered, so as to contain the values of the interaction force $\boldsymbol{h}$ and moment $\boldsymbol{m}$. In other words, the desired position and orientation, together with the measured contact force and moment, are input to the impedance equation which, via a suitable integration, generates the position and orientation of the compliant frame to be used as a reference for the motion controller of the robot end effector.

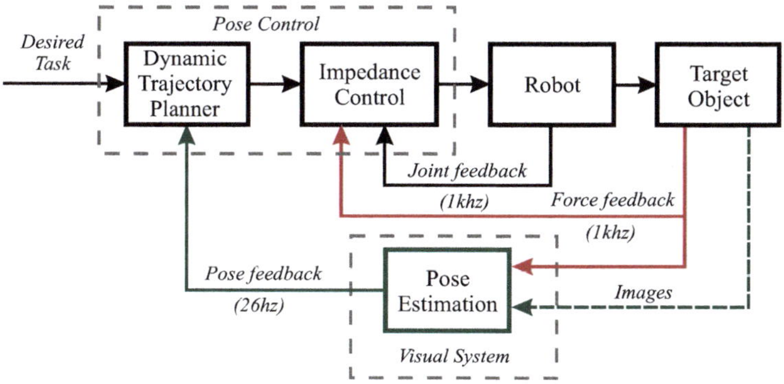

Fig. 3. Position-based visual impedance control

Fig. 4. Experimental setup

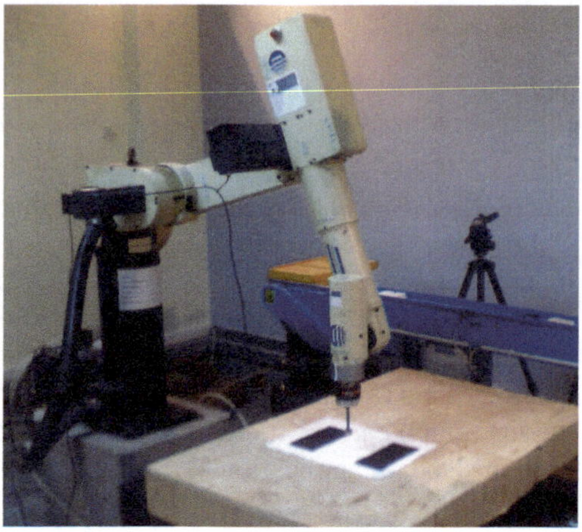

As far as the compliant frame is concerned, the position $\boldsymbol{p}_r$ can be computed via the translational impedance equation

$$\boldsymbol{M}_p \Delta\ddot{\boldsymbol{p}}_{dr} + \boldsymbol{D}_p \Delta\dot{\boldsymbol{p}}_{dr} + \boldsymbol{K}_p \Delta\boldsymbol{p}_{dr} = \boldsymbol{h}, \tag{21}$$

where $\Delta\boldsymbol{p}_{dr} = \boldsymbol{p}_d - \boldsymbol{p}_r$, and $\boldsymbol{M}_p$, $\boldsymbol{D}_p$ and $\boldsymbol{K}_p$ are positive definite matrices representing the mass, damping, and stiffness characterizing the impedance.

The orientation of the reference frame $\boldsymbol{R}_r$ is computed via a geometrically consistent impedance equation similar to Eq. (21), in terms of an orientation error based on the (3×1) vector ${}^r\boldsymbol{\epsilon}_{dr}$, defined as the vector part of the unit quaternion that can be extracted from ${}^r\boldsymbol{R}_d = \boldsymbol{R}_r^T \boldsymbol{R}_d$. The corresponding mass, damping and inertia matrices are $\boldsymbol{M}_o$, $\boldsymbol{D}_o$ and $\boldsymbol{K}_o$ respectively. More details about the geometrically consistent impedance based on the unit quaternion can be found in [10].

It is interesting to notice that the above scheme is able to manage both contact and free-space motion phases. In fact, during free space motion, the position-based impedance control is equivalent to a pose control and the whole scheme corresponds to a classical position-based visual servoing algorithm [19]. For this reason, this scheme can be classified as position-based visual impedance control.

6 Case Studies

The proposed approach has been tested in a number of simulation and experimental case studies. The setup (Figure 4) is composed of a 6-DOF industrial robot Comau SMART-3 S with an open control architecture based on RTAI-Linux operating system. A six-axis force/torque sensor ATI FT30-100 with force range of ± 130 N and torque range of ± 10 N·m is mounted at the arm's wrist, providing readings of six components of generalized force at 1 ms. A visual system composed of a PC equipped with a Matrox GENESIS board and a Sony 8500CE B/W camera is available. The camera is fixed and calibrated with respect to the base frame of the robot.

6.1 Interaction with an Object

Experimental tests have been carried out in order to assess the performance of the impedance control scheme. The environment is a planar wooden horizontal surface, with an estimated stiffness (along ${}^{o}\boldsymbol{n}$) of about 46000 N/m. The object features are 8 landmark points lying on the plane at the corners of two rectangles of 0.1×0.2 m size (as in Figure 4).

The impedance parameters are chosen as: $\boldsymbol{M}_p = 40\boldsymbol{I}_3$, $\boldsymbol{D}_p = 26.3\boldsymbol{I}_3$ and $\boldsymbol{K}_p = 1020\boldsymbol{I}_3$, $\boldsymbol{M}_o = 15\boldsymbol{I}_3$, $\boldsymbol{D}_o = 17.4\boldsymbol{I}_3$ and $\boldsymbol{K}_o = 3\boldsymbol{I}_3$; a 1 ms sampling time has been selected for the impedance and pose control. Notice that the stiffness of the object is much higher that the positional stiffness of the impedance, so that the environment can be considered rigid.

The desired task is planned in the object frame and consists of a straight-line motion of the end effector along the z_o-axis, while keeping a fixed orientation with the stick normal to the $x_o y_o$-plane. The final position is ${}^{o}\boldsymbol{p}_f = \begin{bmatrix} 0 & 0 & 0.05 \end{bmatrix}^T$ m, which is chosen to have a normal force of about 50 N at the equilibrium, with the selected value of the impedance positional stiffness. A trapezoidal velocity profile time-law is adopted, with a cruising velocity of 0.01 m/s. The absolute trajectory is computed from the desired relative trajectory using the current object pose estimation. The final position of the end effector is held for 2 s; after, a vertical motion in the opposite direction is commanded.

In the EKF, the non-null elements of the matrix $\boldsymbol{\Pi}$ have been set equal to $2.5\,\text{pixel}^2$ for $\boldsymbol{f}$, $5 \cdot 10^{-4}$ for $\boldsymbol{n}_h$ and $10^{-6}\,\text{m}^2$ for δ_{hq}. These values have been set on the basis of the calibration errors of the camera. The state noise covariance matrix has been selected so as to give a rough measure of the errors due to the simplification introduced in the model (constant velocity), by considering only velocity disturbance, i.e.

$$\boldsymbol{Q} = \text{diag}\{0,0,0,0,0,0,0,10,10,10,1,1,1\} \cdot 10^{-9}.$$

Notice that the unit quaternion has been used for the orientation in the EKF. Moreover a 38 ms sampling time has been used for the estimation algorithm, corresponding to the typical camera frame rate of 26 Hz.

Two different experiments are presented, to show the effectiveness of the use of force and joint position measurements, besides visual measurements.

In the first experiment only the visual measurements are used. The time history of the measured force and moment in the sensor frame are reported in Figure 5. Notice that the force is null during the motion in free space and becomes different from zero after contact. The impedance control keeps the force limited during the transient while, at steady state, the force component along the z axis reaches a value of about -65 N, which is different from the desired value of -50 N. This is caused by the presence of estimation errors on the position of the plane due to calibration

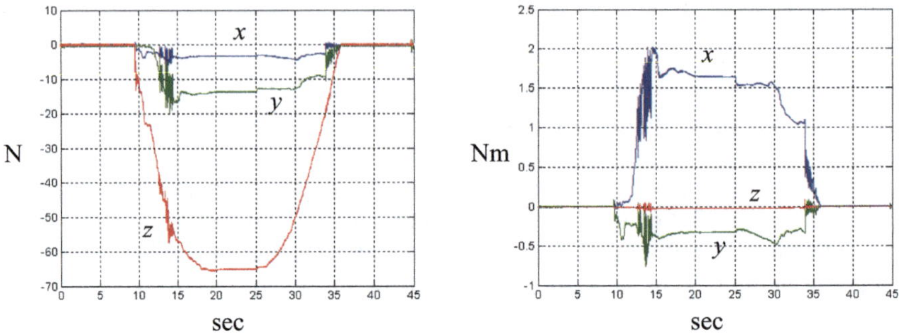

Fig. 5. Measured force (left) and moment (right) in the first experiment

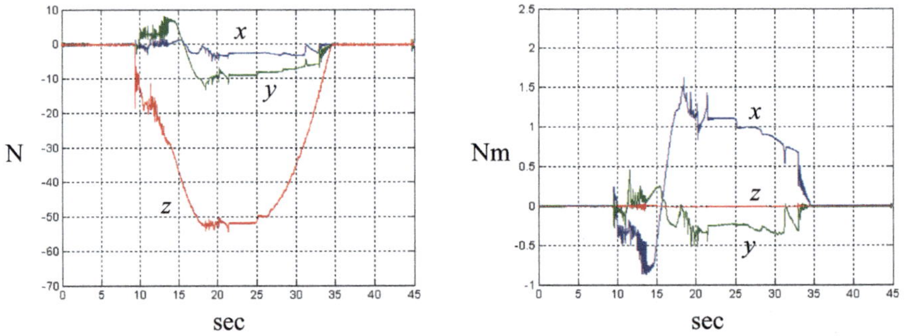

Fig. 6. Measured force (left) and moment (right) in the second experiment

errors of the camera. Moreover, the moment measured by the sensor is different from zero due to the misalignment between the estimated and the real normal direction of the plane.

The same task is repeated using also the contact force and the joint position measurements for object pose estimation; the results are reported in Figure 6. It can be observed that the benefit of using additional measurements in the EKF results in a force along the vertical direction which is very close to the desired value, due to the improved estimation of the position of the plane; moreover, the contact moment is also reduced because of the better estimation of the normal direction of the plane.

6.2 *Vision-Based Head Avoidance*

During a task involving interaction with an object, there is the possibility that a human operator is present in the workspace. In such a case, the robot has to reconfigure in order to avoid the body of the operator, tracked by a camera. In a simple case, it is possible to consider the head and the arms of a person present in the workspace as a source of a repelling elastic force. A volume is created around the head and the

Fig. 7. Trajectory modifications for collision avoidance

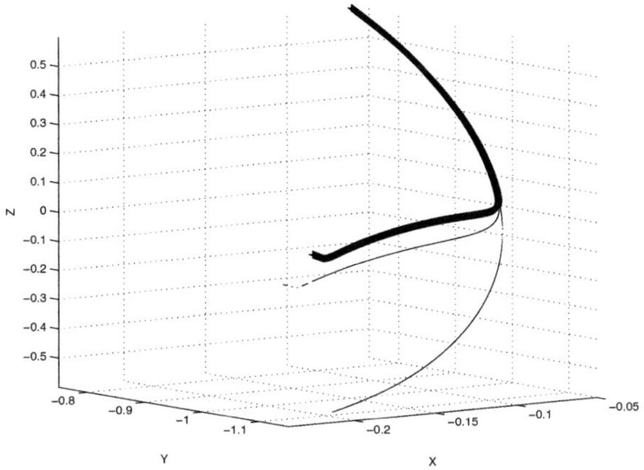

arms: the robot is pushed with continuous reference values given to force or velocity for a point on each link which is the closest to the "safety volume" considered.

Results of an experiment with the Comau SMART 3S industrial robot are reported in Fig. 7. The planned trajectory (dotted line) is abandoned for the presence of the arm (segment parallel to the axis x, with $y = 1$ and $z = 0.5$). The bold trajectory is the path followed with an elastic constant $K = 0.5$ for planning the desired velocity v of the closest points with the formula $v = K(d_{ref} - d)$ for $d < d_{ref}$, where d_{ref} is the radius of the protective sphere and d is the distance between the robot links and the center of such a sphere. The thin path in Fig. 7 is tracked for $K = 0.5$. This simple case study shows the robustness of the skeleton algorithm, which gives continuous references to different inverse kinematics schemes (one for each robot link) in order to push the robot in a certain direction during any kind of operation and with any kind of motion/force controller.

7 Conclusions

The integration of force and visual control to achieve safe human-robot interaction has been presented. A hybrid force/position control scheme and a position-based visual impedance control scheme were introduced, employing a pose estimation algorithm on the basis of visual, force and joint position data. The addition of collision-avoidance facilities with the "skeleton algorithm" gives the opportunity of sharing the workspace with a human operator.

Acknowledgements. The research leading to these results has been supported by the DEXMART Large-scale integrating project, which has received funding from the European Community's Seventh Framework Programme (FP7/2007-2013) under grant agreement ICT-216239. The authors are solely responsible for its content. It does not represent the opinion of the European Community and the Community is not responsible for any use that might be made of the information contained therein.

References

1. De Santis, A., Siciliano, B., De Luca, A., Bicchi, A.: An atlas of physical Human–Robot Interaction. Mechanism and Machine Theory 43, 253–270 (2008)
2. Zinn, M., Khatib, O., Roth, O., Salisbury, J.K.: Playing it safe. IEEE Robotics and Automation Magazine 11(2), 12–21 (2004)
3. Bicchi, A., Tonietti, G.: Fast and soft-arm tactics. IEEE Robotics and Automation Magazine 11(2), 22–33 (2004)
4. Hashimoto, H.: Intelligent interactive spaces — Integration of IT and robotics. In: Proceedings of IEEE Workshop on Advanced Robotics and its Social Impacts, pp. 85–90 (2005)
5. Hosoda, K., Igarashi, K., Asada, M.: Adaptive hybrid control for visual and force servoing in an unknown environment. IEEE Robotics and Automation Magazine 5(4), 39–43 (1998)
6. Nelson, B.J., Morrow, J.D., Khosla, P.K.: Improved force control through visual servoing. In: Proceedings of American Control Conference, pp. 380–386 (1995)
7. Baeten, J., De Schutter, J.: Integrated Visual Servoing and Force Control. The Task Frame Approach. Springer, Heidelberg (2004)
8. Morel, G., Malis, E., Boudet, S.: Impedance based combination of visual and force control. In: Proceedings of IEEE International Conference on Robotics and Automation, pp. 1743–1748 (1998)
9. Olsson, T., Johansson, R., Robertsson, A.: Flexible force-vision control for surface following using multiple cameras. In: Proceedings of IEEE/RSJ International Conference on Intelligent Robots and System, pp. 798–803 (2004)
10. Siciliano, B., Villani, L.: Robot Force Control. Kluwer, Dordrecht (1999)
11. Lippiello, V., Siciliano, B., Villani, L.: A position-based visual impedance control for robot manipulators. In: Proceedings of IEEE International Conference on Robotics and Automation, pp. 2068–2073 (2007)
12. Lippiello, V., Siciliano, B., Villani, L.: Robot force/position control with force and visual feedback. In: Proceedings of European Control Conference, pp. 3790–3795 (2007)
13. De Santis, A., Albu-Schaeffer, A., Ott, C., Siciliano, B., Hirzinger, G.: The skeleton algorithm for self-collision avoidance of a humanoid manipulator. In: Proceedings of IEEE/ASME International Conference on Advanced Intelligent Mechatronics (2007)
14. Hirzinger, G., Albu-Schaeffer, A., Hahnle, M., Schaefer, I., Sporer, N.: On a new generation of torque controlled light-weight robots. In: Proceedings of IEEE International Conference on Robotics and Automation, pp. 3356–3363 (2001)
15. De Santis, A., Pierro, P., Siciliano, B.: The virtual end-effectors approach for human-robot interaction. In: Lenarčič, J., Roth, B. (eds.) Advances in Robot Kinematics. Springer, Heidelberg (2006)
16. Espiau, B., Chaumette, F., Rives, P.: A new approach to visual servoing in robotics. IEEE Transactions on Robotics and Automation 12, 313–326 (1996)
17. Lippiello, V., Villani, L.: Managing redundant visual measurements for accurate pose tracking. Robotica 21, 511–519 (2003)
18. Villani, L., De Schutter, J.: Force control. In: Siciliano, B., Khatib, O. (eds.) Springer Handbook of Robotics. Springer, Heidelberg (2008)
19. Lippiello, V., Siciliano, B., Villani, L.: Position-based visual servoing in industrial multi-robot cells using a hybrid camera configuration. IEEE Transactions on Robotics 23, 73–86 (2007)

3D Automatic Segmentation of the Hippocampus Using Wavelets with Applications to Radiotherapy Planning

Yi Gao, Benjamin W. Corn, Dan Schifter, and Allen Tannenbaum

Abstract. During the past half-century, the cornerstone of treatment for brain metastases has been whole brain irradiation (WBI). WBI has multiple salutary effects including rapid relief of neurological signs and symptoms as well as enhanced local control. Unfortunately, WBI may also engender side effects including memory deficits and decrements in quality of life. Since memory control is thought to be mediated by the hippocampus, attention has been turned to whole brain radiotherapeutic techniques that allow sparing of the hippocampus. In order to be able to minimize dose deposition within the hippocampus, clinicians must be

Yi Gao
School of Biomedical Engineering, Georgia Institute of Technology, Atlanta, GA 30332, USA; (001)404-660-8156
e-mail: yi.gao@gatech.edu

Benjamin W. Corn
Tel-Aviv Sourasky Medical Center, Tel-Aviv Israel
e-mail: bencorn@tasmc.health.gov.il

Dan Schifter
Tel-Aviv Sourasky Medical Center, Tel-Aviv Israel
e-mail: dansch@tasmc.health.gov.il

Allen Tannenbaum
Schools of Electrical & Computer Engineering and Biomedical Engineering, Georgia Institute of Technology, Atlanta, GA, 30332-0250, USA & Department of EE, Technion-IIT, Israel
e-mail: allen.tannenbaum@ece.gatech.edu

J. Angeles et al. (Eds.): Brain, Body and Machine, AISC 83, pp. 17–31.
springerlink.com

able to confidently identify that structure. However, manually tracing out the hippocampus for each patient is time consuming and subject to individual bias. To this end, an automated method can be very useful for such a task. In this paper, we present a method for extracting the hippocampus from magnetic resonance imaging (MRI) data. Our method is based on a multi-scale shape representation using statistical learning in conjunction with spherical wavelets for shape representation. Indeed, the hippocampus shape information is statistically learned by the algorithm and is further utilized to extract a hippocampus from the given 3D MR image. Results are shown on data-sets provided by Brigham and Women's Hospital.

Keywords: Radiotherapy planning, Whole brain irradiation, Hippocampus extraction, Automatic segmentation.

1 Introduction

Brain metastases constitute the most common manifestation of cancer involving the central nervous system. Indeed, it has been estimated that over 170,000 cases of brain metastases are diagnosed annually in the United States alone [1]. During the past half-century, the cornerstone of treatment for this oncologic phenomenon has been whole brain irradiation (WBI) [2]. WBI has multiple salutary effects including rapid relief of neurological signs and symptoms as well as enhanced local control. Moreover, WBI represents an attractive clinical alternative because it can potentially suppress micrometastases that are undetectable with the current degree of resolution by MR imaging. Unfortunately, WBI may also engender side effects including memory deficits and decrements in quality of life [3].

Stereotactic irradiation is an option that has gained popularity in the management of brain metastases. Stereotactic irradiation is appealing because it is of short duration, uses multiple intersecting beams to augment the dose within the tumor volume and provides a rapid dose fall-off thereby optimizing the dosimetric gradient outside the tumor. This rationale allowed Chang *et al.* [4] to mount a trial comparing WBI to stereotactic techniques for patients suffering from brain metastases. The results of the trial indicated less decline in learning and memory function within the stereotactic arm. Yet, others have questioned whether an unacceptable subset of patients (among those treated focally) failed intracranially, albeit outside the radiosurgical treatment volumes.

In order for those receiving WBI to obtain the optimal intracranial control of disease while simultaneously preserving neurocognitive function, it is of strategic importance to recognize that the primary neurocognitive impairments are in memory. Since memory control is thought to be mediated by the hippocampus, attention has been turned to whole brain radiotherapeutic techniques that allow the sparing of the hippocampus. In order to be able to minimize dose deposition within the hippocampus, clinicians must be able to confidently identify that structure. The accuracy and consistency of segmentation can be improved by automating the process and including shape prior knowledge. Also, segmentation is a necessary step prior to registration; segmented structures provide landmarks and can be used to limit the number of free variables in deformable registration, which, in turn, leads to more accurate results.

However, due to typical problems such as imaging noise, low contrast, and partial volume effects, the "bottom-up" segmentation methods that solely utilize image information are not sufficient to obtain satisfying results. Accordingly, higher level shape knowledge is necessary to incorporate the necessary "top-down" information into the segmentation process for better performance. For example, Cootes *et al.* [5] used a parametric representation for the shapes and constructed a point-wise correspondence among all the training shapes. By assuming the distribution of corresponding points to be Gaussian, principle component analysis (PCA) was utilized to extract the eigenmodes of the variations. To extend such an approach to situations where the contour is implicitly represented, Leventon *et al.* performed PCA directly on level set functions [6]. Furthermore, Tsai *et al.* [7] employed PCA as the basis for the learned shape space with the subsequent image segmentation being carried out in a finite dimensional shape space setting. Finally, Rousson *et al.* [8] performed a similar principle component analysis, but then incorporated the shape constraint into a certain segmentation energy functional in order to guarantee that the final shape does not deviate too drastically from those in the learned shape space.

While the previous shape based segmentation methods achieved great success in solving segmentation problems where the image quality is suboptimal, they also have certain limitations. First, the shape variance is limited by the number of training cases. Indeed, when performing PCA, the covariance matrix is of dimension $N - 1$ when there are N training cases. Hence, the shape variance has the upper limit of $N - 1$. More importantly, the shape variances are encoded by the eigenvectors of the covariance matrix, which captures the most global modes of variations of the shape. In other words, small scale variance is very likely to be missed by such analysis because its energy is overwhelmed by a global variance. However, especially for medical applications, global variance may be due to inter-subject differences not related with pathology, whereas a little "bump" on the other hand, may be an important indicator of pathology.

To address such issues of "treating the shape as a whole," a natural way is to decompose a single shape into several "sub-shapes." For example, a surface may be decomposed into linear combination of sinusoidal functions using the Fourier transform [9]. Davatzikos *et al.* [10] treat the coordinates of the points on the curve as 1D functions and perform a wavelet decomposition on them. However, extending such a scheme to higher dimensions and more complicated topologies is not straightforward. More recently, Nain *et al.* extended this type of methodology to 3D spherical shapes [11]. In their work, the surface is parametrically represented and mapped to a sphere using a conformal mapping [12]. By doing so the surface becomes a 3D function defined on the sphere and is then decomposed using spherical wavelets. Nevertheless, not only the spherical topology is a limitation, the conformal step in mapping a surface to a sphere needs the selection of certain landmarks, which introduces a certain arbitrariness into the procedure.

Inspired by Nain *et al.* [11], in this work we perform the wavelet decomposition on the implicit representation of the shape. More explicitly, we use a modified version of the level set function to represent the contour, and the subsequent wavelet decomposition is performed on the resulting level set representation. This effectively removes any topological restrictions on the contour. A nice feature is that by removing the spherical mapping step, the correspondence amongst the various shapes is more naturally handled. Furthermore, with the shape being learned at various scales, the subsequent shape based segmentation is performed by utilizing the image information as well as respecting the shape prior knowledge. In doing so, we provide a general 3D shape-based segmentation method which handles non-spherical topologies and respects fine scale structure of the shape.

The remainder of the paper is organized as follows. In Section 2, we provide the detailed method of extracting hippocampus from MR images using a multi-scale shape prior. More explicitly, the shape is learned in a multi-scale fashion and after that, the learned shape information is utilized to extract the hippocampus. Next, the experiments and results are given in Section 3. Finally, future and ongoing work are discussed in Section 4.

2 Method and Materials

In this section, we provide the details of using the proposed multi-scale shape based segmentation technique to extract the hippocampus region from the MR brain imagery. In general, the method can be divided into an off-line learning step and an on-line segmentation step. In the learning step, a set of MR brain images (called *training images*) along with their segmented label maps for the hippocampus (and the amygdala) are provided for the statistical learning algorithm. The reason for including amygdala is because the visual boundary between hippocampus and amygdala is hardly seen in the images at hand. Hence, in order to obtain an accurate hippocampus segmentation, we first treat the hippocampus-amygdala as a single complex in the shape learning. After the shape of the complex is learned, such shape information is then utilized in extracting the complex from a new MR brain image (called the *raw image*). Next, the hippocampus and the amygdala are separated to form the final result. The details of each step will now be provided.

2.1 Shape Learning

A set of N training MR images: $J_1, J_2, \dots, J_N: \mathbb{R}^3 \to \mathbb{R}$ is provided for the algorithm, along with the corresponding label maps: $U_1, U_2, \dots, U_N: \mathbb{R}^3 \to \{0,1,2\}$ in which the hippocampus and amygdala are labeled by 1 and 2, respectively. For the reasons stated above, the labels of the amygdala and hippocampus are first combined as a single label. That is, we define $\widetilde{U}_i(\boldsymbol{x}) = 1$ if $U_i(\boldsymbol{x}) = 1\ or\ 2$. Then, the shapes of the hippocampus-amygdala complex (HAC) are aligned so that the variances captured by the subsequent statistical learning step are only due to shape, not pose. The aligned shapes are denoted by $V_1, \dots, V_N$. Next, in order to aid the statistical learning and the subsequent segmentation, we represent the aligned shapes as *sparse field level set* (SFLS) functions [13], and denote them by $L_1, \dots, L_N$.

At this point, we are able to formulate the multi-scale shape representation for the HAC's. In general, given a SFLS function L, its wavelet decomposition is written as:

$$W(\boldsymbol{u}) := DWT[L(\boldsymbol{x})] \tag{1}$$

where we assume the domain dimensions of L are all power of 2. Different wavelet bases can be chosen. In fact, in this study we tested the Daubechies wavelets $D4$, $D6$, and $D8$, but the final segmentation results were not significantly affected by the choice. Moreover, the wavelet coefficients are shrunk using the soft-thresholding proposed in [14] to remove the noise on the shape. By slight abuse of notation, the denoised wavelet coefficients are still denoted by W, which are stored in an 3D volumetric array the same size as L. Next, in order to represent the shape in multiple scales, the coefficients are decomposed into multiple bands corresponding to various scales. As is illustrated in Figure 1, the coefficients in each difference scale are concatenated to form a long column vector, called band, and all coefficients are rearranged into different bands.

Formally, we write this operation as:

$$\mathfrak{R}[W] = \{B_1, B_2, \dots, B_K\} \tag{2}$$

where K is the total number of bands. For example, in Figure 1 there are $K = 13$ bands in total. It is noted that later we will need to "inversely rearrange" the wavelet coefficients from the bands and this process is denoted as:

$$\mathfrak{R}^{-1}[\{B_1, B_2, \dots, B_K\}] = W. \tag{3}$$

Up to this point, the boundaries of the N hippocampus-amygdala complexes are now represented by N *band structures*:

$$\mathfrak{B}^{(i)} = \left\{B_1^{(i)}, B_2^{(i)}, \dots, B_K^{(i)}\right\}, \qquad i = 1, \dots, N. \tag{4}$$

A nice property of such multi-scale representation derives from the fact that the shape variance in different spatial locations and frequency scales are now separated in various bands.

In order to statistically learn the shape variances existing in the training shapes, we apply principle component analysis (PCA) to each band in order to compute the statistical model of the shape. More explicitly, the mean band structure is computed as:

$$\bar{\mathfrak{B}} = \{\bar{B}_1, \bar{B}_2, \dots, \bar{B}_K\} = \left\{\frac{1}{N}\sum_{i=1}^N B_1^{(i)}, \frac{1}{N}\sum_{i=1}^N B_2^{(i)}, \dots, \frac{1}{N}\sum_{i=1}^N B_K^{(i)}\right\}. \tag{5}$$

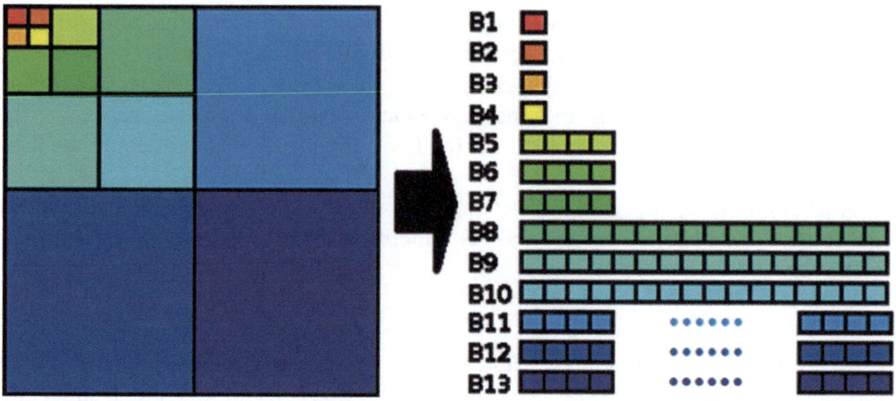

Fig. 1. Decomposition of wavelet coefficients into various bands. The band structure of wavelet coefficients for an image having size 16×16. The coefficients in difference scales are rearranged into different vectors (bands).

The mean shape can be considered as the shape without any variations, regardless of the scale of the variance. Because of this, it is expected that the mean shape computed this way should be consistent with the mean shape from a non-multi-scale shape representation. In fact, it is easy to show that such mean shape, defined as:

$$M := DWT^{-1}\left[\mathfrak{R}^{-1}[\bar{\mathfrak{B}}]\right] \tag{6}$$

is the same as the one computed in [6-7].

While the proposed shape representation does not affect the mean shape, the multi-scale property is beneficial when the shape variances are analyzed. In order to utilize such property, PCA is performed in each band to capture the shape variance. Formally, the mean of each band is first removed as $\tilde{B}_k^{(i)} = B_k^{(i)} - \bar{B}_k, i = 1,2, \dots, N$. The covariance matrix for the k-th band is then formed as:

$$C_k := \frac{1}{N}\sum_{i=1}^{N} \tilde{B}_k^{(i)}\left(\tilde{B}_k^{(i)}\right)^T \in \mathbb{R}^{l_k \times l_k} \tag{7}$$

where $l_k = \left|B_k^{(1)}\right|$ denotes the length of the k-th band. Furthermore, the eigen-decomposition of C_k gives:

$$C_k = Q_k \Lambda_k \mathrm{Q}_\mathrm{k}^\mathrm{T} = \left(\mathbf{q}_1^\mathrm{k}, \dots, \mathbf{q}_{\mathrm{p_k}}^\mathrm{k}\right)\begin{pmatrix} \lambda_1^\mathrm{k} & 0 & \ddots \\ 0 & \ddots & 0 \\ \ddots & 0 & \lambda_{\mathrm{p_k}}^\mathrm{k} \end{pmatrix}\mathrm{Q}_\mathrm{k}^\mathrm{T} \tag{8}$$

where $p_k := \min\left(N-1, l_k\right)$ and the vectors $\boldsymbol{q}_j^k, j = 1, \dots, p_k$ consist of the bases for this band.

In contrast to the PCA scheme used in [5-7], where for N training shapes there are at most $N-1$ bases for the whole learned shape space, here we have bases in each band to record the shape variance at the given scale. Specifically, since the band length is varying, in the k-th band, the number of bases does not exceed p_k. For an 2D image with size 16×16 shown in Figure 1, it can be seen that starting from the band B_{11}, the band lengths $l_k (k \geq 11)$ are greater or equal than 64. So for a moderate size of training shapes, except for the first few bands, we have $p_k = N-1$. Hence the total number of bases is: $\sum_{k=1}^{K} p_k \approx K(N-1)$. In comparison to $N-1$, this greatly enhances the capability of capturing the shape variances at multiple scales.

In order to construct the allowable shape space which restricts the subsequent segmentation, we assume the eigenvalues obtained in Equation (7) are ordered decreasingly. Furthermore, the first Y eigenmodes, $\boldsymbol{q}_1^k, \dots, \boldsymbol{q}_Y^k$ are picked to record the shape variance in this band (scale). In practice, we fixed Y at $\min(6, l_k)$. Moreover, for every $k = 1, \dots, K$, the shape space in that band, $\mathcal{S}_k$ is defined as:

$$\mathcal{S}_k := \left\{ (\boldsymbol{q}_1^k, \dots, \boldsymbol{q}_Y^k) \begin{pmatrix} \sqrt{\lambda_1^k} & 0 & \ddots \\ 0 & \ddots & 0 \\ \ddots & 0 & \sqrt{\lambda_Y^k} \end{pmatrix} \begin{pmatrix} a_1^k \\ \vdots \\ a_Y^k \end{pmatrix}, \; a_i \in [-6,6], \forall\, i \right\}. \tag{9}$$

The above formula defines a hyper- rectangular shaped region centered at the mean in each band, with axes of length 6 times the standard deviation. Equivalently, this imposes a uniformly distributed prior for the shapes. In most of the literature, however, the Gaussian distribution is used. We should note here that in order to maximize the shape likelihood in the Gaussian case, there is always a tendency of evolving towards the mean shape which eliminates the individual shape variance. This is not always a desired property. On the other hand, the uniform distribution does preserve the shape variances. Finally, we note that the "shape space" above is in fact topologically closed but not a mathematical vector space, because it is not closed under the addition.

2.2 *Shape Based Segmentation*

In this section, we describe our multi-scale shape representation based segmentation strategy with its application on extracting the hippocampus from MR images. Briefly, given the image to be segmented, first we use a registration method to estimate the target location in the new image. Next, a data driven segmentation scheme is applied to extract the initial estimate of the shape. Thirdly, the multi-scale shape constraints are used to regularize the previously obtained shape in the learned shape space. The second and the third steps are iterated until arriving at a steady state, which provides the final segmentation. We now give the details.

2.3 Shape Initialization

Given a new image $I: \mathbb{R}^3 \rightarrow \mathbb{R}$, the objective of the present section is to provide an initial shape, which is to be used as the starting point for the subsequent segmentation process presented in the next section. To this end, an atlas-based segmentation approach is adopted [15]. That is, all of the training images are deformably registered to the new image by maximizing mutual information [16-17]. By using mutual information, the proposed method is capable of handling images of multiple modalities. With the optimal deformable transformations being computed and denoted by $T_1, \dots, T_N$, the initial shape $P(\boldsymbol{x})$ is then defined using the "majority rule:"

$$P(\boldsymbol{x}) = \begin{cases} 1, & if \sum_{i=1}^{N} \tilde{U}_i(T_i(\boldsymbol{x})) \geq \frac{N}{2} \\ 0, & otherwise. \end{cases} \tag{10}$$

This is to be used as the initialization for the subsequent segmentation for the new image I.

2.4 Data Driven Segmentation

Given the initial shape P, the proposed segmentation method is an alternating data-driven and multi-scale shape-based process that starts with a data-driven (not shape guided) algorithm. Specifically, we use the energy functional defined in [18]: Denote the initial contour, the level set representation of the initial shape $P(\boldsymbol{x})$ in the previous section, as ϕ. Then, the contour evolves to minimize the energy functional defined as:

$$E(\phi) := \alpha \int \delta\phi(\boldsymbol{x})|\nabla\phi(\boldsymbol{x})|\mathrm{d}\boldsymbol{x} + \int \delta\phi(\boldsymbol{x}) \int \chi(x,y) \Big(\mathcal{H}(\phi(\boldsymbol{y}))(I(\boldsymbol{y}) - u_x) + \big(1 - \mathcal{H}(\phi(\boldsymbol{y}))\big)(I(\boldsymbol{y}) - v_x)\Big) \,\mathrm{d}\boldsymbol{y}\mathrm{d}\boldsymbol{x}. \tag{11}$$

In Equation (11), $\mathcal{H}$ is the smoothed Heaviside function defined in [18], and δ is the derivative of $\mathcal{H}$. Moreover, the χ is the characteristic function for the r-neighborhood around $\boldsymbol{x}$, and α is a positive weighting parameter. One can compute the first variation of E with respect to ϕ and we obtain the corresponding gradient descent flow for ϕ; see [18] for all the details. The evolution of ϕ is only performed for a few steps (not until convergence), and the resulting ϕ is then fed into the shape-filtering processing described in the next section.

2.5 Shape Filtering with Shape Prior

Since physical imagery is never perfect (noise, etc.), the data driven segmentation ϕ performed in the previous section usually suffers from under-segmentation, leakage, or a combination of both. To solve this problem, the extracted shape in the previous step is processed through a "shape filter" so as to fall into the admissible shape space. First of all, the current contour ϕ is registered to the mean shape M in Equation (6) by minimizing the following energy over all $T \in \boldsymbol{T}$, the space of 3D similarity transformations:

$$E_M(T) := \int \left(\phi\left(T(\boldsymbol{x})\right) - M(\boldsymbol{x})\right)^2 \mathrm{d}\boldsymbol{x}. \tag{12}$$

The optimal similarity transformation is denoted by T^* and the registered contour $\phi\left(T^*(\boldsymbol{x})\right)$ is denoted by $\psi(\boldsymbol{x})$. Moreover, $\psi(\boldsymbol{x})$ is sampled on the same discrete domain as $M(\boldsymbol{x})$, so that they have the same dimensions. Then, the wavelet transform is applied to $\psi(\boldsymbol{x})$ to obtain its band structure as:

$$\mathfrak{B}^{\psi} = \left\{B_1^{\psi}, B_2^{\psi}, \dots, B_K^{\psi}\right\}. \tag{13}$$

In a manner similar to that of the learning part of the algorithm, the mean bands are subtracted from $\mathfrak{B}^{\psi}$ to obtain $\widetilde{\mathfrak{B}}^{\psi}$ as:

$$\widetilde{\mathfrak{B}}^{\psi} = \left\{B_k^{\psi} - \bar{B}_k : k = 1, \dots, K\right\} =: \left\{\tilde{B}_k^{\psi} : k = 1, \dots, K\right\}. \tag{14}$$

Next, in each of the K bands, the band $\tilde{B}_k^{\psi}$ is projected to the learned band space. Without loss of generality, we can take the k-th band. Accordingly, we project $\tilde{B}_k^{\psi}$ to the bases vectors in the band $\left\{\boldsymbol{q}_j^k : j = 1, \dots, p_k\right\}$ and we obtain the coefficients:

$$A_k = \left(\frac{\tilde{B}_k^{\psi} \cdot \boldsymbol{q}_j^k}{\sqrt{\lambda_j^k}} : j = 1, \dots, p_k\right)^T =: \left(a_1, a_2, \dots, a_{p_k}\right)^T \in \mathbb{R}^{p_k}. \tag{15}$$

In order to restrict the given novel space to the learned shape space, the novel shape is mapped to its closest shape in the allowable shape space learned in the ***Shape Learning*** part of the scheme. Accordingly, we adjust each A_k as:

$$\tilde{A}_k := \begin{cases} A_k, & if \ \|A_k\| \leq 6 \\ \dfrac{6A_k}{\|A_k\|}, & if \ \|A_k\| > 6\,. \end{cases} \tag{16}$$

After that, the adjusted band structure $\widehat{\mathfrak{B}}^{\psi}$ is obtained as:

$$\widehat{\mathfrak{B}}^{\psi} := \left\{ \left(\boldsymbol{q}_1^k, \dots, \boldsymbol{q}_Y^k\right) \begin{pmatrix} \sqrt{\lambda_1^k} & 0 & \ddots \\ 0 & \ddots & 0 \\ \ddots & 0 & \sqrt{\lambda_Y^k} \end{pmatrix} \cdot \tilde{A}_k + \bar{B}_k : \forall k \right\}, \tag{17}$$

and the filtered shape $\tilde{\psi}$ is then obtained as:

$$\tilde{\psi} = DWT^{-1}\left[R^{-1}\left[\widehat{\mathfrak{B}}^{\psi}\right]\right]. \tag{18}$$

Finally, the filtered shape is transformed back to the new image domain using the inverse of the similarity transformation T^* recorded above. Thus $\tilde{\psi}((T^*)^{-1}(\boldsymbol{x}))$ is then used as the initial segmentation contour for the next-round data driven segmentation. These two steps alternate until convergence. Usually, it only takes two or three iterations for convergence on the data sets we used.

2.6 Separation of the Hippocampus and Amygdala

The previous segmentation provides a binary image $H(\boldsymbol{x}): \mathbb{R}^3 \to \{0,1\}$ representing the hippocampus-amygdala complex, not the hippocampus. In order to form the final segmentation of the hippocampus, the complex needs to be separated. Such separation is achieved in this section using a scheme similar to described in ***Shape initialization*** above. More explicitly, the binary label images $\tilde{U}_i(\boldsymbol{x})$'s are deformably registered to the newly obtained binary image $H(\boldsymbol{x})$. The optimal deformable transformations are denoted by $\mathcal{T}_1, \dots, \mathcal{T}_N: \mathbb{R}^3 \to \mathbb{R}^3$. Then, the final label image $F(\boldsymbol{x})$ is defined as:

$$F(\boldsymbol{x}) := \begin{cases} 1, & if \ \sum_{i=1}^{N} U_i(\mathcal{T}_i(\boldsymbol{x})) \in [N, 2N] \\ 0, & if \ \sum_{i=1}^{N} U_i(\mathcal{T}_i(\boldsymbol{x})) \in [0, N)\,. \end{cases} \tag{19}$$

3 Results

In this section, we report the results of 24 spoiled gradient-recalled (SPGR) coronal-plane volumetric T1-weighted images acquired in the Brigham and Women's Hospital, Harvard Medical School. The ages of the subjects range from 28 to 55 with mean 43.6 and standard deviation 7. The image resolution is $0.938 \times 0.938 \times 1.5mm^3$. Fortunately, we have all the images segmented by experts at the Brigham and Women's Hospital. In order to make full use of all the data, we choose the leave-one-out strategy to perform the shape learning and segmentation. That is, 23 out of 24 images as well as their segmented shapes are used for learning and the image left is segmented and the result is compared with its expert hand segmentation. The segmentation is fully-automatic, so that no human interaction is needed. On an Intel 3.0GHz quad-core machine with 8G memory, segmenting one volumetric image takes about 1 hour.

3.1 Hippocampus and Amygdala Segmentation Results

One important component of the proposed method is to first segment the hippocampus-amygdala complex as a whole, and then separate the two components. In doing so, we successfully solve the segmentation difficulty caused by the fact that the image intensity patterns in amygdala and hippocampus are extremely similar, if not identical. Therefore, in Figure 2 we present the segmented surfaces for both hippocampus and amygdala.

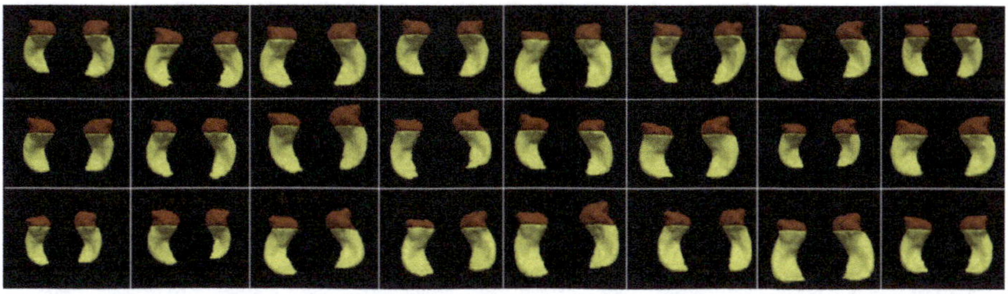

Fig. 2. The segmented surfaces of both hippocampus and amygdala for all the 24 subjects.

Moreover, it is noted that further qualitative and quantitative analysis in the subsequent sections are targeted only on the hippocampus since that is the main objective of present work.

3.2 Segmentation Results Shown by Slices

From here on, we only perform the analysis on the hippocampus. In Figure 3 we present the segmentation of one randomly chosen volume.

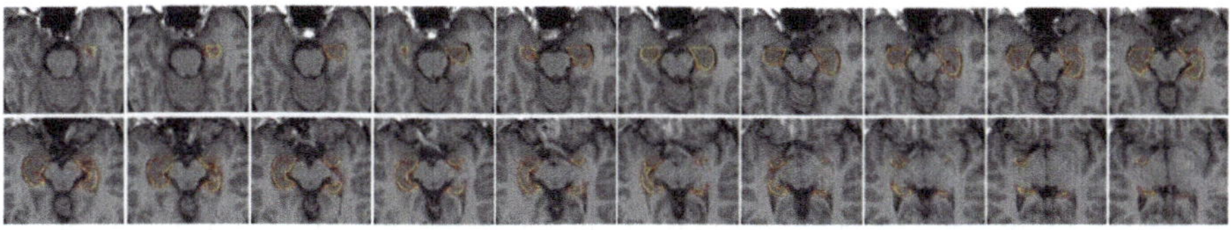

Fig. 3. Axial slices of a segmentation result. The yellow contours are generated by the proposed automatic segmentation algorithm, while the red contours are drawn manually by an expert in BWH.

In this result, the hippocampus boundary generated by the proposed automatic algorithm is colored in yellow whereas the manual boundary, drawn by an expert radiologist at the Brigham and Women's Hospital, is colored in red and used as the reference. As can be observed from the comparison, the two sets of contours match very well in all the slices. Due to space constraints, only one case is shown here. However, it is noted that all the 24 experiments work consistently, and the detailed statistical analysis is given below.

3.3 Distance on Mesh

In order to demonstrate the performance of the algorithm in 3D, in this section we show the surfaces of the segmented hippocampi. In Figure 4, all the 24 hippocampal surfaces are shown.

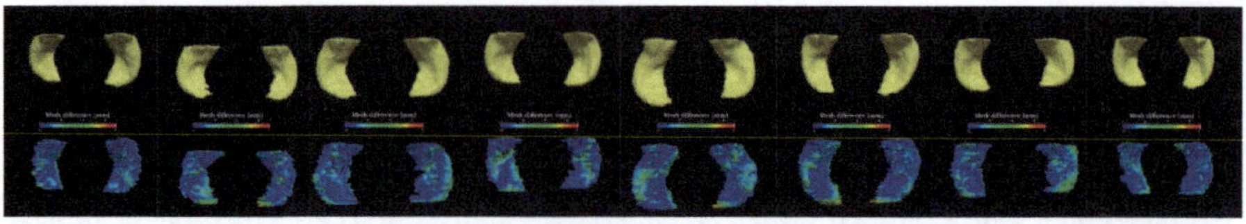

(a)

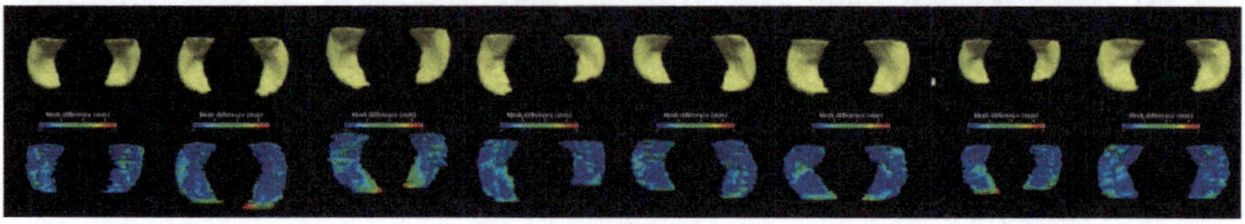

(b)

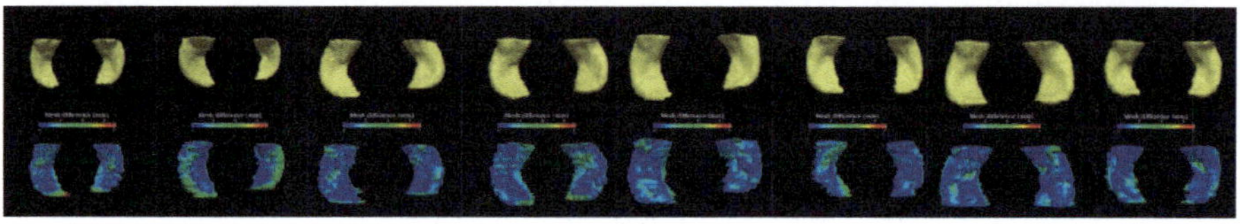

(c)

Fig. 4. The surfaces of all the 24 testing hippocampi. The yellow ones are generated by the proposed method and the ones below are the corresponding manual results. On the manual surfaces, the color shows the difference between the automatic and manual surfaces.

From Figure 4, we can observe that the manual delineated hippocampal surfaces have many discontinuities. This is mainly due to the fact that the contours are traced out slice-by-slice. Thus, although in that process the image information in the adjacent slices is utilized, nevertheless, the resulting surfaces still have a jagged appearance. On the other hand, the proposed 3D segmentation method achieves better continuity and smoothness in all the three spatial dimensions, and the jagged appearance is avoided.

Moreover, in order to quantify the differences between the two sets of results, the difference maps are computed: Given any point on the manual surface, the difference map value is the closest distance to the automatically generated surface. Altogether, they form a scalar function defined on the manual surface. The difference maps are illustrated as the colors on the manual surfaces: Given any location on the surface, the red/blue color means that there exists a relatively large/small difference between the surface generated by expert and the algorithm, respectively. As can be seen in Figure 4, except at few isolated regions, the automatic results are very consistent with the expert manual results.

3.4 Further Quantitative Analysis

To further analyze the results quantitatively, similar to [19], the following parameters were measured: (1) left/right hippocampus volume; (2) volume ratio (automatic/manual) for the left and right hippocampus (ideal value is 1); (3) volume overlap ratio (VOR), i.e., the ratio of the volume of intersection to the volume of union (ideal value is 1); and (4) correctly delineated volume (CDV), i.e., the ratio of the volume of intersection to the manual volume (ideal value is 100%). In addition to these, the Hausdorff distance (HD), which represents the largest distance between the corresponding automatic and manual segmentation surfaces, is also computed. All the results are given in Table 1. It is seen that the algorithm performed correctly and consistently in all the cases.

Table 1. Quantitative analysis for the segmentation results.

		Volume		Volume ratio*	VOR†	HD†† (mm)	CDV§
		Manual(cm^3)	Auto(cm^3)				
Left	Mean	3.80	3.93	1.04	0.81	4.89	86%
	STD	0.64	0.64	0.10	0.048	1.31	4.3%
Right	Mean	3.59	3.78	1.05	0.82	4.97	88%
	STD	0.57	0.62	0.11	0.045	1.27	6.4%

* Volume ratio is computed as automatic/manual.

† VOR: Volume overlap ratio is computed as the ratio of the volume of intersection to the volume of union. Optimal value is 1.0.

†† HD: Hausdorff distance is defined as the largest distance between the corresponding automatic and manual segmentation surfaces. Optimal value is 0.0.

§ CDV: Correctly delineated volume is defiend as the ratio of the volume of intersection to the manual volume. Optimal value is 100%.

4 Discussion

In the present work, we described a multi-scale shape based framework to automatically extract the hippocampus from MR brain imagery. First, we proposed a multi-scale representation for the shape using the wavelet transform. In doing so, we were able to take advantage of the multi-scale property of the wavelet representation. In particular, in learning the shapes, a small-scale shape variance was not overwhelmed by large ones. Moreover, in the segmentation step, we combined the atlas-based scheme and the active contour methods to provide an automatic and robust segmentation algorithm to extract the hippocampus. Finally, the algorithm was tested on 24 T1 MR brain images and the results were analyzed.

Further work includes investigating the number of training shapes needed for learning as well as the atlas-based initialization, in order to achieve optimal efficiency and accuracy. Moreover, in the data driven segmentation step, the local image mean intensity is used as the criteria in separating the hippocampus with its surrounding tissue. However, since the training images are available, one promising alternative would be learning the image information of hippocampus region in the training images, and use such information to drive the segmentation.

This methodology can now be incorporated into existing treatment planning approaches. In this manner, the objective of assessing the merits of whole brain irradiation with hippocampal avoidance can be pursued. With time, the neuro-oncologic community will be able to determine whether control of microscopic disease can be achieved without significant compromise of neurocognitive function.

Acknowledgments. This work was supported in part by grants from NSF, AFOSR, ARO, as well as by a grant from NIH (NAC P41 RR-13218) through Brigham and Women's Hospital. This work is part of the National Alliance for Medical Image Computing (NAMIC), funded by the National Institutes of Health through the NIH Roadmap for Medical Research, Grant U54 EB005149. Information on the National Centers for Biomedical Computing can be obtained from http://nihroadmap.nih.gov/bioinformatics.

References

1. Patchell, R.: The management of brain metastases. Cancer Treat Rev. 29(6), 533–540 (2003)
2. Knisely, J.: Focused attention on brain metastases. Lancet Oncol. 10, 1037–1044 (2009)
3. Aoyama, H., Tago, M., Kato, N., et al.: Neurocognitive function of patients with brain metastasis who received either whole brain radiotherapy plus stereotactic radiosurgery or radiosurgery alone. In: Int. J. Radiat Oncol, vol. 68(5), pp. 1388–1395 (2007)
4. Chang, E., Wefel, J., Hess, K., et al.: Neurocognition in patients with brain metastases treated with radiosurgery or radiosurgery plus whole-brain irradiation: a randomised controlled trial. Lancet Oncology 10(11), 1037–1044 (2009)
5. Cootes, T., Taylor, C., Cooper, D., et al.: Active shape models-their training and application. Comput. Vis. Image Und. 61(1), 38–59 (1995)
6. Leventon, M., Grimson, W., Faugeras, O.: Statistical shape influence in geodesic active contours. In: IEEE Conference on CVPR (2000)
7. Tsai, A., Yezzi, A., Wells, W., et al.: A shape-based approach to the segmentation of medical imagery using level sets. IEEE T. Med. Imaging 22(2), 137–154 (2003)
8. Rousson, M., Paragios, N., Deriche, R.: Implicit active shape models for 3D segmentation in MR imaging. In: Barillot, C., Haynor, D.R., Hellier, P. (eds.) MICCAI 2004. LNCS, vol. 3216, pp. 209–216. Springer, Heidelberg (2004)
9. Staib, L., Duncan, J.: Model-based deformable surface finding for medical images. IEEE T. Med. Imaging 15(5), 720–731 (1996)

10. Davatzikos, C., Tao, X., Shen, D.: Hierarchical active shape models, using the wavelet transform. IEEE T. Med. Imaging 22(3), 414–423 (2003)
11. Nain, D., Haker, S., Bobick, A., et al.: Multiscale 3-D Shape Representation and Segmentation Using Spherical Wavelets. IEEE T. Med. Imaging 26(4), 598–618 (2007)
12. Angenent, S., Haker, S., Tannenbaum, A., et al.: On the Laplace-Beltrami operator and brain surface flattening. IEEE T. Med. Imaging 18(8), 700–711 (1999)
13. Whitaker, R.: A level-set approach to 3D reconstruction from range data. Int. J. Comput. Vision 29(3), 231 (1998)
14. Donoho, D.: De-noising by soft-thresholding. IEEE T. Inform. Theory 41(3), 613–627 (1995)
15. Rohlfing, T., Brandt, R., Menzel, R., et al.: Quo Vadis, Atlas-Based Segmentation? Handbook of Biomedical Image Analysis. In: Registration models, p. 435 (2005)
16. Wells, W., Viola, P., Atsumi, H., et al.: Multi-modal volume registration by maximization of mutual information. Med. Image Anal. 1(1), 35–51 (1996)
17. Rueckert, D., Sonoda, L., Hayes, C., et al.: Nonrigid registration using free-form deformations: application to breast MR images. IEEE T. Med. Imaging 18(8), 712–721 (1999)
18. Lankton, S., Tannenbaum, A.: Localizing Region-Based Active Contours. IEEE T. Image Process 17(11), 2029–2039 (2008)
19. Pasquier, D., Lacornerie, T., Vermandel, M., et al.: Automatic segmentation of pelvic structures from magnetic resonance images for prostate cancer radiotherapy. Int. J. Radiat. Oncol. 68(2), 592–600 (2007)

Rigid Registration of 3D Ultrasound and MRI: Comparing Two Approaches on Nine Tumor Cases

Laurence Mercier, Vladimir Fonov, Rolando F. Del Maestro, Kevin Petrecca, Lasse R. Østergaard, and D. Louis Collins

Abstract. We present a new technique for registering ultrasound and magnetic resonance (MR) images in the context of neurosurgery. It involves generating a pseudo-ultrasound (pseudo-US) from a segmented MR image and uses cross-correlation as the cost function to register with ultrasound. The algorithm's performance is compared to a state-of-the-art technique that uses a median filtered MR images to register with a Gaussian-blurred ultrasound using normalized mutual information (NMI). The two methods are tested on nine tumor cases, including both high- and low-grade gliomas. The pseudo-US method yielded significantly better alignment average than that obtained by NMI ($p = 0.0009$). If one case where NMI failed is excluded, the mean distance obtained by the pseudo-US approach (2.6 mm) is slightly lower than the one obtained by NMI (2.8mm), but not significantly so ($p = 0.16$). We conclude that the pseudo-US method is more robust for these cases.

1 Introduction

Numerous papers have described the use of ultrasound [1-4] and, more recently, tracked ultrasound [5-7], in brain tumor surgery. For brain tumors, fluorescence

Laurence Mercier · Vladimir Fonov · D. Louis Collins
McConnell Brain Imaging Centre, Montreal Neurological Institute,
McGill University, Montreal, Canada
e-mail: `laurence@bic.mni.mcgill.ca`,
`{vladimir.fonov, louis.collins}@mcgill.ca`

Rolando F. Del Maestro · Kevin Petrecca
Brain Tumour Research Center, McGill University, Montreal, Canada
e-mail: `{kevin.petrecca, rolando.delmaestro}@mcgill.ca`

Lasse R. Østergaard
Dept. Of Health Science and Technology, Aalborg University, Denmark
e-mail: `lasse@hst.aau.dk`

J. Angeles et al. (Eds.): Brain, Body and Machine, AISC 83, pp. 33–43.
springerlink.com

imaging [8] and intraoperative MR imaging (iMRI) [9, 10] are other popular intraoperative modalities. Fluorescence is only useful for visualization of certain tumor types and is currently limited to imaging surfaces. The advantage of ultrasound over iMRI is its significantly shorter preparation and acquisition time, while its drawbacks include a limited field of view and the potential difficulty of interpretation. Tracked ultrasound also costs a fraction of the price of an iMRI system. Very few papers have formally compared these modalities. When comparing ultrasound with a high-field iMRI during 26 brain tumor cases, Gerganov et al. [11] found that the image quality before the resection to be similar in both modalities. Ultrasound is therefore useful to visualize the tumor and its boundaries before surgery. As such it can be used for direct navigation during the whole procedure until the end, where it can be used to assess residual tumor or detect a hemorrhage beyond the resection borders. Ultrasound can also be used to estimate brain shift in order to update intraoperative images or to feed in a deformation model.

In image-guided neurosurgery, the preoperative images are converted to patient space through a patient-image registration before the surgery begins. When tracked ultrasound is used, the position of the ultrasound image is also known in patient space because of a tracker fixed on the probe combined with the probe's calibration. Bucholz et al. [12] have shown however that even before opening the dura, a mismatch of varying magnitude is observed between the tracked ultrasound images and preoperative images. This mismatch is evidence of errors in the conversion to patient space of the ultrasound and MRI, and potential brain shift. Hence, the initial ultrasound-to-MRI alignment needs a correction (or re-registration) to be accurately co-aligned.

Registering ultrasound and MR images is challenging because of their dissimilar appearance. In addition, brain tumor cases are among the most complicated pathologies since their intensity, contrast, texture and shape can differ in both modalities. The challenge that we face in the operating room at our institution is to register gadolinium-enhanced T1-weighted (gad-T1w) MRI and B-mode ultrasound images, since the gad-T1w MRI is often the only high resolution scan included in the clinical tumor protocol. To address this specific registration issue, a few automatic registration techniques have been proposed and tested on neurosurgical cases. Roche et al. [13] evaluated a rigid registration technique that uses correlation ratio to match the ultrasound intensity with both the MRI's intensity and gradient on 1 tumor case. Arbel et al. [14, 15] converted an MRI into a pseudo-ultrasound (pseudo-US) to facilitate its registration with ultrasound using cross-correlation on 3 surgical cases including 1 tumor case. Ji et al. [16] pre-processed the MRI and ultrasound before registering them with NMI. They showed a significant reduction of the tumor boundary misalignment in 6 tumor cases. Methods for registering MRI sequences other than a gad-T1w or methods using Doppler ultrasound of the brain have also been proposed [17, 18], however these methods were not applicable in our clinical context.

The main goal of this paper is to compare two rigid registration techniques for aligning preoperative gad-T1w MR images from brain tumor cases with intraoperative 3D B-mode ultrasounds taken on the dura. Given Ji et al.'s [16] success with the NMI approach on six tumor cases, we implemented a similar technique. However, the six tumors selected for Ji et al.'s study included only enhancing (bright) tumors on the gad-T1w MR image with a sharp edge on the ultrasound (mostly meningiomas and metastasis). As one of our objectives was to test the registration methods with more complex tumor cases, the 9 tumor cases in this study include both low-grade (non-enhancing) and high-grade gliomas (often very heterogeneous). The NMI approach is compared to a new pseudo-US technique, with a completely different implementation of the idea first published in [14, 15]. The pseudo-US is generated by segmenting structures from a gad-T1w MRI and remapping their intensities to resemble those of ultrasound.

2 Materials and Methods

2.1 Clinical Data

Ten consecutive cases with B-mode ultrasound acquisition on the dura were considered for analysis in this study. One case was dropped as the reference was moved by the surgeon between the patient-MR registration and the ultrasound acquisition. This error was evident as the ultrasound and MRI did not overlap at all. The remaining nine cases consisted in five high-grade gliomas, three low-grade gliomas and one dysembryoplastic neuroepitheal tumor (DNT). Four tumors were located in the left temporal lobe, three in the left frontal lobe, one in the left occipital lobe, and one in the right parietal lobe. The acquisitions were done on the dura, after the bone flap is removed, but before the resection begins. Ultrasound images were acquired with an HDI 5000 (ATL/Philips, USA) using a P7-4 MHz phased array transducer at a depth setting of 6.5 cm and 8 cm. The 2D images were reconstructed in a 3D volume using a pixel-based method [19], which maps each pixel from each 2D image into a regular 3D grid. Typically, sweeps contain between 200 and 600 frames and are reconstructed at a voxel size of 0.3 x 0.3 x 0.3 mm^3.

2.2 Pseudo-ultrasound Generation

The use of a pseudo-US is inspired from [14, 15], however the implementation here is completely different and more sophisticated than the original method. Differences include: mapping white matter and grey matter to different intensities and adding vessels, sulci and a bright rim to mimic the bone interface. The resulting pseudo-US is very similar to the actual ultrasound, except for the speckle, which does not need to be simulated for registration purposes.

The preoperative MRI that is converted into a pseudo-US is a global gad-T1w MRI (TE = 8 ms, TR = 23 ms, flip angle = 20°, 1mm slice thickness, 0.5x0.5 mm^2 pixel size) acquired on a 1.5 Tesla GE Signa EXCITE® (General Electric, Milwaukee, WI, USA). The MRI is first filtered using a nonlocal means denoising technique [20] with a beta value of 0.5. This step produces a visually subtle denoising, but helps to produce smoother automatic segmentations. The denoised

MRI is then used as input to a discrete classifier [21] in which it is registered to the average nonlinear ICBM152 model [22] to take advantage of standard tissue priors for classification. The classifier first finds a brain mask [23] and then classifies each voxel in the mask as cerebrospinal fluid (CSF), grey matter and white matter. This classifier does not have a lesion class and the tumor was therefore segmented semi-automatically using a publicly available minc tool called Display (packages.bic.mni.mcgill.ca). Any other segmentation tool like the popular ITK-SNAP (www.itksnap.org) can be used. Fully automatic tumor segmentation is challenging as the gliomas in this study are either highly heterogeneous (high-grade gliomas) or almost the same intensity as grey matter (low-grade gliomas).

The next step is the automatic vessel segmentation which is done by: 1) Multi-scale vessel enhancement filtering [24] to intensify vessel-like (tubular) structures. 2) Iterative fuzzy c-means clustering [25] to divide the vessel-filtered data into 3 clusters of which the two higher intensity clusters are selected. In summary, this first segmentation stage classifies each voxel in the denoised MRI as: background, CSF, white matter, grey matter, vessel or tumor. The sulci category is obtained by automatically removing the ventricles from the CSF mask. Finally, to simulate the appearance of the inner border of the skull, a gradient filter is applied on the blurred (FWHM = 2 mm) brain mask.

The next stage consists in assigning an intensity value to each of the classes found above. All intensities are mapped in the 0-100 interval as follows: background (0), CSF (4), white matter (12), grey matter (32), tumor (50), sulci (70) and vessels (100). Even if tumors can be highly heterogeneous on gad-T1w MRIs, they are simplified as a homogenous mass because no direct intensity correspondence was found between MR and ultrasound for the different tumor components. For example, necrosis is dark on MR, but on ultrasound its appearance varies depending if the necrosis is in a liquid, solid or mixed state. Figure 1 shows images from each stage of image processing: starting from the gad-T1w MRI (Fig. 1A), to the classified MRI with the tumor (Fig. 1B), adding sulcal CSF (Fig. 1C), the skull (Fig. 1D) and the vessels (Fig. 1E). In order to smooth the shape of the objective function, the final pseudo-US is blurred using a Gaussian kernel with a 2mm FWHM (Fig. 1F).

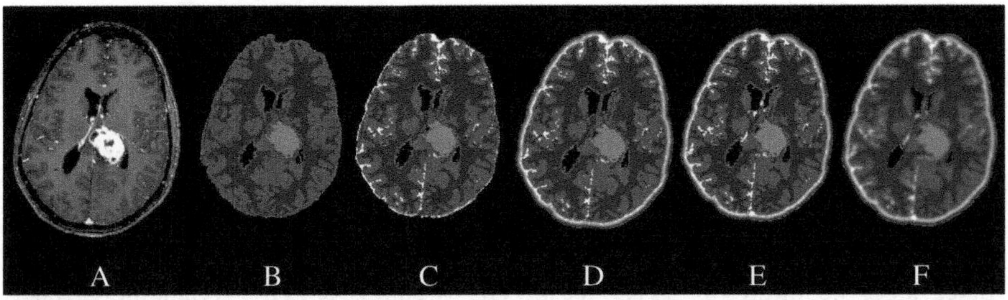

Fig. 1. Processing pipeline for the creation of a pseudo-ultrasound. Initial gadolinium-enhanced T1-weighted MRI (A), intermediate steps (B-E), final blurred pseudo-ultrasound (F).

2.3 Normalized Mutual Information Technique

An NMI-based method inspired from the work of Ji et al. [3] was implemented with two important differences: 1) a downhill simplex optimization algorithm is used instead of Powell's algorithm and 2) the reconstructed 3D ultrasound acquisitions described above were used as opposed to a few 2D ultrasound images as in the original paper. These changes were made because of constraints in the current version (0.99.6) of the registration program (minctracc) available with the MINC tools. However, using 3D ultrasound instead of 2D ultrasound was suggested as a potential improvement by the authors. On the other hand, using a simplex instead of Powell's optimization could make our implementation less robust. Aside from these differences, similar preprocessing steps were applied to the MRI and the ultrasound. However, because of the important differences from the version presented by Ji et al, the method will simply be referred to as "NMI" in this paper. The NMI approach was chosen because its robustness has been proven in the clinical context with brain tumor images and at the time of this study, was the state-of-the-art technique for that specific application.

2.4 Rigid Body Registration

All rigid body registrations are computed with a MINC tool called *minctracc*. Many initial simplex sizes were tested, ranging from 1.25 to 20. A simplex of 2 was used to initialize the downhill simplex algorithm because it produced the best results on our 9 data sets with both registration techniques. In minctracc, a simplex of 2 corresponds to an initial translation of 2mm in each direction and an initial rotation of 2.0 degrees around each axis. The pseudo-US technique is evaluated using the cross-correlation (xcorr) cost functions. The source image is the MRI converted into a pseudo-US as previously described. Because the pseudo-US is only defined in the brain region, a brain mask is used when registering. The target is a 3D ultrasound on which a median filter (radius = 2mm) was applied in order to remove some speckle and obtain a smoother objective function. The purpose of the ultrasound's mask is to exclude the black background which contains no data. The background comes from the reconstruction stage in which the 2D ultrasound images are masked to remove all annotations, scale bars, etc. around the image. For the NMI approach, the source is a median filtered MRI (radius = 2mm). Initially, the same brain mask was used for pseudo-US and NMI technique, but results were more robust when the MRI was not masked for the NMI technique. The target is blurred by a Gaussian kernel (FWHM = 5mm). The ultrasound mask for NMI is smaller: it only contains areas of higher ultrasound intensities. This mask is obtained by removing the background of the blurred ultrasound with Otsu's method [26] and then dilating the foreground mask by a 5mm radius kernel, similar to Ji et al. Fig. 2 below shows an example of the original ultrasound and MRI (left), the processed MRI and ultrasound for the NMI technique (center) and for registering with the pseudo-US approach (right).

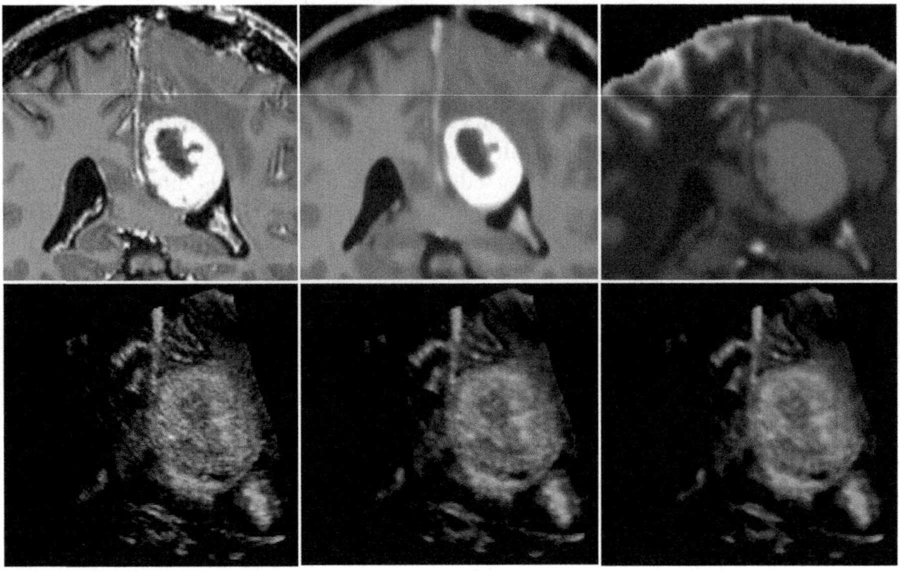

Fig. 2. Image processing for case #4. : Left: Original gadolinium-enhanced T1-weighted MRI (top) and original ultrasound (bottom). Center: processed MRI (top) and ultrasound (bottom) for the NMI registration techniques. Right: processed MRI (top) and ultrasound (bottom) for pseudo-US registration technique.

2.5 Registration Validation

To evaluate the magnitude of the ultrasound-MRI misalignment, it was necessary to estimate an ideal ultrasound-MR registration as a baseline for comparison. As no gold standard is available for clinical data, our bronze standard [27] was defined by selecting corresponding anatomical landmarks in both modalities. For each of the nine cases, a senior neurosurgeon (R.D.M) chose seven to eleven corresponding anatomical features in both the ultrasound and MR volumes. The distance between the corresponding points enables a quantitative estimation of the accuracy of the alignment between both modalities. As reported in other similar papers [17, 28], the mean distance between the ultrasound and MRI was computed as the mean absolute Euclidian distance between the corresponding points in both modalities.

3 Results

Table 1 below shows the quantitative results for each clinical case. The estimated initial misalignment (after the standard clinical skin-landmark registration) between the ultrasound and MR in this series is: 6.1±3.4mm; after bronze standard re-registration, 1.3±0.9; after NMI re-registration, 4.9±5.8 and after a pseudo-US re-registration, 3.0±1.7.

Table 1. Quantitative results for each clinical case. The values given in the table are the Euclidian mean distance in mm with its associated standard deviation (SD).

No	Initial misalignment	Bronze standard re-registration	NMI re-registration	Pseudo-US re-registration
1	7.8 ± 0.9	0.9 ± 0.9	2.8 ± 0.9	1.7 ± 0.7
2	10.5 ± 0.5	1.1 ± 0.5	20.0 ± 1.9	5.8 ± 1.4
3	10.9 ± 1.8	1.7 ± 0.9	3.3 ± 1.6	2.8 ± 1.4
4	2.6 ± 0.9	1.4 ± 0.8	2.0 ± 1.0	2.4 ± 1.0
5	3.2 ± 0.6	1.2 ± 0.9	1.8 ± 0.8	2.1 ± 0.7
6	2.4 ± 1.8	1.4 ± 1.0	4.6 ± 2.5	4.0 ± 1.8
7	3.6 ± 1.8	1.4 ± 1.0	4.6 ± 2.5	4.0 ± 1.8
8	6.9 ± 2.6	1.4 ± 1.2	2.7 ± 1.6	3.7 ± 1.7
9	7.3 ± 0.8	1.0 ± 0.5	2.7 ± 0.7	1.4 ± 0.8
mean	6.1 ± 3.4	1.3 ± 0.9	4.9 ± 5.8	3.0 ± 1.7

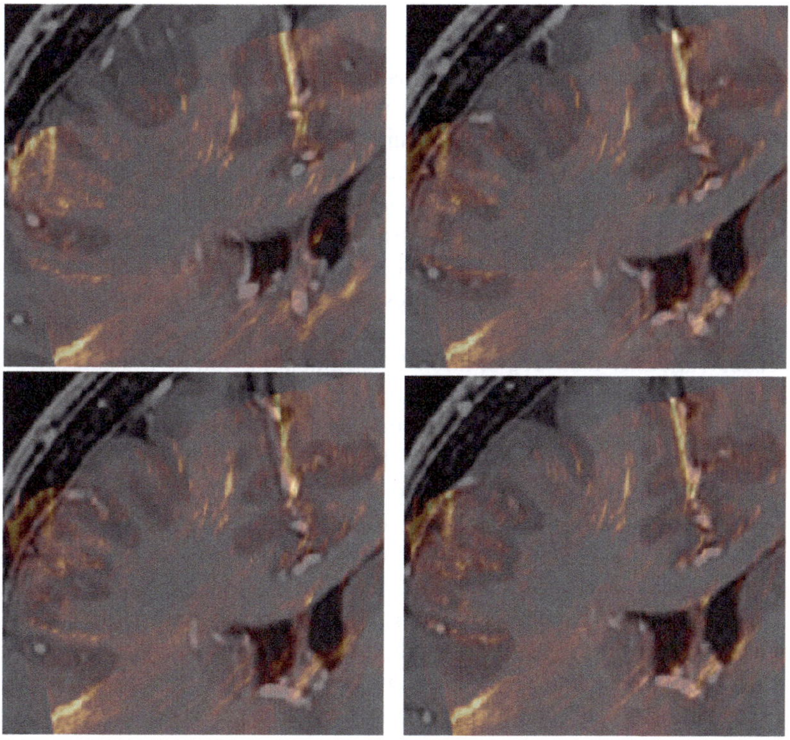

Fig. 3. Visual results for case #9. The 2D reslice of the 3D ultrasound (in orange tones) is overlaid on a gad-T1w (in grayscale). The ultrasound plane is constant and the associated MR varies depending on the registration technique. top left: initial alignment, top right: after the bronze standard re-registration, bottom left: after the NMI re-registration and bottom right: after the pseudo-US re-registration. Notice how the falx and sulci are better aligned in bottom right (pseudo-US) compared to bottom left (NMI).

A multi-factorial analysis of variance (ANOVA) analysis was performed on all 75 point pairs (7 to 11 corresponding points for each of the 9 patients). The ANOVA was computed with both repeated measures for the 4 registration methods and a between group design for the 9 patients. It revealed the existence of a significant difference between methods ($F(9,3) = 383.3$, $p < 0.001$). To determine which of the methods were different, multiple pair-wise t-tests were done using a Bonferroni correction (significance level = 0.05/6 = 0.0083). The statistical tests revealed that the pseudo-US significantly improved the alignment of ultrasound and MRI when compared to the initial state ($p < 0.0001$) and when compared to the NMI technique ($p = 0.0009$). This is in great part due to case #2, a large very heterogeneous high-grade glioma, for which the NMI technique completely failed (distance > 20mm). If this outlier is removed and the statistics recomputed, the pseudo-US registration (mean = 2.6) is not significantly different from NMI (mean = 2.8) ($N = 66$, $p = 0.16$). Figure 3 shows a visual example of the results for case #9.

4 Discussion

The estimated initial ultrasound-MRI misalignment in this series (6.1 ± 3.4 mm) is similar to what is found in the literature for neurosurgical cases. Reinertsen et al. [18] measured the co-registration accuracy on 5 patients using 10 manually selected landmarks from Doppler ultrasound on the dura and the preoperative MR angiography after registration and found a mean initial misalignment of 7.34 ± 2.1 mm. Lindseth et al. [29] reported a mismatch of 5.33 ± 2.43 mm on 12 cases. Hartov et al. [30] found a mean of 4.01 ± 2.5 mm on 35 cases. It was however measured by comparing tumor boundaries after applying a 3D translation only (no rotations).

Ji et al. validated their registration by segmenting tumor boundaries in both modalities and computing the 2D distance between certain edges of the tumors before and after registration. This validation technique does not necessarily reflect the quality of the fit in the rest of the image, which is important if the intraoperative ultrasound's purpose is to update the patient-image registration at the beginning of surgery. Its disadvantage also lies in its 2D nature. The attempt here was to find a quantitative, 3D measure covering more than only the tumor region. However the bronze standard chosen here is not perfect as the manual point selection is associated to a certain imprecision and might not detect subtle differences between the registration methods. Figure 3 shows an example of a case where the pseudo-US technique (D) produced a slightly better alignment than the bronze standard (B). This can be seen by looking at the sulci on the left side of the image. Our next step will be to add the qualitative assessment of a neurosurgeon (visual ranking and ratings) to detect potential subtle differences between techniques.

The NMI approach has the obvious advantage of requiring minimal prior knowledge, but might be less robust in cases of large heterogeneous tumors. In these cases the tumor can occupy the majority of the ultrasound image and have different internal divisions. The pseudo-US approach appears more robust to the differences in tumor appearance in both modalities, but requires a careful tumor segmentation as input. The classifier used in this study to segment the brain in order to generate the pseudo-US was very robust, even with the highly atypical brains and the very distorted anatomy of some of the cases included here.

5 Conclusions

In conclusion, we have presented a new MRI–ultrasound registration technique, which is based on converting a gad-T1w MR image into a pseudo-US image. We compare its performance with that of a state-of-the-art NMI-based method similar to the one described by Ji et al. [16]. Both registration techniques were tested on nine tumor cases, which included both low grade and high-grade gliomas. Both techniques significantly improved the MRI-US registration when compared to the initial alignment, and the pseudo-US method yielded significantly better alignment on average than the NMI. If we exclude from the comparison the one case where NMI failed, the mean distance obtained by the pseudo-US approach is slightly smaller than that obtained by NMI, but not significantly. We conclude that the pseudo-US method is more robust for these cases.

References

[1] van Velthoven, V.: Intraoperative ultrasound imaging: comparison of pathomorphological findings in US versus CT, MRI and intraoperative findings. Acta Neurochir Suppl. 85, 95–99 (2003)

[2] Woydt, M., et al.: Correlation of intra-operative ultrasound with histopathologic findings after tumour resection in supratentorial gliomas. A method to improve gross total tumour resection. Acta Neurochir (Wien) 138, 1391–1398 (1996)

[3] LeRoux, P.D., et al.: A comparison between preoperative magnetic resonance and intraoperative ultrasound tumor volumes and margins. J. Clin. Ultrasound 22, 29–36 (1994)

[4] Erdogan, N., et al.: Ultrasound guidance in intracranial tumor resection: correlation with postoperative magnetic resonance findings. Acta Radiol. 46, 743–749 (2005)

[5] Tirakotai, W., et al.: A novel platform for image-guided ultrasound. Neurosurgery 58, 710–718 (2006) (discussion 710–718)

[6] Unsgaard, G., et al.: Ability of navigated 3D ultrasound to delineate gliomas and metastases–comparison of image interpretations with histopathology. Acta Neurochir (Wien) 147, 1259–1269 (2005) (discussion 1269)

[7] Unsgaard, G., et al.: Neuronavigation by intraoperative three-dimensional ultrasound: initial experience during brain tumor resection. Neurosurgery 50, 804–812 (2002) (discussion 812)

[8] Stummer, W., et al.: Fluorescence-guided surgery with 5-aminolevulinic acid for resection of malignant glioma: a randomised controlled multicentre phase III trial. Lancet. Oncol. 7, 392–401 (2006)

[9] Hatiboglu, M.A., et al.: Impact of Intraoperative High-Field Magnetic Resonance Imaging Guidance on Glioma Surgery: A Prospective Volumetric Analysis. Neurosurgery 64, 1073–1081 (2009)

[10] Nimsky, C., et al.: Volumetric assessment of glioma removal by intraoperative high-field magnetic resonance imaging. Neurosurgery 55, 358–370 (2004) (discussion 370-371)

[11] Gerganov, V.M., et al.: Reliability of intraoperative high-resolution 2D ultrasound as an alternative to high-field strength MR imaging for tumor resection control: a prospective comparative study. J. Neurosurg. 111, 512–519 (2009)

[12] Bucholz, R.D., et al.: The correction of stereotactic inaccuracy caused by brain shift using an intraoperative ultrasound device. In: Troccaz, J., Mösges, R., Grimson, W.E.L. (eds.) CVRMed-MRCAS 1997, CVRMed 1997, and MRCAS 1997. LNCS, vol. 1205, pp. 459–466. Springer, Heidelberg (1997)

[13] Roche, A., et al.: Rigid registration of 3-D ultrasound with MR images: a new approach combining intensity and gradient information. IEEE Trans. Med. Imaging 20, 1038–1049 (2001)

[14] Arbel, T., et al.: Automatic non-linear MRI-ultrasound registration for the correction of intra-operative brain deformations. Comput. Aided Surg. 9, 123–136 (2004)

[15] Arbel, T., et al.: Automatic Non-linear MRI-Ultrasound Registration for the Correction of Intra-operative Brain Deformations. In: Niessen, W.J., Viergever, M.A. (eds.) MICCAI 2001. LNCS, vol. 2208, pp. 913–922. Springer, Heidelberg (2001)

[16] Ji, S., et al.: Mutual-information-based image to patient re-registration using intraoperative ultrasound in image-guided neurosurgery. Med. Phys. 35, 4612–4624 (2008)

[17] Coupe, P., et al.: A Probabilistic Objective Function For 3d Rigid Registration Of Intraoperative Us And Preoperative MR Brain Images. Presented at the ISBI (2007)

[18] Reinertsen, I., et al.: Validation of vessel-based registration for correction of brain shift. Med. Image Anal. 11, 374–388 (2007)

[19] Solberg, O.V., et al.: Freehand 3d Ultrasound Reconstruction Algorithms-a Review. Ultrasound Med. Biol., May 17 (2007)

[20] Coupe, P., et al.: An optimized blockwise nonlocal means denoising filter for 3-D magnetic resonance images. IEEE Trans. Med. Imaging 27, 425–441 (2008)

[21] Zijdenbos, A.P., et al.: Morphometric analysis of white matter lesions in MR images: method and validation. IEEE Trans. Med. Imaging 13, 716–724 (1994)

[22] Janke, A.L., et al.: MNI and Talairach space: Everything you wanted to know but were afraid to ask. In: HMB, Florence, Italy (2006)

[23] MacDonald, D., et al.: Automated 3-D extraction of inner and outer surfaces of cerebral cortex from MRI. Neuroimage 12, 340–356 (2000)

[24] Frangi, A.F., et al.: Multiscale vessel enhancement filtering. In: Wells, W.M., Colchester, A.C.F., Delp, S.L., et al. (eds.) MICCAI 1998. LNCS, vol. 1496, pp. 130–137. Springer, Heidelberg (1998)

[25] Bezdek, J.C., et al.: Review of MR image segmentation techniques using pattern recognition. Med. Phys. 20, 1033–1048 (1993)

[26] Otsu, N.: A threshold selection method from gray-level histograms. IEEE Trans. Sys. Man. Cyber. 9, 62–66 (1979)

[27] Jannin, P., et al.: Validation of medical image processing in image-guided therapy. IEEE Trans. Med. Imaging 21, 1445–1449 (2002)

[28] Reinertsen, I., et al.: Clinical validation of vessel-based registration for correction of brain-shift. Medical Image Analysis 11, 673–684 (2007)

[29] Lindseth, F., et al.: Image fusion of Ultrasound and MRI as an Aid for assessing Anatomical Shifts and improving Overview and Interpretation in Ultrasound Guided Neurosurgery. In: CARS 2001, pp. 247–252 (2001)

[30] Hartov, A., et al.: A comparative analysis of coregistered ultrasound and magnetic resonance imaging in neurosurgery. Neurosurgery 62, 91–99 (2008)

A New Approach to Virtual Mirroring for View Integration

Carmen E. Au and James J. Clark

Abstract. In this paper, we present an improvement to an image integration technique called Virtual Mirroring (VM). The basis for this approach is formed on the notion that humans are familiar with seeing multiple views of a scene through reflections off various reflective surfaces. VM overlays secondary views of a scene onto virtual mirrors that can be seen by a primary view of the scene, thereby providing more views of a given scene to the observer. We refer to the cameras that capture the secondary and primary views as Secondary (SC) and Principal (PC) Cameras respectively. In our previous approach, the camera setup was constrained to a specific relative position between the cameras. Moreover, the virtual mirrors we created did not respect the laws of reflection. In this new approach, we assume an object plane on the scene in order to be able to determine the world position of each SC image point and subsequently, the corresponding points of reflection through a virtual mirror. By using an object plane, the camera placement is not constrained and the virtual mirrors are created using equations that adhere to the laws of reflection.

1 Introduction

In the recent past, a number of image integration techniques have been developed. The purpose of image integration is to combine a collection of images of a common scene into a single composite image. One such group of image integration techniques is called image stitching. In image stitching, images are pieced together into a single larger image, such as panoramas. There are different approaches to solving the image stitching problem. Brown and Lowe categorize the methods into two main categories: direct and feature-based methods [1]. In direct methods [2, 3], the camera parameters are iteratively estimated by minimizing the difference in pixel intensity in the area of overlap. In feature-based methods [4, 5, 6, 7], a correspondence

Carmen E. Au · James J. Clark
McGill University, 3480 University St., Montreal, Quebec, Canada
e-mail: {au, clark}@cim.mcgill.ca

J. Angeles et al. (Eds.): Brain, Body and Machine, AISC 83, pp. 45–55.
springerlink.com

Fig. 1. Example "paparazzi" images with virtual mirroring: (a) image from principal camera, (b) image from secondary camera, and (c) image from principal camera with image from secondary camera overlaid onto a virtual mirror. Image created using an image editing program.

between points in the images is established using either low-level features, such as corners and edges, or high level features such as parts of objects. Most of these methods, however, are not robust to changes in illumination, zoom and rotation. In order to develop techniques that are in fact robust to the aforementioned changes, a type of features called invariant features was developed [8, 9]. Most notably, Brown and Lowe developed an image stitching technique which uses Lowe's Scale Invariant Feature Transform (SIFT) [1, 10]. For these image stitching techniques, most of the image acquisition is done by translating the camera in a direction parallel to the image plane. In fact all of the aforementioned image stitching techniques excluding one notable exception [8], require that the camera be moved along the same direction as the image plane. Since all the acquired images are either on, or approximately on the same plane, the resulting panoramic images created do not create the feeling of looking into a three-dimensional environment. Even in Baumberg's approach [8], which allows for viewpoint changes and rotations, the maximum amount of rotation before the results begin to degrade is around 51 degrees. In many real world applications, either the cameras are mobile or fixed to maximize the coverage of a scene as with surveillance scenarios; it would be remiss to limit the amount of allowable viewpoint change.

Beyond image stitching, a few papers have described forays into dynamic panorama mosaicing, which is the creation of panoramic videos. Many of the techniques employ systems that have either specialized panoramic cameras such as catadioptric cameras, or use a mirroring system to generate the panoramas [11, 12, 13, 14, 15, 16, 17]. Agarwala *et al.* created Panoramic Video Textures (PVTs) with a single panning video camera [18]. Likewise, Rav-Acha *et al.* [19] created Dynamosaics using a single camera. In both cases, since only a portion of the scene has been imaged at any given time, the panorama must contain images from different time slices in the video sequence. Thus, the PVTs and Dynamosaics are limited to scenes with repetitive motion, such as swaying trees or waterfalls. Neither technique can create dynamic panoramas for most real world scenes, which have highly aperiodic motion.

In this paper, we describe an improvement to a technique called Virtual Mirroring (VM) for integrating images [20]. This approach has the advantage over previous efforts of being able to integrate images without being constrained by the viewpoint. In many real world situations, cameras are not always placed in such a way that viewpoints overlap. On the contrary, situations such as video surveillance, paparazzi photographers or fans taking pictures of a rock concert are all scenarios in which cameras are not necessarily placed to allow for view integration with the aforementioned approaches. VM does in fact account for such real world situations. The name is derived from the use of virtual mirrors to integrate these views with widely differing viewpoints, into a single image. In Figure 1 we show an example, created using an image editing program, of how VM works. Two images, (a) and (b), are combined by overlaying one image (b) onto a virtual mirror that was placed in (a). The resulting composite image is shown in (c). This paper extends our earlier work by introducing an approach that allows for dynamic cameras. Moreover, our previous effort did not employ true mirror reflection properties to determine the pixel mapping between images.

1.1 Virtual Mirroring Technique

The goal of this research is to integrate images acquired by several camera phones of a common scene, and to display them on a 2D display for remote viewing. To

Fig. 2. The coordinate frame for CE, {CE}, is shown in relation to the coordinate frame for CS, {CS}, which is aligned with the world coordinate frame, {W}

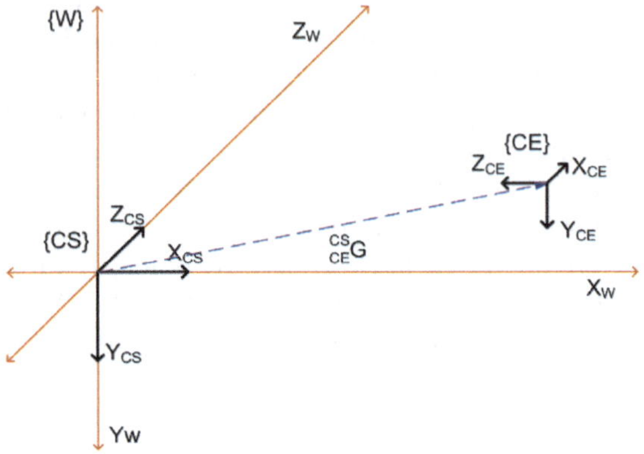

that end, existing techniques, do not provide a natural way of compositing images from vastly differing viewpoints. For example, if two people, who are standing on either side of the room, are photographing the same person, one person might capture an image of the front side of the person, while the other person would capture an image of the back side of the person. The question then begs as to how to combine images from opposing sides of a person into one image in a natural way. The answer lies in the rather commonplace phenomenon of reflections. In everyday interactions, humans are accustom to seeing two views of objects that are placed between themselves and a mirror, or any reflective surface. For that matter, the first view is observed by looking at the object directly, and the second view is observed by looking at the object's reflection. Because of the familiarity that humans have with observing multiple views via mirrors, we adopted a virtual mirroring technique, whereby virtual mirrors are placed on the scene to provide additional views to the viewer.

The notion of displaying images captured by a camera on a virtual mirror is a recent idea. In 2002, François *et al.* explored the idea of creating a handheld virtual mirror, which combined a camera and a flat LCD screen to form a mirror [21]. One major issue they tackled was that of reconciling the different perspectives between the camera and the user. Kojima *et al.* developed an augmented reality system called *NaviView* [22]. This system is designed to serve as an aid to drivers for when they encounter "blind" intersections, which are intersections with limited or no visibility. A virtual mirror is overlaid onto a Windshield Display device that can show images on a windshield. Images from surveillance cameras, which generally have a better view of the intersection from which the driver is "blinded", are displayed on the virtual mirrors. By doing so, drivers are provided another view that is otherwise unavailable to them without the use of a virtual mirror. Bichlmeier *et al.* also made use of virtual mirrors when they designed an augmented reality system for situations that require *in-situ* visualization [23]. They likened their system to a dentist's mirror, where the mirror is used to visualize the oral cavity. Dentists move small handheld mirrors around to establish better views while the patient mostly remains still. Similarly, Bichlmeier *et al.*'s virtual handheld mirrors could be moved around, so that the 3D objects being visualized could remain stationary. This method for visualizing

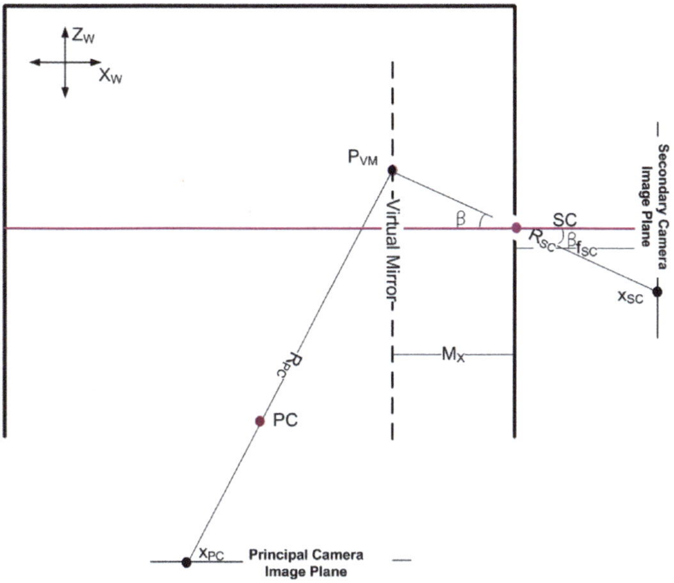

Fig. 3. Diagram of Pixel Mapping. Points projected onto the X-Z plane, to determine the horizontal components of the pixel mapping. The point x_{SC} on the secondary camera (SC) image plane, is mapped to the point P_{VM} on the virtual mirror plane, which is in turn mapped to the point x_{PC} on the principal camera (PC) image plane.

Fig. 4. Composite images using old approach to VM.

objects would be useful in many applications, particularly in computer-aided surgeries, where the patients should not be moved.

In our 2008 paper, we describe how we adopted the use of virtual mirroring for view integration. While the technique can be extended to multiple cameras, we describe how to map two cameras to one another. We call these cameras the "principal" (PC) and "secondary" (SC) camera. The SC is mapped to a virtual mirror and then correspondingly to the PC. If we refer again to Figure 1, (a) and (b) are images from the PC and SC respectively. The virtual mirror can be seen by the PC, and SC is mapped onto it. In the resulting composite image, both the images from PC and SC are visible. In our initial implementation of the technique, videos taken from two calibrated cameras in McGill University's *Shared Reality Environment* [24] (SRE) were used. In Figure 2, the coordinate frame for CE, {CE}, is shown in relation to the coordinate frame for CS, {CS}, which is aligned with the world coordinate frame, {W}. The transformation which describes {CE} in relation to {CS} is defined as:

$$ {}^{CS}_{CE}G = \begin{bmatrix} {}^{CS}_{CE}R & {}^{CS}_{CE}T \\ 0\ \ 0\ \ 0 & 1 \end{bmatrix} \tag{1} $$

As can be seen in Figure 3, the unique setup of the SRE resulted in the optical axes of the cameras being orthogonal to one another, allowing for simple trigonometric functions to solve for the camera mapping. The resulting composite images are shown in Figure 4.

There are two major disadvantages to the described approach to VM. Firstly, the approach is only valid due to the unique setup which caused the optical axes of the cameras to be orthogonal to one another. The resulting equations cannot be extended to other camera setups. Secondly, while the approach is called Virtual Mirroring, the equations used to solve for the camera mapping do not adhere to mirror properties, and thus, the resulting "mirrors" are not true mirrors. In the following section, we describe our new approach for VM and how it is an improvement on the previous approach.

2 New VM Approach

Our new VM approach corrects the two major flaws of the previous approach. The new approach is no longer constrained by a specific camera setup and the created virtual mirrors respect the laws of reflection.

2.1 Making a True Virtual Mirror

The first issue we addressed was the issue that the previous approach did not adhere to true mirror properties, and the virtual mirror is not in fact a mirror. The law of reflection states that angle of incidence is equal to the angle of reflection. Thus, in order to overlay the SC image onto the virtual mirror in such a way that upheld the law of reflection, we would require *a priori* knowledge of the depth of all the points in the scene imaged by the SC. By so doing, the angle of incidence can be determined. Rather than attempting to determine such a depth map, since it is a non-trivial task, we assume that all points lie on a single object plane. We expect that any distortion that may result from this assumption will be negligible to the human observer.

The placement of the object plane is determined by the user, with two constraints. To minimize distortion, the object plane should be as close to the middle of the scene as possible. Secondly, in order to simplify the mapping from one camera to the other considerably, we make the assumption that the object plane is parallel to the SC image plane. In so doing, we can take advantage of the fact that the spherical angles between a point on the SC image plane and SC are equal to the angles to the corresponding point on the object plane and SC. Figure 5 shows an example of how the object plane would be placed for the given configuration of SC and PC.

As with the previous approach, the virtual mirror is placed close to the SC and also parallel to the SC image plane. Thus, when the SC image is mapped onto the virtual mirror, the resulting image appears as though the mirror is reflecting the contents of the scene. If instead the mirror were placed elsewhere, the mapping can still be computed, but the reflection would not appear natural.

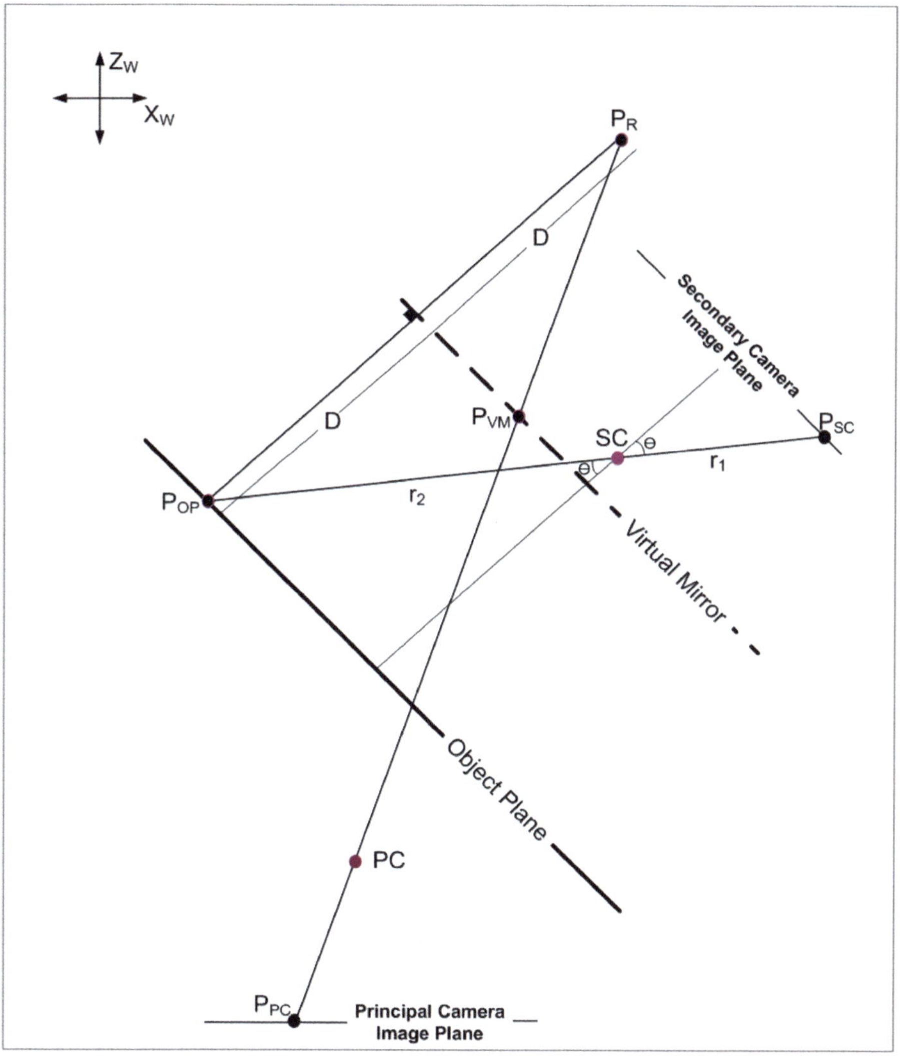

Fig. 5. Diagram for new approach. P_{SC} and P_{PC} are points on the SC and PC image planes respectively. P_{OP} and P_R are the corresponding points on object plane and the reflection to that point.

2.2 Algorithm for Generalized Technique

In this section, we present the algorithm for our new approach, which does not constrain the SC and PC placement to an orthogonal one as the previous approach did.

For each point on the SC image plane, P_{SC}, find the corresponding point on the object plane, P_{OP}. Placing the camera, SC, at the origin, (0,0,0), the spherical coordinates of P_{SC} can be determined using the following equations:

$$\theta = \arctan(\sqrt{x^2+y^2}/z) \tag{2}$$

$$\phi = \arctan(y/z) \tag{3}$$

$$r = \sqrt{x^2+y^2+z^2} \tag{4}$$

where x, y, and z are the Cartesian coordinates of P_{SC}. A ray passing through SC and P_{SC} intersects the object plane at point, P_{OP}. Since the object plane is assumed to be parallel to the SC image plane, the spherical angles of P_{OP} with respect to the origin are the same as those of P_{SC}. The radial distance, r, can be determined using similar triangles.

$$r_2 = r_1 * distOP/z \tag{5}$$

where r_2 is the radial distance of P_{OP}, r_1 is the radial distance of P_{SC}, and distOP is the assigned distance of the object plane from SC. The spherical coordinates of P_{OP} can easily be converted back into Cartesian coordinates using the following equations:

$$P_{OP}(x) = r2 * \sin(\theta) * \cos(\phi) \tag{6}$$

$$P_{OP}(y) = r2 * \sin(\theta) * \sin(\phi) \tag{7}$$

$$P_{OP}(z) = r2 * \cos(\theta) \tag{8}$$

For each point on the object plane, P_{OP}, the corresponding point on the PC image plane, P_{PC} can be found. Thus far, the coordinates are in the SC coordinate frame, to facilitate computation, we convert the coordinates into the PC coordinate frame by multiplying by the transformation matrix, G, which maps the two coordinates.

$$P_{OP_1} = G * P_{OP_0} \tag{9}$$

Once we have the point on the object plane represented in PC coordinate frame, we can calculate the corresponding point of reflection. The virtual mirror is user defined as being parallel to the SC image plane. The distance of P_{OP} from VM can be calculated.

$$Dm = (Pop \bullet normVM)/\left|normVM\right| \tag{10}$$

where *normVM* is a normal to the virtual mirror. Given the distance, D, of P_{OP} from the virtual mirror, the corresponding point of reflection, P_R, can be determined. Knowing that a point and its reflection are equidistant from a plane mirror, and the ray between P_R and P_{OP} are in the same direction as the normal to the virtual mirror, *normVM*, P_R can be found with the following equation:

$$P_R = 2 * D * (-normVM) + P_{OP} \tag{11}$$

The corresponding point on the PC image plane, P_{PC}, is the intersection of the ray from P_R to PC and the PC image plane. Using the equation to determine a line-plane intersection we can determine P_{PC}.

$$P_{PC} = (normPC \bullet (P - PC))/(normPC \bullet (PC - P_R) \tag{12}$$

where *normPC* is a normal to the PC image plane and P is a point on that plane. We note that at this point, P_{PC} is in world coordinates, and we would like to represent them in image coordinates. To do so, we convert to spherical coordinates using Equations 2, 3, and 4 and finally back into PC image coordinates using Equations 6 and 7.

3 Resulting Composite Images for New VM Approach

In Figure 6 we show the result of integrating three images of the same scene. The images were acquired by a camera stage with three camera each placed at the vertex of an equilateral triangle, and each facing the center of the triangle. As can be seen, three views of the same scene can be viewed quite naturally in the same composite

Fig. 6. Integrated view of three views of the same scene using new VM approach

image. The placement of the virtual mirrors are such that scale and relative orientation are maintained. Thus, the resulting composite image inherently provide the viewer with information about the camera positions, and subsequently, likely with more information about the scene than if the three images were not integrated. Finally, we note that if there is any distortion as a result of using an assumed object plane, it is not perceptible in the resulting image.

4 Limitations and Future Works

The new VM approach allows for the integration of images from unconstrained cameras with widely differing viewpoints. Future work should include a method for handling occlusion. In this new approach, we have not improved upon the rudimentary background subtraction method used in the previous approach. Moreover, the current approach assumes the transformation between SC and PC are known. Therefore, future works can include an auto-calibration step, which determines the relative position of the input cameras.

References

1. Lowe, D.: Distinctive image features from scale-invariant keypoints. International Journal of Computer Vision 20, 91–110 (2004)
2. Beis, J., Lowe, D.: Shape indexing using approximate nearest-neighbor search in highdimensional spaces. In: Conference Computer Vision Pattern Recognition, pp. 1000–1006 (1997)
3. Irani, M., Anandan, P.: About direct methods. In: Proceedings of the International Workshop on Vision Algorithms, London, UK, pp. 267–277. Springer, Heidelberg (2000)
4. Capel, D., Zisserman, A.: Automated mosaicing with super-resolution zoom. In: IEEE Computer Society Conference on Computer Vision and Pattern Recognition (1998)
5. Debevec, P., Yu, Y., Boshokov, G.: Efficient view-dependent image-based rendering with projective texture-mapping. Technical Report CSD-98-1003, University of California at Berkeley, 20 (1998)
6. Kanazawa, Y., Kanatani, K.: Image mosaicing by stratified matching. In: Workshop on Statistical Methods in Video Processing (2002)
7. Rankov, V., Locke, R.J., Edens, R.J., Barber, P.R., Vojnovic, B.: An algorithm for image stitching and blending. In: Proceedings of SPIE, vol. 5701, pp. 190–199 (2005)
8. Baumberg, A.: Reliable feature matching across widely separated views. In: IEEE Computer Society Conference on Computer Vision and Pattern Recognition, pp. 774–781 (2000)
9. Schmid, C., Mohr, R.: Local grayvalue invariants for image retrieval. IEEE Transactions on Pattern Analysis and Machine Intelligence 19(5), 530–535 (1997)
10. Lowe, D.G.: Object recognition from local scale-invariant features. In: International Conference on Computer Vision, pp. 1150–1157 (1999)
11. Chahl, J.S., Srinivasan, M.V.: Reflective surfaces for panoramic imaging. Applied Optics 36, 8275–8285 (1997)

12. Chahl, J.S., Srinivasan, M.V.: A complete panoramic vision system, incorporating imaging, ranging, and three dimensional navigation. In: OMNIVIS 2000: Proceedings of the IEEE Workshop on Omnidirectional Vision, Washington, DC, USA, p. 104. IEEE Computer Society, Los Alamitos (2000)
13. Ishiguro, H., Ng, K.C., Capella, R., Trivedi, M.M.: Omnidirectional image-based modeling: three approaches to approximated plenoptic representations. Machine Vision Applications 14(2), 94–102 (2003)
14. Nayar, S.K.: Catadioptric omnidirectional camera. In: IEEE Computer Society Conference on Computer Vision and Pattern Recognition, Washington, DC, USA, p. 482. IEEE Computer Society, Los Alamitos (1997)
15. Svoboda, T., Pajdla, T.: Epipolar geometry for central catadioptric cameras. International Journal on Computer Vision 49(1), 23–37 (2002)
16. Tan, K., Hua, H., Ahuja, N.: Multiview panoramic cameras using mirror pyramids. IEEE Transaction on Pattern Analysis and Machine Intelligence 26(7), 941–946 (2004)
17. Zimmermann, S., Kuban, D.P.: A video pan/tilt magnify/rotate system with no moving parts for motion simulation and telepresence. In: Proceedings of the 3rd Annual Virtual Reality Conference and Exhibition on VR becomes a Business, Westport, CT, USA, pp. 181–189. Meckler Corporation (1993)
18. Agarwala, A., Zheng, K.C., Pal, C., Agrawala, M., Cohen, M., Curless, B., Salesin, D., Szeliski, R.: Panoramic video textures. ACM Transactions on Graphics 24(3), 821–827 (2005)
19. Rav-Acha, A., Pritch, Y., Lischinski, D., Peleg, S.: Dynamosaics: Video mosaics with non-chronological time. In: IEEE Computer Society Conference on Computer Vision and Pattern Recognition, Washington, DC, USA, pp. 58–65. IEEE Computer Society, Los Alamitos (2005)
20. Au, C.E., Clark, J.J.: Multiple view integration and display using virtual mirrors. In: 5th Canadian Conference on Computer and Robot Vision, Ottawa, Canada, pp. 286–293. CRV (2008)
21. François, A., Kang, E., Malesci, U.: A handheld virtual mirror. In: SIGGRAPH 2002: ACM SIGGRAPH 2002 Conference Abstracts and Applications, pp. 140–140. ACM, New York (2002)
22. Kojima, K., Sato, A., Taya, F., Kameda, Y., Ohta, Y.: Naviview: visual assistance by virtual mirrors at blind intersection. In: Proceedings of Intelligent Transportation Systems, pp. 592–597. IEEE, Los Alamitos (September 2005)
23. Bichlmeier, C., Sielhorst, T., Navab, N.: The tangible virtual mirror: New visualization paradigm for navigated surgery. In: AMIARCS - The Tangible Virtual Mirror: New Visualization Paradigm for Navigated Surgery, Copenhagen, Denmark, MICCAI Society (October 2006)
24. Cooperstock, J.: Shared reality environment, `http://www.cim.mcgill.ca/sre/`

Designing a Metric for the Difference between Gaussian Densities

Karim T. Abou–Moustafa, Fernando De La Torre, and Frank P. Ferrie

Abstract. Measuring the difference between two multivariate Gaussians is central to statistics and machine learning. Traditional measures based on the Bhattacharyya coefficient or the symmetric Kullback–Leibler divergence do not satisfy metric properties necessary for many algorithms. This paper proposes a metric for Gaussian densities. Similar to the Bhattacharyya distance and the symmetric Kullback–Leibler divergence, the proposed metric reduces the difference between two Gaussians to the difference between their parameters. Based on the proposed metric we introduce a symmetric and positive semi-definite kernel between Gaussian densities. We illustrate the benefits of the proposed metric in two settings: (1) a supervised problem, where we learn a low-dimensional projection that maximizes the distance between Gaussians, and (2) an unsupervised problem on spectral clustering where the similarity between samples is measured with our proposed kernel.[1]

1 Introduction

The Gaussian distribution plays a crucial role in multivariate statistics in general, and in discrimination theory in particular [1]. A direct realization of this fact is to

Karim T. Abou–Moustafa · Frank P. Ferrie
Centre of Intelligent Machines (CIM), McGill University, 3480 University street, Montreal, QC, H3A 2A7, Canada
e-mail: {karimt, ferrie}@cim.mcgill.ca

Fernando De La Torre
The Robotics Institute, Carnegie Mellon University, 5000 Forbes Avenue, Pittsburgh, PA 15213, USA
e-mail: ftorre@cs.cmu.edu

[1] This is the main McGill CIM Technical Report (#TR–CIM–10–05) for our research describing a measure for the difference between Gaussian densities. This technical report was presented at the *International Symposium on Brain, Body and Machine* – Nov. 10–12, 2010 – which was held to celebrate the 25th anniversary of McGill Centre of Intelligent Machines (CIM). The first and third authors are affiliated with CIM. The proceedings of the symposium have been published in this Springer Series on *Advances in Intelligent and Soft Computing.*

J. Angeles et al. (Eds.): Brain, Body and Machine, AISC 83, pp. 57–70.
springerlink.com

note how Gaussian densities are pervasive in statistical machine learning. A major aspect in discrimination theory, and consequently in statistical learning, is to reflect how two probability distributions are *close to*, or *far away* from each other; or more formally, quantify the separation or similarity/dissimilarity between probability distributions. Recently, there has been an increasing interest in defining dissimilarity measures on probability distributions to tackle problems involving structured data and/or objects not in vectorized form, when locally represented by generative models or probability distributions [11, 26, 17, 20, 12, 19]. If $\mathscr{X} = \{X_1, \ldots, X_n\}$ is the input space of such data points or objects (images, documents, proteins, variable length sequences of audio or video frames, etc.), and $\mathscr{P}$ is the space of a certain parametric family of probability distributions, then handling this type of data is usually done by mapping each datum from $\mathscr{X}$ to a probability distribution in $\mathscr{P}$. Hence, defining a dissimilarity measure on $\mathscr{P}$ in fact induces a dissimilarity measure on $\mathscr{X}$.

Our contribution in this paper is three–fold. Due to the importance of the Gaussian distribution, we define a separation or dissimilarity measure for the family of d-dimensional Gaussian distributions $\mathscr{G}_d$, such that the measure, among other requirements, should be a *full metric*; i.e. satisfy the three metric axioms: positivity, symmetry and obey the triangle inequality. Based on the three metric axioms satisfied by our metric, (1) we propose a kernel between Gaussian densities and show that it is symmetric and positive semi–definite (PSD), and (2) define an embedding for the objects in $\mathscr{X}$ into a low dimensional subspace $\mathbb{R}^{d_0}$ where $d_0 \ll n$. As it will be shown here, (1) and (2) can not be achieved if any of the three metric axioms are not satisfied.

Our proposed measure is in many ways very similar to the closed form expressions of the Bhattacharyya divergence [3] and the symmetric Kullback–Leibler (KL) divergence [16] between two multivariate Gaussian densities. However, unlike those measures of divergence that are positive, symmetric, and violate the triangle inequality [13], our proposed metric meets the three metric axioms. As will be discussed below, all measures of divergence for probability distributions are positive by definition of the divergence and can be symmetrized [1, 6]. However, very few of them meet the triangle inequality axiom.

Since our proposed measure is a full metric (by definition) on $\mathscr{G}_d$, then mapping from $\mathscr{X}$ to $\mathscr{G}_d$ yields interesting consequences for various learning algorithms. **First**; most classification and clustering algorithms assume that $\mathscr{X} \subseteq \mathbb{R}^d$ and hence, they rely on the Euclidean measure to define distances/similarities between points. If objects in $\mathscr{X}$ are complex structured data – variable length time series data or not in vectorized form – it becomes very difficult to apply these algorithms on such data. However, mapping these objects from $\mathscr{X}$ to $\mathscr{G}_d$ and using our proposed metric alleviates this difficulty by using these algorithms on their images in $\mathscr{G}_d$. **Second**; there have been some serious advances recently in speeding up the k–means algorithm by avoiding many distance computations between points and cluster centres [8]. This was possible to achieve through the triangle inequality property of the Euclidean metric to compute upper and lower bounds on these distances. It is straight forward that our proposed metric can use these same bounds to speed up clustering in $\mathscr{G}_d$. **Third**; by exponentiating the negative value of the metric, one directly obtains a kernel $K_{\mathscr{G}} : \mathscr{G}_d \times \mathscr{G}_d \to \mathbb{R}$ that, as will be shown here, is symmetric and PSD [5, 10]. This allows a smooth extension for all kernel based methods [23] to be applied on objects mapped to $\mathscr{G}_d$.

The triangle inequality axiom, in addition, allows us to consider a more general aspect of our proposed measure. If $\boldsymbol{D}_{\mathscr{G}} \in \mathbb{R}^{n \times n}$ is a symmetric matrix with zero diagonal elements (self distances) and filled with the mutual distances between the n objects in $\mathscr{X}$ using our proposed metric on $\mathscr{G}_d$, and $\tilde{\boldsymbol{D}}_{\mathscr{G}}$ is the centralized[2] distance matrix of $\boldsymbol{D}_{\mathscr{G}}$, then : (1) $\boldsymbol{G} = -\frac{1}{2}\tilde{\boldsymbol{D}}_{\mathscr{G}}$ is a PSD matrix that defines a dot product (or a gram) matrix in a Hilbert space, (2) there exists a matrix $\boldsymbol{X} \in \mathbb{R}^{n \times d_0}$ s.t. $\boldsymbol{G} = \boldsymbol{X}\boldsymbol{X}^\top$ that provides for the objects in $\mathscr{X}$ an embedding in $\mathbb{R}^{d_0}$, and the dimensionality d_0 is the rank of the matrix $\boldsymbol{G}$, and (3) for the case of $n = 3$, that $\boldsymbol{G}$ is PSD is equivalent to the triangular inequality relation between the three points. These results are credited to Young and Householder [27] (and recently by Roth *et al.* [22]) who establish the equivalence between the triangle inequality axiom of a metric and the positive semi–definiteness of the gram matrix $\boldsymbol{G}$.

Our research work starts by analyzing the closed form expressions for the Bhattacharyya divergence and the symmetric KL divergence between two multivariate Gaussian densities. We note that both have very similar properties and structure with regard to their closed form expression. Next, we propose our dissimilarity metric for Gaussian densities. Using this proposed metric, we introduce a kernel for Gaussian densities and show that it is symmetric and PSD. Finally, using preliminary experiments, we validate the proposed metric in two settings; (1) supervised, where the metric is maximized to learn a lower dimensional subspace for discriminant analysis, and (2) unsupervised, where our proposed kernel is used with spectral clustering to measure the similarity between images.

2 Related Work

Earlier work on dissimilarity measures for probability distributions started with kernels for generative models in order to plug them in discriminative models such as SVMs. This includes the Fisher kernel by Jaakkola and Haussler [11] and then the TOP kernel by Tsuda *et al.* [26].

Lafferty and Lebanon [17] study the statistical manifold of probability distributions and arrive at heat or diffusion kernels. In particular, they find closed form expressions for Gaussian distributions with spherical covariances and for multinomial distributions, where the latter is used for text documents classification. Moreno *et al.* [20] exponentiate the negative symmetric KL divergence (2) to define a kernel between Gaussian distributions for an SVM classifier: $K_{KL} = \exp\{-\alpha d_{KL} + \beta\}$, where α and β are scaling and shifting parameters respectively. Since K_{KL} is not PSD, the authors raise the need for α and β to scale and shift the kernel until it gets closer to a PSD one. Jebara and Kondor [12] define the probability product kernel (PPK) as a general symmetric and PSD kernel that can be applied to various

[2] $\tilde{\boldsymbol{D}}_{\mathscr{G}} = \boldsymbol{Q}\boldsymbol{D}_{\mathscr{G}}\boldsymbol{Q}$, where $\boldsymbol{Q} = \boldsymbol{I} - \frac{1}{n}\boldsymbol{1}\boldsymbol{1}^\top$; i.e. the row sum and column sum of $\tilde{\boldsymbol{D}}_{\mathscr{G}}$ is zero.

probability distributions and generative models: $K(P_1,P_2) = \int_{\mathscr{X}} p_1^{\alpha}(\mathbf{x})p_2^{\alpha}(\mathbf{x}) = \langle p_1^{\alpha}, p_2^{\alpha}\rangle_{L_2}$, where α is a positive constant. They consider two cases for α; 1) $\alpha = 1/2$, where they arrive at the Bhattacharyya affinity ρ, and 2) $\alpha = 1$, where they arrive at the expected likelihood kernel. More recently, Martins *et al.* [19] expand the set of kernels based on information theoretic measures by incorporating nonextensive information measures as similarity measures between probability distributions.

In independent and different research paths, Roth *et al.* [22] develop a formal treatment for correcting dissimilarity measures that do not satisfy the triangle inequality based on the results of Young and Householder [27]. The kernel community, in another direction, has recently developed a framework for distances between probability distributions based on a Hilbert space embedding of these distributions without explicit density estimation. They further plug this distance it in a variety of problems arising in statistics such as homogeneity tests, independence measurement and feature selection. Please refer to Sriperumbudur *et al.* [25] for recent advances and results in this direction.

3 Divergences and Distances for Probability Distributions

In statistics and information theory, dissimilarity measures of probability distributions are known as coefficients of divergence, Ali–Silvey distances [1], or f–divergence according to Csiszar [6]. If $P_1,P_2 \in \mathscr{P}$ are two probability distributions defined over the same domain of events $\mathscr{E}$, then the divergence of P_2 from P_1 is defined as $d_f(P_1,P_2) = \mathbb{E}_{p_1}\{C(\phi)\} = \int_{\mathscr{E}} p_1(x)C(\phi(x))\, dx$, where $d_f(P_1,P_2) \in [0,\infty)$, p_1, p_2 are the probability density functions of P_1 and P_2 respectively, $\phi(x) = p_1(x)/p_2(x)$ is the likelihood ratio[3], and C is a continuous convex function on $(0,\infty)$.

The divergence, according to Ali & Silvey or Csiszar, has to satisfy certain requirements. The most relevant to our discussion is that it should be zero when $P_1 = P_2$ and as large as possible when P_1 and P_2 are farthest apart. This is exactly the first axiom of a metric. However, the divergence by definition, is not symmetric and need not to obey the triangle inequality. Although, any divergence can be transformed to a symmetrized measure by summing $d_f(P_1,P_2)$ and $d_f(P_2,P_1)$, it is neither trivial nor obvious how to satisfy the triangle inequality. For the purpose of our discussion, we shall consider the symmetric KL divergence [16] and Chernoff's measure for discriminatory information [4] which yields the Bhattacharyya coefficient [3] and consequently the Bhattacharyya divergence [3] and the Hellinger distance [21]. Note that all these measures can be directly derived from $d_f(P_1,P_2)$ (see [1] for more details).

[3] The original definition of ϕ is the generalized Radon–Nikodym derivative of P_1 with respect to P_2.

3.1 Distances and Divergences for Gaussian Densities

Before proceeding, we need to introduce our notation for multivariate Gaussian densities. Let $\{\mathscr{N}_j(\mathbf{x};\boldsymbol{\mu}_j,\boldsymbol{\Sigma}_j) \in \mathscr{G}_d \mid \mathbf{x},\boldsymbol{\mu} \in \mathbb{R}^d,\ \boldsymbol{\Sigma}_j \in \mathbb{S}_{++}^{d\times d},\ j=1,2\}$ be two Gaussian densities where:

$$\mathscr{N}_j(\mathbf{x};\boldsymbol{\mu}_j,\boldsymbol{\Sigma}_j) = |2\pi\boldsymbol{\Sigma}_j|^{-\frac{1}{2}} \exp\{-\tfrac{1}{2}(\mathbf{x}-\boldsymbol{\mu}_j)^\top \boldsymbol{\Sigma}_j^{-1}(\mathbf{x}-\boldsymbol{\mu}_j)\}, \tag{1}$$

$|\cdot|$ is the determinant, $\boldsymbol{\mu}_j$ is the mean vector, $\boldsymbol{\Sigma}_j$ is the covariance matrix, and $\mathbb{S}_{++}^{d\times d}$ is the space of real symmetric PSD matrices.

The symmetric KL divergence is based on Kullback's measure of discriminatory information: $I(P_1,P_2) = -\int_{\mathscr{E}} p_1 \log(p_1/p_2)dx$. Kullback realizes the asymmetry of $I(P_1,P_2)$ and describes it as the *directed divergence*. To achieve symmetry, Kullback defines the divergence as $I(P_1,P_2)+I(P_2,P_1)$ and notes that it is positive and symmetric but violates the triangle inequality [16] (p. 6,7). Hence, it can not define a metric structure. The closed form expression for the symmetric KL divergence between $\mathscr{N}_1$ and $\mathscr{N}_2$ can be written as:

$$d_{KL}(\mathscr{N}_1,\mathscr{N}_2) = \tfrac{1}{2}\mathbf{u}^\top(\Sigma_1^{-1}+\boldsymbol{\Sigma}_2^{-1})\mathbf{u} + \tfrac{1}{2}\mathrm{tr}(\boldsymbol{\Sigma}_1^{-1}\boldsymbol{\Sigma}_2 + \boldsymbol{\Sigma}_2^{-1}\boldsymbol{\Sigma}_1 - 2\boldsymbol{I}), \tag{2}$$

where tr is the matrix trace, $\mathbf{u} = (\boldsymbol{\mu}_1 - \boldsymbol{\mu}_2)$, and $\boldsymbol{I}$ is the identity matrix. Equation (2) describes d_{KL} as a sum of two components, one due to the difference in means weighted by the covariance matrices, and the other due to the difference in variances and covariances [16] (p. 6,7). If $\boldsymbol{\Sigma}_1 = \boldsymbol{\Sigma}_2 = \boldsymbol{\Sigma}$, then d_{KL} expresses the difference in means which is the exact form of the Mahalanobis distance: $(\boldsymbol{\mu}_1-\boldsymbol{\mu}_2)^\top\boldsymbol{\Sigma}^{-1}(\boldsymbol{\mu}_1-\boldsymbol{\mu}_2)$. However, if $\boldsymbol{\mu}_1 = \boldsymbol{\mu}_2 = \boldsymbol{\mu}$, then d_{KL} expresses the difference, or the dissimilarity between covariance matrices $\boldsymbol{\Sigma}_1$ and $\boldsymbol{\Sigma}_2$:

$$d_{KL}(\mathscr{N}_1,\mathscr{N}_2) = \tfrac{1}{2}\mathrm{tr}(\boldsymbol{\Sigma}_1^{-1}\boldsymbol{\Sigma}_2 + \boldsymbol{\Sigma}_2^{-1}\boldsymbol{\Sigma}_1 - 2\boldsymbol{I}). \tag{3}$$

The Bhattacharyya divergence, on the other hand, is a special case of Chernoff's [4] measure of discriminatory information: $d_{Ch}(P_1,P_2) = -\ln(\inf_{0<t<1}\int_{\mathscr{E}} p_1^t p_2^{1-t}dx)$. Setting $t = 1/2$, although not the infimum but still within the valid range, yields the Bhattacharyya divergence [3]: $d_B(P_1,P_2) = -\ln\int_{\mathscr{E}}\sqrt{p_1p_2}dx = -\ln\rho(P_1,P_2)$, where ρ is the Bhattacharyya coefficient. Note that $0 \le d_B \le \infty$ and $0 \le \rho \le 1$. The coefficient ρ can define another distance: $d_H(P_1,P_2) = \sqrt{1-\rho(P_1,P_2)}$, $0 \le d_H \le 1$, which is known as the Hellinger distance [21]. Kailath [13] carefully studied d_B and d_H and notes that d_B is positive and symmetric but violates the triangle inequality, while d_H meets all axioms that define a metric. Here, we also note the work of Jebara and Kondor [12] who arrive to the Bhatacharyya coefficient ρ via the probability product kernel (PPK). They define ρ as the Bhattacharyya affinity and confirm through the PPK definition that ρ is a PSD kernel.

The closed form for the Bhattacharyya coefficient ρ between $\mathscr{N}_1$ and $\mathscr{N}_2$ can be written as follows:

$$\rho(\mathcal{N}_1, \mathcal{N}_2) = \frac{|\boldsymbol{\Sigma}_1|^{\frac{1}{4}} |\boldsymbol{\Sigma}_2|^{\frac{1}{4}}}{|\frac{1}{2}\boldsymbol{\Sigma}_1 + \frac{1}{2}\boldsymbol{\Sigma}_2|^{\frac{1}{2}}} \exp\{-\tfrac{1}{8}\mathbf{u}^\top (\tfrac{1}{2}\boldsymbol{\Sigma}_1 + \tfrac{1}{2}\boldsymbol{\Sigma}_2)^{-1}\mathbf{u}\}. \tag{4}$$

The closed form of the Hellinger distance between $\mathcal{N}_1$ and $\mathcal{N}_2$ is directly obtained from the Bhattacharyya coefficient, however the expression for the Bhattacharyya divergence has a more interesting compact form:

$$d_B(\mathcal{N}_1, \mathcal{N}_2) = \tfrac{1}{8}\mathbf{u}^\top (\tfrac{1}{2}\boldsymbol{\Sigma}_1 + \tfrac{1}{2}\boldsymbol{\Sigma}_2)^{-1}\mathbf{u} + \tfrac{1}{2}\ln \frac{\frac{1}{2}|\boldsymbol{\Sigma}_1 + \boldsymbol{\Sigma}_2|}{|\boldsymbol{\Sigma}_1|^{\frac{1}{2}} |\boldsymbol{\Sigma}_2|^{\frac{1}{2}}}. \tag{5}$$

Similar to d_{KL} in Equation (2), d_B in Equation (5) is expressed as the sum of two components, one due to the difference in means, and the other due to the difference in covariance matrices. If $\boldsymbol{\Sigma}_1 = \boldsymbol{\Sigma}_2 = \boldsymbol{\Sigma}$, then d_B is equal to the Mahalanobis distance up to a scaling factor ($\frac{1}{8}$), and if $\boldsymbol{\mu}_1 = \boldsymbol{\mu}_2 = \boldsymbol{\mu}$, then d_B will express the dissimilarity between the matrices $\boldsymbol{\Sigma}_1$ and $\boldsymbol{\Sigma}_2$:

$$d_B(\mathcal{N}_1, \mathcal{N}_2) = \tfrac{1}{2}\ln \frac{\frac{1}{2}|\boldsymbol{\Sigma}_1 + \boldsymbol{\Sigma}_2|}{|\boldsymbol{\Sigma}_1|^{\frac{1}{2}} |\boldsymbol{\Sigma}_2|^{\frac{1}{2}}}. \tag{6}$$

3.2 A Close Look at d_{KL} and d_B

We note that when the Bhattacharyya divergence and the symmetric KL divergence were applied to $\mathcal{N}_1$ and $\mathcal{N}_2$, they factored the difference between the distributions in terms of the difference between their first and second order statistics. In other words, the difference between two Gaussian densities was reduced to the difference between their parameters. Note also that d_{KL} and d_B in Equations (2) and (5) respectively have the same structure; a sum of two components, one due to the difference in means (represented as a Mahalanobis distance), and the other due to the difference in covariance matrices. More precisely, the first component in d_{KL} is the sum of two Mahalanobis distances, while the first component in d_B is a variant of the Mahalanobis distance that uses the inverse of an average covariance matrix. Note that this explanation for the meaning of each term is due to Kullback [16] (p. 6,7). The Mahalanobis distance comprising the first component of d_{KL} and d_B meets the three metric axioms. However, since Equations (2) & (5) are positive and symmetric but violate the triangle inequality, then the reason for the deficiency in meeting the triangle inequality is due to the second component in d_{KL} and d_B which measures the dissimilarity between covariance matrices, i.e. Equations (3) and (6). This observation implies that the measures for the difference between the PSD covariance matrices in Equations (3) and (6) do not define proper metrics for covariance matrices on the manifold $\mathbb{S}_{++}^{d\times d}$.

4 Designing a Metric for Gaussian Densities

The discussion above suggests that if there is a distance measure for covariance matrices that defines a metric on the manifold $\mathbb{S}_{++}^{d\times d}$, then it is possible to design a new separation measure specifically for Gaussian densities. The designed measure however, should meet certain requirements: 1) it should satisfy all the metric axioms, 2) reduce to the Euclidean distance when $\boldsymbol{\Sigma}_1 = \boldsymbol{\Sigma}_2 = \boldsymbol{I}$, 3) reduce to the Mahalanobis distance when $\boldsymbol{\Sigma}_1 = \boldsymbol{\Sigma}_2 = \boldsymbol{\Sigma}$, and 4) reduce to a metric for covariance matrices, satisfying all metric axioms, when $\boldsymbol{\mu}_1 = \boldsymbol{\mu}_2 = \boldsymbol{\mu}$.

Note that requirements 2 & 3 are in the same spirit of d_{KL} and d_B in Equations (2) and (5) which provide an intuitive meaning and explanation for the metric in any of these special cases. Although the Hellinger distance d_H is a true metric by definition, it does not yield such properties, however d_H has a defined range ($d_H \in [0,1]$) which might be desirable in certain contexts.

Similar to d_{KL} and d_B, the new designed metric will comprise two components, a measure for the difference in means, and a measure for the difference in covariance matrices. However, unlike these measures, the second component will be a true metric for real symmetric PSD matrices on the manifold $\mathbb{S}_{++}^{d\times d}$.

4.1 A Metric for Symmetric and Positive Semi–definite Matrices

Förstner and Moonen [9] proposed a metric measure for covariance matrices that is derived from a canonical invariant Riemannian metric on the manifold $\mathbb{S}_{++}^{d\times d}$. The measure is a full metric, invariant under affine transformations of the coordinate system, and invariant to inversion. For two matrices $\{\boldsymbol{A},\boldsymbol{B} \in \mathbb{S}_{++}^{d\times d}\}$ the distance measure between them is:

$$d_{FM}(\boldsymbol{A},\boldsymbol{B}) = \sqrt{\mathrm{tr}\{\ln^2 \boldsymbol{\Lambda}(\boldsymbol{A},\boldsymbol{B})\}}, \tag{7}$$

where $\boldsymbol{\Lambda}(\boldsymbol{A},\boldsymbol{B}) = \mathrm{diag}(\lambda_1,\ldots,\lambda_d)$ is the solution of a generalized eigenvalue problem (GEP): $\boldsymbol{AV} = \boldsymbol{\Lambda BV}$. The proof that d_{FM} defines a metric on the manifold $\mathbb{S}_{++}^{d\times d}$ and that it satisfies all the axioms of a metric can be found in [9]. The basic idea of comparing covariance matrices is to reflect the deviations in variances in all directions. In d_{FM}, these deviations are evaluated as the ratio of variances for all dimensions. The ln in d_{FM} measures these deviations as factors, while squaring guarantees that deviations by a factor of f and $1/f$ will be equally penalized.

4.2 The Proposed Metric $d_{\mathscr{G}}$

Our metric is designed based on the first component of the Bhattacharyya distance for the difference in means, and on the metric d_{FM} in Equation (7) for covariance matrices. For two Gaussian densities $\mathscr{N}_1$ and $\mathscr{N}_2$, the proposed metric $d_{\mathscr{G}}$ is defined as follows:

$$d_{\mathscr{G}}(\mathscr{N}_1,\mathscr{N}_2) = \left(\mathbf{u}^\top \boldsymbol{S}^{-1}\mathbf{u}\right)^{\frac{1}{2}} + \left(\sum_{k=1}^{d} \ln^2 \lambda_k(\boldsymbol{\Sigma}_1,\boldsymbol{\Sigma}_2)\right)^{\frac{1}{2}}. \tag{8}$$

Except for the invariance to inversion property of d_{FM}, $d_{\mathscr{G}}(\mathscr{N}_1, \mathscr{N}_2)$ inherits all the properties of its constituting components: 1) it is invariant to affine transformations, 2) it is a full metric, and 3) it fulfils the special cases of requirements 2, 3 & 4 mentioned above.

Similar to the Bhattacharyya distance (5) and the symmetric KL divergence (2), the proposed metric $d_{\mathscr{G}}$ reduces the difference between $\mathscr{N}_1$ and $\mathscr{N}_2$ to the difference between their parameters. Moreover, it has the exact same structure as d_{KL} and d_B, where the first term measures the difference in means, while the second term measures the difference between two covariance matrices (i.e. two symmetric and PSD matrices). In other words, each term in $d_{\mathscr{G}}$ has a clear meaning and measures a well defined quantity.

4.3 A Kernel Based on $d_{\mathscr{G}}$

We can define a kernel $K_{\mathscr{G}} : \mathscr{G}_d \times \mathscr{G}_d \to \mathbb{R}$ for two Gaussian densities based on $d_{\mathscr{G}}$ as follows:

$$K_{\mathscr{G}}(\mathscr{N}_1, \mathscr{N}_2) = \exp\{-d_{\mathscr{G}}(\mathscr{N}_1, \mathscr{N}_2)\}. \tag{9}$$

The kernel $K_{\mathscr{G}}$ is an exponential function of a distance measure that is not an Euclidean norm. Genton in [10] studies different classes of kernels with their properties and points to [5] for a formal treatment for the case of $K_{\mathscr{G}}$. In particular, Christakos and Papanicolaou [5] set conditions for the class of exponential kernels when the distance that defines the kernel is not an Euclidean metric on $\mathbb{R}^d$. To show that $K_{\mathscr{G}}$ is a PSD kernel, we rewrite $d_{\mathscr{G}}(\mathscr{N}_1, \mathscr{N}_2)$ in Equation (8) as follows:

$$\begin{aligned} d_{\mathscr{G}}(\mathscr{N}_1, \mathscr{N}_2) &= (\mathbf{u}^\top \boldsymbol{\Phi}^\top \boldsymbol{\Gamma}^{-1} \boldsymbol{\Phi} \mathbf{u})^{\frac{1}{2}} + \left(\sum_{i=1}^{d} \ln^2 \lambda_i \right)^{\frac{1}{2}} \\ &= \left(\sum_{j=1}^{d} \gamma_j (\mathbf{u}^\top \boldsymbol{\phi}_j)^2 \right)^{\frac{1}{2}} + \left(\sum_{i=1}^{d} \omega_i \ln^2 \lambda_i \right)^{\frac{1}{2}}, \text{ and hence} \end{aligned}$$

$$K_{\mathscr{G}}(\mathscr{N}_1, \mathscr{N}_2) = \exp\left\{ -\left(\sum_{j=1}^{d} \gamma_j (\mathbf{u}^\top \boldsymbol{\phi}_j)^2 \right)^{\frac{1}{2}} \right\} \exp\left\{ -\left(\sum_{i=1}^{d} \omega_i \ln^2 \lambda_i \right)^{\frac{1}{2}} \right\}, \tag{10}$$

where $\omega_i = 1$, for $1 \le i \le d$, $\boldsymbol{\Phi} = [\boldsymbol{\phi}_1 \dots \boldsymbol{\phi}_d]$ is a column matrix with the eigenvectors of $\boldsymbol{S}$, and $\boldsymbol{\Gamma} = \text{diag}(\gamma_1, \dots, \gamma_d)$ is the diagonal matrix of its eigenvalues. To show that $K_{\mathscr{G}}$ is a PSD kernel, one has to show that each term in the right hand side (RHS) of (10) is a PSD kernel since the multiplication of two PSD kernels is another PSD kernel [10] (p. 300). Let the lag vectors $\mathbf{h}_1$ and $\mathbf{h}_2$ and the weight vectors $\mathbf{w}_1$ and $\mathbf{w}_2$ be respectively defined as follows:

$$\mathbf{h}_1 = [\mathbf{u}^\top \boldsymbol{\phi}_1, \ldots, \mathbf{u}^\top \boldsymbol{\phi}_d]^\top, \quad \mathbf{h}_2 = [\ln \lambda_1, \ldots, \ln \lambda_d]^\top, \tag{11}$$

$$\mathbf{w}_1 = [\gamma_1, \ldots, \gamma_d]^\top, \quad \mathbf{w}_2 = [\omega_1, \ldots, \omega_d]^\top. \tag{12}$$

It is shown in [5] (p. 475) that if an exponential kernel K is of the form:

$$K(\mathbf{h}) = \exp\{-(w_1|h_1|^p + \cdots + w_n|h_n|^p)^{\frac{1}{2}}\}, \tag{13}$$

for a lag vector $\mathbf{h} \in \mathbb{R}^n$ and a weight vector $\mathbf{w} \in \mathbb{R}^n$, then K is a PSD kernel if and only if $0 < p \leq 2$. Setting $p = 2$ and using the definition of $\mathbf{h}_1$, $\mathbf{h}_2$, $\mathbf{w}_1$ and $\mathbf{w}_2$ from (11) and (12), then $K_{\mathscr{G}}$ can be written as:

$$\begin{aligned} K_{\mathscr{G}}(\mathscr{N}_1, \mathscr{N}_2) = \exp\Big\{ &- \big(w_1^1 (h_1^1)^2 + \cdots + w_d^1 (h_d^1)^2\big)^{\frac{1}{2}} \Big\} * \\ \exp\Big\{ &- \big(w_1^2 (h_1^2)^2 + \cdots + w_d^2 (h_d^2)^2\big)^{\frac{1}{2}} \Big\}, \end{aligned} \tag{14}$$

where each term on the RHS of (14) has the exact same structure of (13), and hence each term defines a PSD kernel. Consequently, it follows that $K_{\mathscr{G}}$ is a PSD kernel.

Note that the definition of $K(\mathbf{h})$ in Equation (13) allows the introduction of a kernel parameter $\sigma > 0$ in $K_{\mathscr{G}}$ that controls the affinity between the Gaussian densities (in fact there parameters) without loosing its PSD property. Therefore, the final form of our proposed kernel is : $\exp\{-d_{\mathscr{G}}(\mathscr{N}_1, \mathscr{N}_2)/\sigma\}$ and $\sigma > 0$.

Finally, if $\boldsymbol{W}_{\mathscr{G}} \in \mathbb{R}^{n \times n}$ is the kernel or gram matrix obtained from $K_{\mathscr{G}}(\mathscr{N}_i, \mathscr{N}_j)$, for $1 \leq i, j \leq n$, then $\boldsymbol{W}_{\mathscr{G}}$ meets with the gram matrix $\boldsymbol{G} = -\frac{1}{2}\tilde{\boldsymbol{D}}_{\mathscr{G}}$ (defined earlier) in that both are derived from $\boldsymbol{D}_{\mathscr{G}}$ and both are symmetric and PSD. Therefore, using the results in [27, 22], there exist matrices $\boldsymbol{X}_1 \in \mathbb{R}^{n \times d_1}$ and $\boldsymbol{X}_2 \in \mathbb{R}^{n \times d_2}$ s.t. $\boldsymbol{G} = \boldsymbol{X}_1 \boldsymbol{X}_1^\top$ and $\boldsymbol{W}_{\mathscr{G}} = \boldsymbol{X}_2 \boldsymbol{X}_2^\top$ that provide an embedding for the objects in $\mathscr{X}$ into a lower dimensional space $\mathbb{R}^{d_1}$ and $\mathbb{R}^{d_2}$, and the dimensionality d_1 and d_2 is the rank of the matrices $\boldsymbol{G}$ and $\boldsymbol{W}_{\mathscr{G}}$ respectively. This establishes the relation between the triangle inequality of our metric and the positive semi–definitness of $\boldsymbol{G}$ and $\boldsymbol{W}_{\mathscr{G}}$ (please refer to Section 1).

5 Experimental Results

In the experimental results, we validate the proposed metric in two different learning settings. First, we consider a supervised learning problem in the context of learning a linear transformation for dimensionality reduction. Second, we investigate the problem of unsupervised learning via spectral clustering algorithms where each entry in the affinity matrix uses the proposed kernel to measure the similarity between samples.

Table 1. Specifications of the data sets used in the discriminant analysis experiments where number of classes, size and the number of attributes are denoted by c, n and d respectively.

Data set	c	n	d	Data set	c	n	d
UCI Bupa	2	345	6	UCI Monks–III	2	554	6
UCI HouseVotes	2	435	16	UCI Pima	2	768	8
UCI Monks–I	2	556	6	UCI TicTacToe	2	958	9
UCI Monks–II	2	601	6				

5.1 Supervised Discriminative Dimensionality Reduction

Fisher/Linear discriminant analysis (FDA/LDA) seeks a low dimensional subspace where d_{KL} in Equation (2) is maximized [16]. For a 2–class/multi–class problem, FDA/LDA model each class as a Gaussian distribution under the assumption that all classes have equal covariance matrices. In this case, d_{KL} reduces to the Mahalanobis distance and FDA/LDA reduces to a GEP. To extend this framework when the covariance assumption does not hold, De La Torre and Kanade [7] proposed MODA that searches for a low dimensional subspace that explicitly maximizes the objective function in Equation (2). Our objective here is to use our separation measure in the same context and compare it to other discriminant analysis techniques.

Here, we only consider 2–class problems and model each class, C_1 and C_2 as a Gaussian distribution; $\mathcal{N}_1(\,\cdot\,;\boldsymbol{\mu}_1,\boldsymbol{\Sigma}_1)$ and $\mathcal{N}_2(\,\cdot\,;\boldsymbol{\mu}_2,\boldsymbol{\Sigma}_2)$ respectively. Similar to FDA/LDA and MODA, we search for a linear transformation $\boldsymbol{B} \in \mathbb{R}^{d\times k}$ with $k < d$ such that it maximizes $d_{\mathscr{G}}(\mathcal{N}_1,\mathcal{N}_2)$ in the lower dimensional space. The linear transformation $\boldsymbol{B}$ can have any number of bases k such that $1 \leq k \leq \min(d-1, n-1)$. This is unlike FDA/LDA which can only define subspaces of dimensionality $k \leq \min(c-1, d-1)$, where c is the number of classes.

Let the distance between $\mathcal{N}_1$ and $\mathcal{N}_2$ under the linear transformation $\boldsymbol{B}$ be defined as follows:

$$d_{\mathscr{G}}(\mathcal{N}_1,\mathcal{N}_2;\boldsymbol{B}) = \underbrace{\mathrm{tr}\{(\boldsymbol{B}^\top \boldsymbol{S}\boldsymbol{B})^{-1}(\boldsymbol{B}^\top \boldsymbol{U}\boldsymbol{B})\}}_{\mathrm{I}(\boldsymbol{B})} + \underbrace{\mathrm{tr}\{\log^2\{(\boldsymbol{B}^\top \boldsymbol{\Sigma}_1\boldsymbol{B})^{-1}(\boldsymbol{B}^\top \boldsymbol{\Sigma}_2\boldsymbol{B})\}\}}_{\mathrm{II}(\boldsymbol{B})}, \tag{15}$$

where $\boldsymbol{U} = \mathbf{u}\mathbf{u}^\top$ and $\boldsymbol{S} = (\frac{1}{2}\boldsymbol{\Sigma}_1 + \frac{1}{2}\boldsymbol{\Sigma}_2)$. Maximizing Equation (15) with respect to $\boldsymbol{B}$ yields a basis $\boldsymbol{B}^*_{\mathscr{G}}$ that is optimal, in terms of separation, for classes C_1 and C_2. Since there is no closed form solution for the maximum of Equation (15), we use an iterative procedure based on gradient ascent:

$$\boldsymbol{B}^{t+1} = \boldsymbol{B}^t + \eta\frac{\partial d_{\mathscr{G}}(\mathcal{N}_1,\mathcal{N}_2;\boldsymbol{B})}{\partial \boldsymbol{B}} = \boldsymbol{B}^t + \eta\frac{\partial \mathrm{I}(\boldsymbol{B})}{\partial \boldsymbol{B}} + \eta\frac{\partial \mathrm{II}(\boldsymbol{B})}{\partial \boldsymbol{B}},$$

where

$$\frac{\partial \mathrm{I}(\boldsymbol{B})}{\partial \boldsymbol{B}} = 2\boldsymbol{U}\boldsymbol{B}\boldsymbol{\Theta} - 2\boldsymbol{S}\boldsymbol{B}\boldsymbol{\Theta}(\boldsymbol{B}^\top \boldsymbol{U}\boldsymbol{B})\boldsymbol{\Theta}, \quad \boldsymbol{\Theta} = (\boldsymbol{B}^\top \boldsymbol{S}\boldsymbol{B})^{-1},$$

$$\frac{\partial \mathrm{II}(\boldsymbol{B})}{\partial \boldsymbol{B}} = [2\log \boldsymbol{L}\{2(\boldsymbol{B}^\top \boldsymbol{\Sigma}_2 \boldsymbol{B})^{-1}\boldsymbol{B}^\top \boldsymbol{\Sigma}_2 - 2(\boldsymbol{B}^\top \boldsymbol{\Sigma}_1 \boldsymbol{B})^{-1}\boldsymbol{B}^\top \boldsymbol{\Sigma}_1\}]^\top,$$

with $\boldsymbol{L} = \mathrm{diag}(\ell_1, \ldots, \ell_k)$ is the eigenvalue matrix of $(\boldsymbol{B}^\top \boldsymbol{\Sigma}_1 \boldsymbol{B})^{-1}(\boldsymbol{B}^\top \boldsymbol{\Sigma}_2 \boldsymbol{B})$, and η is the step length. The gradient ascent procedure starts with a reasonable step length and it is decreased by 50% if it increases the value of the objective function. Other strategies such as line search are possible but this simple method has provided good preliminary results. Similar to MODA, the objective function in Equation (15) is non–convex and any gradient ascent procedure can be trapped into local minima. Therefore, we typically start the algorithm with multiple initializations and select the solution $\boldsymbol{B}^*_{\mathscr{G}}$ with the lowest training error.

The error considered here is the error of a quadratic classifier in the lower dimensional space. Since each class is modelled as a Gaussian distribution, a sample $\mathbf{x}$ with an unknown label y is assigned the label of its closest class, where closeness is based on the Mahalanobis distance between the sample $\mathbf{x}$ and the class C_j: $(\boldsymbol{\mu}_j - \mathbf{x})^\top \boldsymbol{\Sigma}_j^{-1}(\boldsymbol{\mu}_j - \mathbf{x})$, $j = 1, 2$.

Table (1) shows seven data sets from the UCI ML Repository that are used in this experiment. The empirical error (with standard deviation) was averaged over 10 folds nested cross validation for three different projection dimensions $k = 1, 2, 3$. Table (2) shows the empirical error for LDA, PCA+LDA, PCA, RCA [2], MODA and $\boldsymbol{B}^*_{\mathscr{G}}$ on the UCI data sets for projection dimension $k = 1$. It is clear that linear transformation $\boldsymbol{B}^*_{\mathscr{G}}$ yields very competitive results with standard discriminant analysis techniques and with more recent approaches such as RCA and MODA.

5.2 *Unsupervised Clustering of Images*

In the second experiment, we consider an unsupervised learning problem where our main objective is to compare different distance and divergence measures between Gaussian densities in the context of clustering. Our hypothesis is that full metric

Table 2. Empirical error (with standard deviation) for discriminant analysis experiments using a projection dimension $k = 1$. Due to space limitations, please see supplementary material for $k = 2, 3$.

Data set	LDA	PCA+LDA	PCA	RCA	MODA	$\boldsymbol{B}^*_{\mathscr{G}}$
Bupa	44.7 (5.1)	37.3 (4.8)	45.5 (6.0)	37.9 (34.1)	34.1 (8.1)	**32.0** (6.2)
HouseVotes	11.1 (5.5)	4.5 (3.4)	12.6 (5.8)	4.5 (3.4)	4.5 (3.4)	**4.2** (3.6)
Monks–I	33.3 (8.6)	36.1 (10.5)	33.3 (8.6)	36.1 (10.5)	34.4 (10.5)	**33.1** (12.2)
Monks–II	37.9 (4.8)	33.7 (4.2)	43.9 (5.4)	35.0 (4.9)	32.5 (4.5)	**31.3** (4.7)
Monks–III	**18.8** (10.2)	22.4 (6.7)	34.2 (9.1)	22.4 (6.7)	23.3 (8.4)	21.1 (8.0)
pima	37.2 (5.1)	**24.0** (4.6)	39.6 (5.3)	**24.0** (4.8)	27.8 (4.5)	28.5 (5.9)
TicTacToe	38.9 (11.0)	**1.4** (4.6)	54.9 (9.6)	**1.4** (4.6)	**1.4** (4.5)	**1.5** (4.8)

Table 3. *Col. 1* The three data sets used in the spectral clustering experiments. *Col. 2* Number of classes, size and attributes for each data set. *Col. 3* The accuracy of spectral clustering using for the four similarity matrices.

Data set	c	n	d	$\boldsymbol{W}_B$	$\boldsymbol{W}_{KL}$	$\boldsymbol{W}_H$	$\boldsymbol{W}_{\mathscr{G}}$
Yale–A face data set	15	165	32×32	51.0	57.7	**64.9**	59.5
KTH TIPS grey scale textures	10	810	200×200	56.9	56.1	**60.5**	**60.2**
USPS handwritten digits	10	7291/2007	16×16	56.2	57.0	55.2	**59.1**

measures such as $d_{\mathscr{G}}$ and d_H will yield better clustering results than d_{KL}, and very comparable to ρ.

Here, we adopt the same conceptual framework of Kondor and Jebara [15] that models each image as a bag of pixels (BOP). In this framework, instead of directly modelling each BOP as a Gaussian distribution, Kondor and Jebara map each BOP to a high dimensional feature space $\mathscr{H}$ using kernel PCA (KPCA) [23] in order to capture more nonlinear relations between the pixels. For KPCA, they use a Gaussian kernel with kernel width r. Next, they model each BOP in $\mathscr{H}$ as a Gaussian distribution with a full covariance and regularization parameter ε. That is, each image is finally represented as a Gaussian distribution in $\mathscr{H}$. Finally, they use SVMs with the Bhattacharyya kernel to classify the images. Please refer to [15] for more details. Here, we apply spectral clustering (SC) [18] on the Gaussian distributions in $\mathscr{H}$ instead of using SVMs.

Four similarity measures are used to construct the similarity or (adjacency) matrix for SC : $\boldsymbol{W}_B = \rho(\mathscr{N}_i, \mathscr{N}_j)$ – the Bhatacharyya kernel of [15], $\boldsymbol{W}_{KL} = \exp\{-d_{KL}(\mathscr{N}_i, \mathscr{N}_j)/\sigma\}$ – the KL kernel of [20], $\boldsymbol{W}_H = \exp\{-d_H(\mathscr{N}_i, \mathscr{N}_j)/\sigma\}$, and $\boldsymbol{W}_{\mathscr{G}} = \exp\{-d_{\mathscr{G}}(\mathscr{N}_i, \mathscr{N}_j)/\sigma\}$, where $\sigma > 0$ and $1 \leq i, j \leq n$. Note that, in total, there are three parameters to optimize in this framework; r, ε and σ. According to the recommendation in [18, pp. 6], we use the normalized SC of Shi and Malik [24]: $\boldsymbol{L}_{rw} = \boldsymbol{I} - \boldsymbol{D}^{-1}\boldsymbol{W}$, where $\boldsymbol{W}$ can be replaced by any of the previously defined similarity matrices, $\boldsymbol{L}_{rw}$ is the normalized (as a random walk) Laplacian, and $\boldsymbol{D} = \text{diag}(\boldsymbol{W}\mathbf{1}_{n\times n})$. To proceed with SC, we find the c eigenvectors corresponding to the c smallest eigenvalues of the GEP: $\boldsymbol{L}_{rw}\mathbf{v} = \omega\boldsymbol{D}\mathbf{v}$ and form the matrix $\boldsymbol{V} = [\mathbf{v}_1 \ldots \mathbf{v}_c] \in \mathbb{R}^{n\times c}$, where c is the number of clusters. Now each Gaussian distribution in $\mathscr{H}$ (corresponding to one image) is mapped to a row vector in $\boldsymbol{V}$. Finally, we cluster the rows of $\boldsymbol{V}$ using the k–means algorithm – with multiple initializations – and select the clustering configuration with minimum distortion.

Three image data sets are used in these experiments and shown in Table (3). Due to the large size of the USPS data set, the first 100 digits of each class are considered as our data set. The number of clusters is assumed to be known and its equal to the number of classes in each data set. To measure the clustering accuracy, we adopt the technique in [28] that uses a $c \times c$ confusion matrix $\boldsymbol{C}$ and the Hungarian algorithm [14] to solve the following optimization problem: $\max \text{tr}\{\boldsymbol{CP}\}$, where $\boldsymbol{P}$ is a permutation matrix, and the result is divided by the number of data points to be clustered.

Table (3) shows the results of SC using the four different similarity measures. Due to the difficulty of clustering these images with a general and simple representation such as BOP, and due to sensitivity of this framework to the choice of parameter values as acknowledged by Kondor and Jebara, the accuracy is generally low for all the data sets. Nevertheless, we note the difference between ρ and d_{KL} on one hand, and d_H and $d_{\mathscr{G}}$ on the other hand. Counter to our hypothesis, the Bhattacharyya affinity ρ did not perform as good as the similarities induced by $d_{\mathscr{G}}$ and d_H.

The triangle inequality plays an important role for $\boldsymbol{W}$ and consequently for clustering. In the GEP of SC, $\boldsymbol{L}_{rw}$ should be PSD, $\boldsymbol{D}$ should be positive definite, and hence $\boldsymbol{W}$ should be PSD as well. If the triangle inequality is not satisfied, $\boldsymbol{W}$ will be non–definite and the GEP will yield an inaccurate embedding. It follows that the row vectors of $\boldsymbol{V}$ will have inaccurate coordinates, and consequently k–means will yield an inaccurate clustering. The amount of inaccuracy is tightly related to how far is $\boldsymbol{W}$ from a PSD matrix. This is where parameter σ comes into play for $\exp\{-d_{KL}/\sigma\}$ for instance, where it helped improve the positive semi–definiteness of $\boldsymbol{W}_{KL}$ thereby improving the clustering accuracy. A deeper and a more formal investigation is currently undergoing in this direction.

Concluding remarks. We have designed a metric that measures the separation or difference between two Gaussian densities. The measure has interesting properties and consequences for various learning algorithms and showed promising preliminary results in two different learning settings. Also, we have considered the importance of the triangle inequality axiom for metrics and divergence measures, and its relation to the PSD property of the gram matrix derived from these measures. Although our metric is a designed measure, an important and legitimate question to ask is, what is the original divergence measure between P_1 and P_2 such that when plugging in $\mathscr{N}_1$ and $\mathscr{N}_2$ yields our metric $d_{\mathscr{G}}$? The right answer is to generalize the analysis presented here using various divergence measures from the class of Aly–Silvey distances to the general form of the exponential family of probability distributions. On the one hand, it allows us to study which divergence measures factorize the difference in the exponential family in terms of difference in their statistics, and on the other hand, study which of these divergence measures satisfy the three metric axioms or yield symmetric PSD gram matrices. This analysis can result in a very rich set of measures that have different properties and characteristics, however this remains to be explored.

References

1. Ali, S.M., Silvey, S.D.: A general class of coefficients of divergence of one distribution from another. J. of the Royal Statistical Society. Seris B 28(1), 131–142 (1966)
2. Bar-Hillel, A., Hertz, T., Shental, N., Weinshall, D.: Learning a Mahalanobis metric from equivalence constraints. J. of Machine Learning Research 6, 937–965 (2005)
3. Bhattacharyya, A.: On a measure of divergence between two statistical populations defined by their probability distributions. Bull. Calcutta Math. Soc. 35, 99–109 (1943)

4. Chernoff, H.: A measure of asymptotic efficiency for tests of a hypothesis based on the sum of observations. Annals of Mathematical Statistics 22, 493–507 (1952)
5. Christakos, G., Papanicolaou, V.: Norm–dependent covariance permissibility of weakly homogeneous spatial random fields and its consequences in spatial statistics. Stochastic Environmental Research and Risk Management 14, 471–478 (2000)
6. Csiszár, I.: Information–type measures of difference of probability distributions and indirect observations. Studia Scientiarium Mathematicarum Hungarica 2, 299–318 (1967)
7. De La Torre, F., Kanade, T.: Multimodal oriented discriminant analysis. In: ACM Proc. of ICML, pp. 177–184 (2005)
8. Elkan, C.: Using the trinagle inequality to accelerate k–means. In: ACM Proc. of ICML, pp. 147–153 (2003)
9. Förstner, W., Moonen, B.: A metric for covariance matrices. Tech. rep., Dept. of Geodesy and Geo–Informatics, Stuttgart University (1999)
10. Genton, M.: Classes of kernels for machine learning: A statistics perspective. J. of Machine Learning Research 2, 299–312 (2001)
11. Jaakkola, T., Haussler, D.: Exploiting generative models in discriminative classifiers. In: NIPS 11, pp. 487–493. MIT Press, Cambridge (1999)
12. Jebara, T., Kondor, R., Howard, A.: Probability product kenrels. J. of Machine Learning Research 5, 819–844 (2004)
13. Kailath, T.: The divergence and Bhattacharyya distance measures in signal selection. IEEE Trans. on Communication Technology 15(1), 52–60 (1967)
14. Knuth, D.E. (ed.): The Stanford graphbase. Springer, New York (1988)
15. Kondor, R., Jebara, T.: A kernel between sets of vectors. In: ACM Proc. of ICML 2003 (2003)
16. Kullback, S.: Information Theory and Statistics – Dover Edition. Dover, New York (1997)
17. Lafferty, J., Lebanon, G.: Information diffusion kernels. In: NIPS 14, MIT Press, Cambridge (2002)
18. Luxburg, U.v.: A tutotrial on spectral clustering. Tech. Rep. TR–149, Max Plank Institute for Biological Cybernetics (2006)
19. Martins, A., Smith, N., Xing, E., Aguiar, P., Figueiredo, M.: Nonextensive information theoretic kernels on measures. J. of Machine Learning Research 10, 935–975 (2009)
20. Moreno, P., Ho, P., Vasconcelos, N.: A Kullback–Leibler divergence based kernel for svm classification in multimedia applications. In: NIPS 16 (2003)
21. Rao, C.: Use of Hellinger distance in graphical displays. In: Titt, E., Kollo, T., Niemi, H. (eds.) Multivariate Statistics and Matrices in Statistics, pp. 143–161 (1995)
22. Roth, V., Laub, J., Buhmann, J.: Optimal cluster preserving embedding of nonmetric proximity data. IEEE Trans. PAMI 25(12), 1540–1551 (2003)
23. Schölkopf, B., Smola, A.: Learning with kernels. MIT Press, Cambridge (2002)
24. Shi, J., Malik, J.: Motion segmentation and tracking using normalized cuts. In: IEEE Proc. of ICCV, pp. 1154–1160 (1998)
25. Sriperumbudur, B., Gretton, A., Fukumizu, K., Schölkopf, B., Lanckriet, G.: Hilbert space embeddings and metrics on probability distributions. J. of Machine Learning Research 11, 1517–1561 (2010)
26. Tsuda, K., Kawanabi, M., Ratsch, G., Sonnenburg, S., Muller, K.R.: A new discriminative kernel from probability distributions. Neural Computation 14, 2397–2414 (2002)
27. Young, G., Householder, A.: Discussion of a set of points in temrs of their mutual distances. Psychometrika 3(1), 19–22 (1938)
28. Zha, H., Ding, C., Gu, M., He, X., Simon, H.: Spectral relaxation for k–means clustering. In: NIPS 13, MIT Press, Cambridge (2001)

Physical Asymmetries and Brightness Perception

James J. Clark

Abstract. This paper considers the problem of estimating the brightness of visual stimuli. A number of physical asymmetries are seen to permit determination of brightness that is invariant to certain manipulations of the sensor responses, such as inversion. In particular, the light-dark range asymmetry is examined and is shown to result, over a certain range, in increased variability of sensor responses as scene brightness increases. Based on this observation we propose that brightness can be measured using variability statistics of conditional distributions of image patch values. We suggest that a process of statistical learning of these conditional distributions underlies the Stevens effect.

1 Introduction - Is It Dark or Bright?

Suppose one is viewing a scene, such as looking out onto a busy street on a bright sunny day, or looking around your moonlit kitchen for a midnight snack with the lights turned off. In this paper we will be concerned with the perception of how bright a viewed scene is, and consider the question "what makes one scene appear bright while the other appears dark?". The term *brightness* denotes the subjective perception of the *luminance* of a visual stimulus, where luminance is a photometric (i.e. perceptually weighted) measure of the intensity of light (either reflecting from a surface or being emitted from a light source) per unit area traveling in a particular direction.

A naive answer to the question of what determines the perception of brightness would be to simply associate "dark" with low sensor signal values and "light" with high sensor signal values, as depicted in figure 1.

There are a number of problems with this naive approach, however. To begin with, the sign of the variation in sensor signals with the luminance of the visual

James J. Clark
Centre for Intelligent Machines, McGill University, 3480 University Street, Montreal, Quebec, Canada
e-mail: clark@cim.mcgill.ca

J. Angeles et al. (Eds.): Brain, Body and Machine, AISC 83, pp. 71–82.
springerlink.com

Fig. 1. Is the perception of brightness based on the firing rates of neurons in the brain?

stimulus is an arbitrary convention. One could just as easily have sensors whose signals are high when the luminance of the stimulus is low and vice-versa. For example, in the human retina both types of sensors are found, where bipolar cells either respond to the presence (ON-cells) or to the absence (OFF-cells) of incident light [1].

Some sensors respond to spatial or temporal contrasts (or derivatives). To take a specific example, consider that signals to the visual cortex from the retina are in the form of ON-Center/OFF-Surround and OFF-Center/ON-Surround signals. These could be integrated to recover the luminance, but the ambiguity in the sign remains. This also implies that the sensor signal may depend on the spatial and temporal "surround".

The issues just mentioned suggest that brightness is perceived in a way that involves more than just the raw signal levels from the image sensors. This paper describes a possible approach for doing this, one that is based on consideration of physical asymmetries that reveal differences between light and dark.

2 Physical Asymmetries Underlying Brightness Perception

The most fundamental asymmetry that we will look at is the so-called *light-dark range asymmetry*. This asymmetry can be understood by noting that there is a wider range of sensor values possible in a bright scene than in a dark one. Suppose, for argument's sake, that we have a strictly increasing monotonic visual sensor with an infinite dynamic range. That it, its response is a strictly increasing function of the intensity of the incident light. If this sensor views a scene consisting of a single non-luminous textured convex Lambertian object, illuminated by a single point light source having illuminance L, there will be a finite maximum value that this sensor could produce. This maximum value will depend on the sensitivity of the sensor, σ, the maximum albedo of the object, and the illuminance L of the object surface. The range of albedo values for a non-luminous object must be in the range $[0,1]$. Thus the range of sensor values will be $[0, L\sigma]$. As the surface illuminance L increases, so does the range of possible sensor values. This increase of the range would persist even if the sensor was instead taken to have a strictly decreasing response (corresponding to a negative σ) or had a constant offset (so that the sensor had a non-zero response to a zero incident intensity).

The analysis is more involved, but the light-dark range asymmetry will also be present for more complicated scenes, with multiple non-Lambertian objects and multiple distributed illuminants. Singularities such as caustics created by mirrors and lenses can create infinite intensities, but only over vanishingly small areas. Sensors with finite extent will have a finite response to such caustics, and this response will be scaled by the illuminance of the light source.

2.1 *Breakdown of the Light-Dark Range Asymmetry Due to Saturation*

Practical physically realizable sensors will saturate beyond some range of incident light intensity, at both the low and high ends of the sensor's range. The saturation on low end implies that the sensor will be insensitive to scene brightness changes below a certain level. The saturation on the high end, however, will not remove all sensitivity to brightness changes. This is because, in a scene which contains shadowing, or a range of surface albedos that includes zero albedo surfaces, there will be parts of the scene which result in sensor responses below the high-end saturation limit. Figure 2 shows the histogram of sensor values for different scene illuminances given an assumption of uniform distribution of object albedos. We can see the breakdown of the light-dark asymmetry due to saturation. At very low scene illuminances the histogram contains a single impulse at the minimum response value of the sensor. As the scene illuminance increases, some of the values rise above the minimum level, up to a value that scales with the scene illuminance. The height of the impulse at the minimum level drops as fewer sensor responses are below the minimum value. As the scene illuminance increases further, some of the incident light has an intensity above the sensors high-end saturation level. Thus an impulse at this level begins to form. As the scene illuminance increases further, there will always be some responses in the operational range of the sensor but these will become a smaller and smaller fraction of the total. Thus the histogram becomes more and more concentrated in the impulse at the high-level saturation value. Thus we can see that the histogram for the very high and very low scene illuminances are symmetric. It is only for intermediate illuminances, where the sensor does not saturate significantly, that the light-dark range asymmetry is present.

An automatic gain control, such as that provided by the pupil in the human eye, can extend the range of validity of the light-dark range asymmetry. A perfect gain control would seem to obviate the possibility of brightness perception, since the

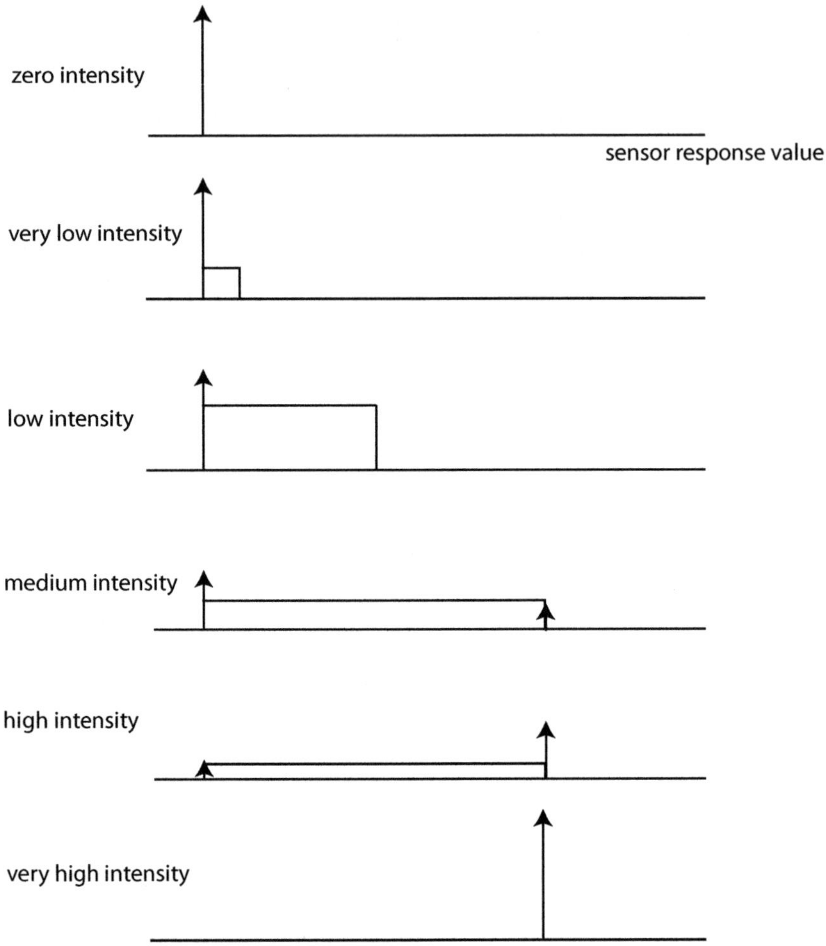

Fig. 2. Sensor saturation causes a breakdown of the light-dark range asymmetry at the extremes of scene brightness.

sensor response would always be the same. However, the gain control signal itself can be used as the brightness measure since the gain control mechanism will necessarily exhibit a light-dark asymmetry. Even if one inverted the sensor signal, the gain control signal would not be inverted (e.g. a camera's aperture would still need to be closed down as the scene illuminance increased).

2.2 Other Asymmetries

The light-dark asymmetry is not the only sort of physical asymmetry that permits differentiating between light and dark. There are also important sensori-motor asymmetries (such as what happens when you close your eyes, or turn off lights, or occlude objects) that can be used to distinguish between dark and light scenes. Shadowing is an asymmetric process. Black patches are often found in shadowed areas, whereas white patches are rarely found there. Specular highlights are very bright compared with other areas, and never darker. A strong asymmetry arises through surface inter-reflection. For example, white patches can illuminate nearby patches, while black patches do not. Langer [2] points out that shadows and inter-reflections are in some sense symmetric with each other, as an intensity inversion transforms shadows into areas that look like inter-reflections and vice-versa. The symmetry is not exact, however, and the shadows and inter-reflections that are produced are often unlikely to be observed. There are other reasons for the lack of an exact symmetry. One reason is that all white surfaces illuminate nearby objects while only some black surfaces are shadow regions. Another is that the white patches 'cause' the illumination of nearby surfaces, while shadows are caused by other surfaces. So the intensity inversion must also imply a causal inversion, as the shadow regions now become illuminating regions and vice-versa.

In color images, there are additional asymmetries to be found. As Myin [3] points out, color is just a multidimensional intensity measure, and the asymmetries associated with intensity transfer to color as well. A commonly considered transformation is spectral inversion. There are many forms of this, but the most common is the independent inversion of each channel in an RGB image (e.g. $R' = R_{max} - R, \quad G' = G_{max} - G, \quad B' = B_{max} - B$). White/Black patches are desaturated, and this persists under spectral inversion. Mid-tones are often highly saturated, and this also persists under spectral inversion. Shadow areas are always desaturated, while illuminated areas can be highly saturated. This asymmetry is reversed through spectral inversion, as dark areas appear colored and light areas (which now correspond to shadowed or dark areas in the world) appear desaturated. Thus low saturation values can indicate shadowed areas in the real-world, no matter whether the RGB values are inverted or not.

3 Statistical Measures of Scene Brightness

Figure 2 suggests that one could obtain a measure of brightness by looking at statistics of the sensor response distributions. In the range over which the light-dark range asymmetry exists, as the scene illuminance increases the sensor response distribution becomes more spread out. There are many difference statistics that could be used to capture this spreading out. For example, one could use the variance of the distribution or its entropy. Figure 3 shows the entropy of the sensor value distribution of the situation associated with figure 2. It can be seen that the light-dark range asymmetry results in a rising entropy value as long as the scene illuminance is relatively low. Beyond a certain point the high-end saturation of the sensor comes into play and begins to reduce the entropy with further increases in scene illuminance.

So far we have been considering global measures applicable to entire scenes. We could narrow our focus to look at small scene or image patches and ask whether we can find measures of patch brightness that are in some sense invariant to the specifics

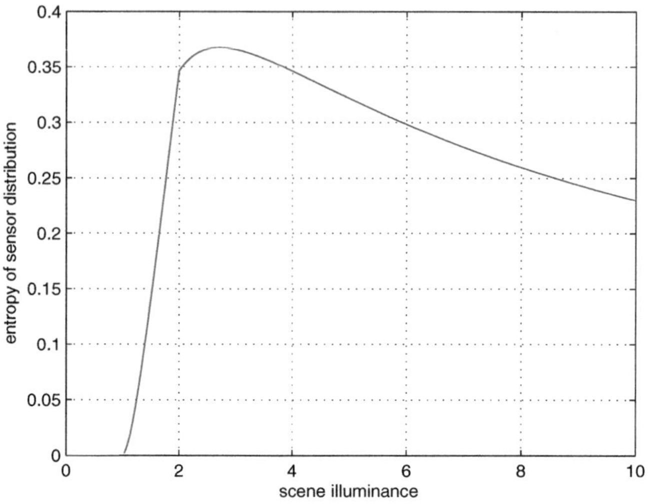

Fig. 3. Entropy of the distribution of sensor values as a function of scene illuminance for a simple scene having a uniform distribution of object albedo. The effect of the light-dark range asymmetry is evident as well as its breakdown at the extremes of scene brightness.

of the sensing process. One extension of the ideas discussed earlier is to apply the statistical measures such as variance or entropy to small patches in the image. The idea here being that bright patches would have a higher contrast measure (such as variance or entropy) than dark patches.

There is some psychological evidence for such an approach. In a study that produced the effect bearing his name, Stevens [4] found that subjects viewing a gray patch in a white surround perceived the contrast between the patches to increase as the intensity of the illumination (see figure 4 for an example of this effect). The background brightness was perceived to increase via a power law, with exponent 0.33, with respect to its luminance. The brightness of the gray patches, on the other hand, had a variable exponent, which became negative for darker patches. Overall, the effect is that the perceived contrast increased with the illumination intensity. Hunt [5] observed an analogous effect in the perception of colored patches - he found that as overall intensity increased so did the perceived colorfulness.

It has long been informally conjectured that a form of "inverse-Stevens effect" exists. That is, *perceived intensity increases with image contrast* (see figure 5 for an example of this effect). As Fairchild [6] points out, photographers often underexpose a high contrast scene (e.g. a dim indoor scene) and overexpose a low contrast scene (e.g. a bright outdoor scene). Fairchild did a psychophysical study to investigate this conjecture [6]. His results were inconclusive, however, showing a wide inter-subject variability. Some subjects had the supposed contrast-intensity relation while others had no relation, and still others had a relation in the direction opposite to that supposed.

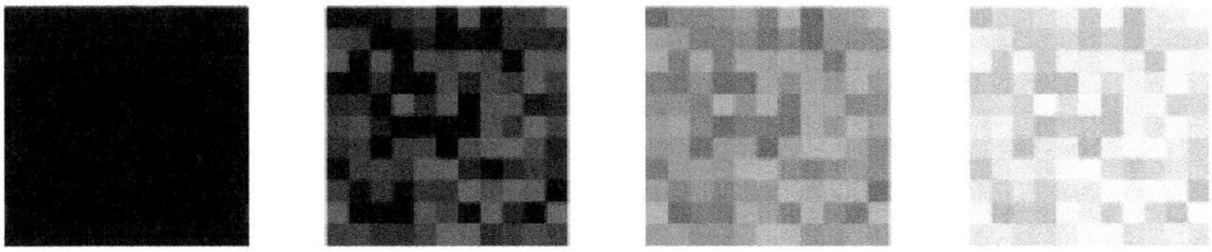

Fig. 4. The Stevens Effect: Shown are image patches with constant contrast and increasing mean intensity. Human observers usually perceive the contrast to increase along with the mean intensity of the patch.

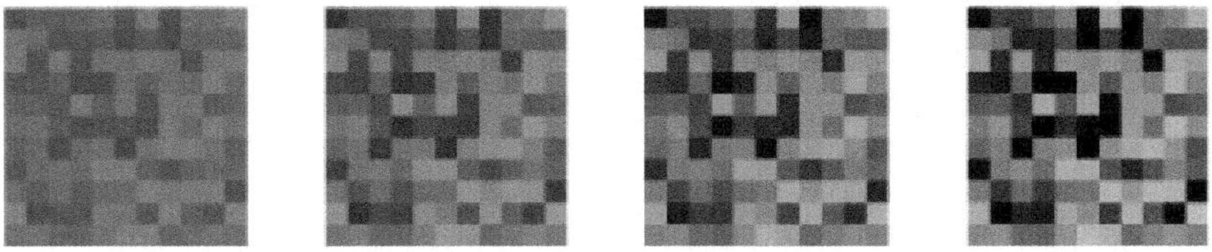

Fig. 5. Inverse Stevens Effect: Shown are image patches with constant mean intensity and increasing contrast. Some observers perceive the intensity to increase along with the contrast.

A straightforward implementation of this concept would be to measure the contrast (or entropy) of the histogram of pixel values in an image patch patch. In general, however, doing this produces only a weak dependance on intensity, and mainly produces a result similar to an *edge detection* operation. Indeed, entropy has frequently been employed as a feature in edge detection systems (e.g. [7]. In addition, as illustrated by figure 6, there can be instances of image patches that are bright but have low contrast or patch variability and others that are dark but have relatively high patch variability. To remedy these problems, we propose that the variability (entropy) of learned *conditional distributions* based on many image patches observed over time should be used. The idea is that, while in a given image patch there may only be a loose correlation between patch brightness and patch variability, a stronger correlation may be observed over a large database of image patches associated with a particular central image intensity. Thus, given a pixel or small image region with a particular intensity value, the entropy of the learned distribution of pixel values conditioned on this immediate value can be used as a measure of patch brightness.

This suggests an explanation of the Stevens effect. The idea is that, through visual experience, an observer learns an association between surface brightness and the entropy of the surface patch intensity values. Our thinking is motivated by the ideas of Norwich [8] who suggests that perception arises through reduction of uncertainty. In his view, a more intense stimulus has more uncertainty, and hence higher entropy. Furthemore, he proposes that the *subjective* impression of the intensity of a stimulus is related to the entropy of the stimulus. This leads to the hypothesis that the subjective impression of increased contrast with brighter images that comprises

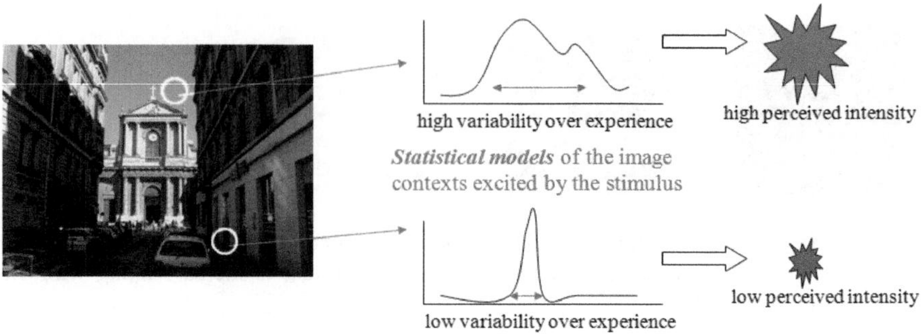

Fig. 6. Simple image patch variability measures cannot be used to measure brightness since some bright patches can have low patch variability and some dark patches can have relatively high patch variability. Instead, the variability of previously experienced image patches associated with a given central value can be used.

the Stevens Effect is a result of a learned association between contrast and entropy. That is, high contrast patches in natural images will statistically tend to have higher entropies than low contrast image patches.

4 Surround Entropy in Natural Images

To test our hypothesis that patch brightness can be related to the entropy of conditional distributions, we carried out an empirical study of the conditional statistics of surrounds in a database of natural images. For our study we used a set of 136 images from the van Hateren database [9]. Each of these images had a size of 1024x1536 pixels. Figure 7 shows four of the images that were used in our study.

The raw image values were scaled by calibration factors provided with the database. These factors account for variations in sensitivity caused by aperture settings and shutter speeds, and permit us, to some extent, to compare intensities across images in the database. Details of the image acquisition can be found in [9]. The image pixels we used were 12-bits each, with a linear intensity scale. These were obtained from the 8-bit van Hateren images which were compressed using a non-linear quantization scheme. The 12-bit linear pixels were obtained using the look-up table provided by van Hateren. We smoothed the scaled images with a 5x5 averaging kernel before computing entropies. This is to remove the residual effects of the non-uniform quantization scheme which would otherwise create a relative decrease in entropy with intensity, due to the greater spread between quantization levels at high intensities than at low. We also eliminated images which exhibited noticeable saturation at the high intensity end, as indicated by examination of the images' intensity histograms. Saturation results in an excessive number of pixels having the same value, which would reduce the compute entropy values, especially at high intensity levels. Although the images that were used in the study did not appear to exhibit any saturation, examination of the conditional histograms (in figure 8) show

Fig. 7. Samples of the images used in the empirical study. These images were taken from the van Hateren image database (van Hateran and van der Schaaf, 1998).

that there is still at least a low level of saturation, as indicated by the blip on the high end of the highest intensity conditional histogram.

To construct the conditional histograms for a given central value I, we searched the database images for pixels with values in the range $[I, I+\Delta I]$. Then we computed the histogram of the pixels in an 11x11 neighborhood centered on these pixels. These individual surround histograms were then summed to give the overall conditional histogram for the value I.

The conditional entropy is shown in figure 9. It is seen to rise, almost linearly, for low intensities, and then flatten out, and finally to drop once again.

The curves shown in figures 3 and 9 have a similar shape, and it is tempting to claim that the empirical result can be completely explained by the light-dark range asymmetry with sensor saturation. However, as mentioned earlier, we deliberately omitted images from our test set which exhibited noticeable sensor saturation, and examination of the conditional distributions shown in figure 8 reveals very little saturation, if any. The conditional variation of entropy must be due, at least in part, to other effects. Our view is that the situation is complicated, as there are many factors which act to determine the surround distribution and hence its entropy. Not all of these are dependent on the surround brightness. It may be that the human visual system is able to factor out these various contributions and isolate those that are related to intensity. The paper [10] developed statistical models for the surround using a Maximum Entropy approach and looked at the effects of three ecological processes: shadowing, occlusion, and inter-reflection. These processes all introduce

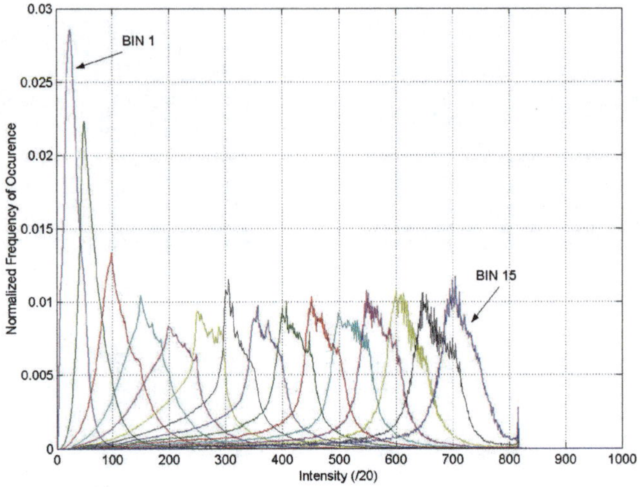

Fig. 8. The conditional histograms of the surround pixel values given the central pixel value (for 15 different ranges of central pixel values).

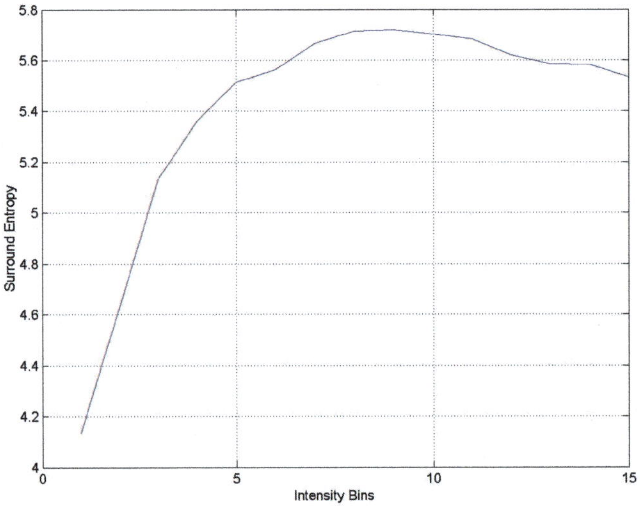

Fig. 9. The entropy of the conditional distributions of surround values given the central value, for 15 different central value ranges.

asymmetries with respect to brightness which can act together with the light-dark range asymmetry to shape the dependance of patch brightness with patch variability.

It is an open question whether humans can adapt to changes in the sensing apparatus which do not alter the physical asymmetries we have discussed. For example, one could photometrically invert the image presented to the visual system. Such an adaptive capability would be predicted by our theory. However, at least one study has been done which shows that humans are not able to re-invert inverted intensity, at least not without a long time for adaptation [11]. In particular, detection of shadows seems to rely crucially on the shadows being darker than the illumination regions [12]. This is not a fatal blow to our theory as it could be that to adapt to such inversions requires an extensive learning process in order to develop a new set of conditional statistical models.

A recent study Mante et al [13] used the van Hateren database and found no correlation between patch intensity and contrast. The patches they used were quite large, however, being circular with a diameter of 64 pixels. Also, they only looked at correlations of luminance and contrast within individual images, and then averaged these correlations across all of the images in the database. Our results are based on small patches and statistics were gathered over all images.

5 Summary

This paper suggests that various physical asymmetries, such as the light-dark range asymmetry, provide a means for measuring scene brightness that is invariant to some manipulations, for example inversion, of the sensor responses. We propose that brightness perception is mediated, not by the level of the raw photoreceptor signals, but by a statistical learning of the association between a particular localized sensor signal and the variability of a learned distribution of surround sensor values conditioned on the localized sensor signal value. Following the ideas of Norwich we propose that humans learn to associate the entropy of a stimulus with the intensity of that stimulus. We examined a small set of natural images and observed that there was, indeed, a relationship between an image pixel value and the entropy of the distribution of surrounding pixel values which would explain the Stevens effect - the perceived increase of contrast as scene intensity increases.

References

1. Hartline, H.K.: The response of single optic nerve fibers of the vertebrate eye to illumination of the retina. American Journal of Physiology 121, 400–415 (1938)
2. Langer, M.S.: A model of how interreflections can affect color appearance. Technical Report No. 70, Max-Planck Institute for Biological Cybernetics, Tuebingen, Germany (1999)
3. Myin, E.: Color and the duplication assumption. Synthese 129, 61–77 (2001)
4. Stevens, S.S.: To Honor Fechner and Repeal His Law. Science 133, 80–86 (1961)
5. Hunt, R.W.G.: The Reproduction of Color, 5th edn. Fountain Press, Kingston-upon-Thames (1995)
6. Fairchild, M.D.: A Victory for Equivalent Background – On Average. In: IS&T/SID 7th Color Imaging Conference, Scottsdale, pp. 87–92 (1999)
7. Shiozaki, A.: Edge extraction using entropy operator. Computer Vision, Graphics, and Image Processing 36, 1–9 (1986)
8. Norwich, K.: The psychophysics of taste from the entropy of the stimulus. Perception and Psychophysics 35(3), 269–278 (1984)

9. van Hateren, J.H., van der Schaaf, A.: Independent component filters of natural images compared with simple cells in primary visual cortex. Proc. R. Soc. Lond. B 265, 359–366 (1998)
10. Clark, J.J., Hernandez, D.: Surround statistics and the perception of intensity and color. In: 3rd International Workshop on Statistical and Computational Theories of Vision, Nice (October 2003)
11. Anstis, S.M.: Visual adaptation to a negative, brightness-reversed world: Some preliminary observations. In: Carpenter, G.A., Grossberg, S. (eds.) Neural Networks for Vision and Image Processing, pp. 1–14. MIT Press, Cambridge (1992)
12. Cavanagh, P., Leclerc, Y.G.: Shape from shadows. J. Exp. Psychol. Hum. Percept. Perform 15(1), 3–27 (1989)
13. Mante, V., Bonin, V., Geisler, W.S., Carandini, M.: Independence of luminance and contrast in natural scenes and in the early visual system. Nature Neuroscience 8, 1690–1697 (2005)

A Learning-Based Patient Repositioning Method from Limited-Angle Projections

Chen-Rui Chou, C. Brandon Frederick, Sha X. Chang, and Stephen M. Pizer

Abstract. This paper presents a novel patient repositioning method from limited-angle tomographic projections. It uses a machine learning strategy. Given a single planning CT image (3D) of a patient, one applies patient-specific training. Using the training results, the planning CT image, and the raw image projections collected at the treatment time, our method yields the difference between the patient's treatment-time postition and orientation and the planning-time position and orientation. In the training, one simulates credible treatment-time movements for the patient, and by regression it formulates a multiscale model that expresses the relationship giving the patient's movements as a function of the corresponding changes in the tomographic projections. When the patient's real-time projection images are acquired at treatment time, their differences from corresponding projections of the planning-time CT followed by applications of the calculated model allows the patient's movements to be estimated. Using that estimation, the treatment-time 3D image can be estimated by transforming the planning CT image with the estimated movements,

Chen-Rui Chou
Department of Computer Science, University of North Carolina at Chapel Hill, 201 South Columbia Street, Chapel Hill, NC 27599-3175 USA
e-mail: cchou@email.unc.edu

C. Brandon Frederick
Department of Biomedical Engineering, University of North Carolina at Chapel Hill, 152 MacNider Hall, Chapel Hill, NC 27599-3175 USA
e-mail: cbfreder@email.unc.edu

Sha X. Chang
Department of Radiation Oncology, University of North Carolina at Chapel Hill, 101 Manning Drive, Chapel Hill, NC 27599-3175 USA
e-mail: sxchang@med.unc.edu

Stephen M. Pizer
Departments of Computer Science and Radiation Oncology, University of North Carolina at Chapel Hill, 201 South Columbia Street, Chapel Hill, NC 27599-3175 USA
e-mail: pizer@cs.unc.edu

J. Angeles et al. (Eds.): Brain, Body and Machine, AISC 83, pp. 83–94.
springerlink.com

and from this, changes in the tomographic projections between those computed from the transformed CT and the real-time projection images can be calculated. The iterative, multiscale application of these steps converges to the repositioning movements. By this means, this method can overcome the deficiencies in limited-angle tomosynthesis and thus assist the clinician performing an image-guided treatment. We demonstrate the method's success in capturing patients' rigid motions with subvoxel accuracy with noise-added projection images of head and neck CTs.

1 Introduction

Approximately 40% of people with cancer have radiotherapy as part of their treatment. While damaging cancer cells in the tumor, radiation can also affect surrounding healthy cells if it is not directed with a high degree of accuracy. In order to attain high accuracy, a real-time imaging device can be employed to guide the radiation treatment so that the patient's treatment region is rigidly well-aligned with that region at planning time. This approach is known as image-guided radiotherapy (IGRT). To support IGRT, traditional imaging approaches, such as computed tomography (CT), are not very appropriate. Despite CT's fine 3D spatial resolution in imaging tissue, the expense of having a CT scanner in each treatment room, the long image acquisition time and the intense dose of ionizing radiation make its direct application in real-time imaging impractical. As described in section 2, recent tomographic imaging advances, like nanotube stationary tomosynthesis (NST) [1] and cone-beam CT (CBCT), are designed to decrease the number of imaging sources or limit the range of x-ray projection orientations to achieve higher image acquisition speed, lowered dose and lowered cost. However, with the deficiency in geometry, traditional tomographic reconstruction methods, like filtered-back projection (FBP) or the simultaneous algebraic reconstruction technique (SART) [2], cannot reconstruct a 3D image with full volumetric information but rather induce many streaking artifacts. (See Fig.1)

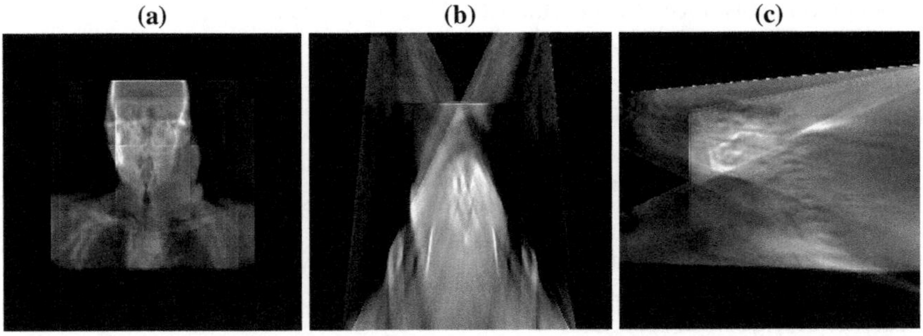

Fig. 1. **(a)** Coronal, **(b)** axial and **(c)** sagittal views of the head and neck image reconstructed from NST's 52 projections. The reconstruction method is SART with 20 iterations.

One way to overcome this geometric deficiency is to use the patient's planning CT image as prior information. With this prior information, one can utilize image registration techniques to register the prior image to either the 2D raw projections [3] or the reconstructed 3D image [4, 5, 6, 7]. In [5, 6, 7], mutual information was used to register the planning CBCT to a 3D image reconstructed from a set of limited-angle projections (80 projections in 45° DTS [5, 6, 7].) Their registration quality varied with the landmark objects used for MI calculations: with bony structures, the mean shift error is about 2 mm [5]; while with soft tissues, there are up to 1.7 mm "one-dimensional" shift errors [7]. Apparently, even when high contrast bony structures are present, their methods require many projections to detect patient's movements. In [8], they register the prior CT (3D) images to several X-ray projection (2D) images by maximizing the similarity between the X-ray projection images and those in the CT projection space. However, almost half of their test cases failed to converge to the correct positions when the initial shifts are more than 1 cm.

Our method, on the other hand, can capture patient's rigid motions yielding displacements in the treatment region up to 3 cm. Moreover, our method requires very few projections (e.g., 2 for both 5° CBCT and NST) but yields patient's treatment-time 3D motions with sub-voxel accuracy in simulated oblique-angle NST and simulated limited-angle CBCT for the head and neck.

Our method incorporates the image-residuals-to-shape-parameters learning strategy in Active Appearance Models (AAMs) [9]. It involves two stages: training and treatment. In the training stage we sample from a range of patient movements, and for each such movement we generate 2D projections[1] by transforming and reprojecting the patient's planning CT. We compute a linear regression between the patient movements and the difference between the projections of the moved CT and those of the CT in the planning position. In the treatment stage, the learned regression model is applied iteratively to the successive residues between the real-time projections and those of the moving CT transformed by the previously predicted parameters. This iteration yields the predicted transformations with high accuracy.

This paper is organized as follows: In section 2, we sketch the imaging arrangements of NST and of limited-angle CBCT. In section 3, we explicitly describe our method with three parts: training, treatment and an implementation design. In section 4, we demonstrate the numerical results and analyze upon them. Finally, we conclude our paper, make some interesting observations and set future work in section 5.

2 Imaging Geometries

While neither CBCT and NST are not presently available for our experiments, we simulate their IGRT projection images by reprojecting the planning-time CT based on the corresponding IGRT geometries. Therefore we describe the imaging geometries as below.

[1] 2D projection is summation of intensities along each ray from a putative x-ray source to a pixel.

2.1 Nanotube Stationary Tomosynthesis (NST)

NST [1] is a stationary imaging system that can perform imaging without interfering with treatment delivery. As illustrated in Fig. 2 (a), it consists of an arrangement of radiation sources arrayed around the treatment portal, together with a planar detector. Firing the sources in sequence produces a different 2D projection image per source. Each projection image requires ~200 ms. The imaging dose is very low relative to CT.

2.2 Limited-Angle Cone-Beam CT (CBCT)

CBCT is a rotational imaging system with a single radiation source and a planar detector. This pair rotates by a limited angle during IGRT, taking projection images during traversal (Fig. 2 (b)). The limited-angle scan provides the shortened imaging time and the lowered imaging dose, e.g., for 5° rotation, it takes less than one second.

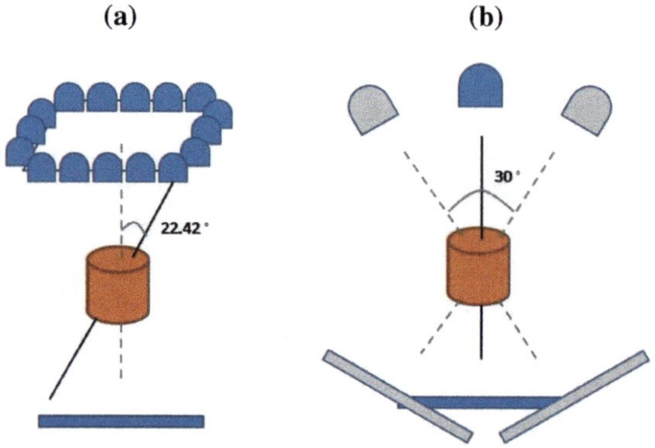

Fig. 2. **(a)** The NST geometry: stationary sources arrayswith max tomographic angle=22.42° **(b)** The 30° CBCT geometry: rotational imaging system with tomographic angle=15°.

3 Method

Our method is made up of two stages: training (sec. 3.1) and treatment (sec. 3.2). A particular form of training, which operates hierarchically by scale, has turned out to yield a good implementation (sec. 3.3).

3.1 The Training Stage

The training stage requires a patient's planning 3D image with full volumetric information, e.g., a CT image: CT_{plan}, prior to the radiotherapy. The planning image, CT_{plan}, is used as a reference to generate N possible treatment-time images, $\mathbf{CT}_{trans}$, by a collection $\mathbf{C}$ of credible rigid transformations C_i. The rigid transformations are sampled from a space Ω represented by the six rigid-motion parameters t_x, t_y, t_z for translation; r_x, r_y, r_z for rotation (Euler angles):

$$\mathbf{CT}_{transf} = \{CT^i_{transf} = Transform(CT_{plan}, C_i) \mid \bigcup_{i=1}^{N} C_i \in \mathbf{C}\} \tag{1}$$

$$\mathbf{C} = span(t_x, t_y, t_z, r_x, r_y, r_z) \in \Omega \tag{2}$$

At training time our method computes the residuals between 2D projections of CT_{plan} and the corresponding projections of $\mathbf{CT}_{trans}$. The tomographic projections are generated by reprojecting the 3D images, with operator $\mathbf{P}$, based on the geometry of the target machine, e.g. NST or limited-angle CBCT. The residuals array $\mathbf{R}$ is a concatenation, over the projections of the differences between corresponding projection pairs. In the work described here the difference between two projections is simply the pixel-by-pixel tuple of intensity differences.

$$\mathbf{R}_{i,j} = P_j(CT^i_{transf}) - P_j(CT_{plan}), \tag{3}$$

where P_j is the j-th projection in $\mathbf{P}$ and $1 \leqslant j \leqslant M$.

$$\mathbf{R} = \mathbf{P}(\mathbf{CT}_{transf}) - \mathbf{P}(CT_{plan}) \tag{4}$$

$$= \begin{bmatrix} P_1(CT^1_{transf}, C_1) - P_1(CT_{plan}) & \cdots & P_M(CT^1_{transf}, C_1) - P_M(CT_{plan}) \\ P_1(CT^2_{transf}, C_2) - P_1(CT_{plan}) & \cdots & P_M(CT^2_{transf}, C_2) - P_M(CT_{plan}) \\ \vdots & \ddots & \vdots \\ P_1(CT^N_{transf}, C_N) - P_1(CT_{plan}) & \cdots & P_M(CT^N_{transf}, C_N) - P_M(CT_{plan}) \end{bmatrix}$$

We calculate the matrix β that when multiplied by the residuals array $\mathbf{R}$ minimizes squared errors in the prediction of the transformation parameters $\mathbf{C}$ (5). That is, our estimate $\hat{\beta}$ is calculated by a deterministic linear regression (6).

$$\mathbf{C} = \mathbf{R} \cdot \beta + \varepsilon \tag{5}$$

$$\hat{\beta} = (\mathbf{R}^{\mathbf{T}}\mathbf{R})^{-1}\mathbf{R}^{\mathbf{T}}\mathbf{C} \tag{6}$$

Fig. 3 (a) is the flow chart of the training procedure.

3.2 *The Treatment Stage*

At treatment time the objective of each iteration i is to produce a geometric transformation C_i to be applied to the present best estimate of the treatment CT, as well as a new estimate of the treatment CT. Once the patient's real-time projections $\mathbf{P}_{real-time}$ are collected, we apply an iterative procedure, (7)-(9), to transform and reproject the planning 3D image such that its tomographic projections, $\mathbf{P}(CT_{moving})$, are consistent with the real-time ones. The transformation C_i is estimated by the projection

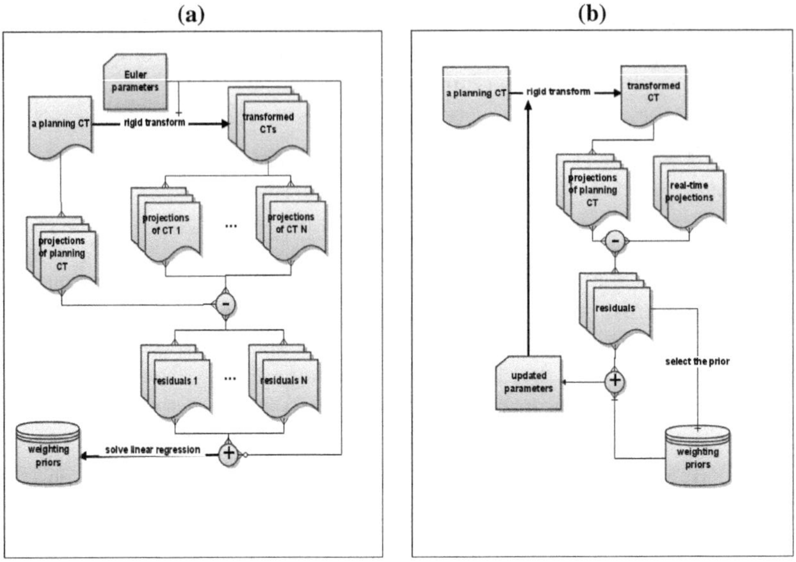

Fig. 3. Flow chart of **(a)** the training procedure and **(b)** the treatment procedure.

residuals **R** and the weightings $\hat{\beta}$ calculated in the training stage. The reprojection process is implemented on GPUs to make it run within 10 ms.

$$\mathbf{R} = \mathbf{P}_{real-time} - \mathbf{P}(CT_{moving}) \tag{7}$$

$$C_i = \mathbf{R} \cdot \hat{\beta} \tag{8}$$

$$CT_{moving} = Transform(CT_{moving}, C_i) \tag{9}$$

Because the composite changes of multiple rigid-motion parameters may be linearly dependent with other composite changes, instead of using the full transformation computed by (8), at each iteration of (7)-(9) we transform the moving image by the one rigid-motion parameter in C_i that decreases the magnitude of the residual $||\mathbf{R}||_2$ most. The method is taken to have converged if the magnitude of the residual is below the threshold. Fig. 3 (b) is the flow chart of the treatment procedure.

3.3 Hierarchical Trainings

It is unrealizable to enumerate many combinations in **C**. Therefore, we used a piecewise approximation scheme called "hierarchical training":

In the training stage, K hierarchical weighting matrices $\hat{\beta}_1, \hat{\beta}_2, \cdots, \hat{\beta}_K$ are generated from large to small scales of training. At the k-th level of training, the transformation parameters are collected from the combinations of translations with $\pm\mu_k, 0$ (cm) and rotations with $\pm\nu_k, 0$ (degrees). In order to have accurate estimations in the whole training domain, the selection of μ_k and ν_k depends on the interpolation accuracy of $\hat{\beta}_k$ calculated in (6) such that the union of each level's confidence intervals, $\tau^k_{translation}$ and $\tau^k_{rotation}$, covers the whole training domain. (See 1D example in Fig. 4)

$$\bigcup_{k=1}^{K} \tau^k_{translation} \supseteq [-Tr_{limit}, Tr_{limit}], \bigcup_{k=1}^{K} \tau^k_{rotation} \supseteq [-Ro_{limit}, Ro_{limit}] \qquad (10)$$

where $\pm Tr_{limit}$ and $\pm Ro_{limit}$ are the training limits for translations and rotations, respectively. For each level of training, the training values in **C** are formed by all combinations, over the six rigid motion parameters, of $\pm Tr_{limit}$ and $\pm Ro_{limit}$, respectively.

In the treatment stage the calculated hierarchical weighting matrices are applied sequentially, from $\hat{\beta}_1$ to $\hat{\beta}_K$, to give new reposition directions when our optimizer fails to decrease the magnitude of the residual by all of the six rigid-motion parameters. In our current implementation, we applied a four-level hierarchy ($K = 4$) with $\mu_k = (K-k+1) \cdot Tr_{limit}/K$, $\nu_k = (K-k+1) \cdot Ro_{limit}/K$. As a result, after applying the hierarchical training, our optimizer can almost surely converge to the global minimum in the residual space.

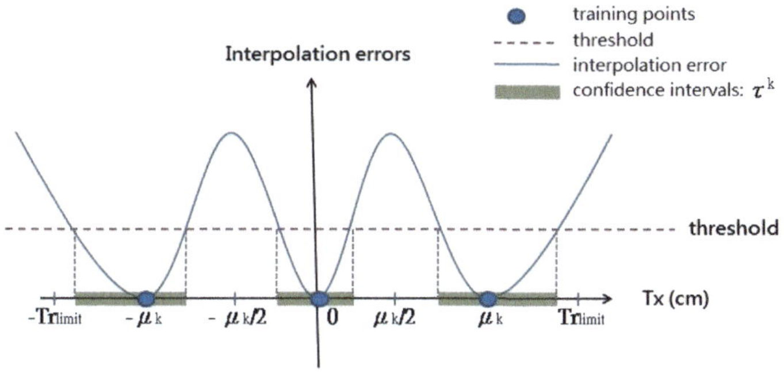

Fig. 4. An illustration of the confidence intervals of the translations along x-axis in the k-th level of training. $\pm\mu_k$, 0 are the training points. Confidence intervals τ^k are calculated by a threshold that outputs the limits of the tolerable wrong estimations. $\pm Tr_{limit}$ are the training limits of translations.

4 Results

4.1 Test Environments

We tested our method using three patients' (P5202, P5203, P5204) head and neck CTs with a voxel size of 0.12 cm lateral $\times$ $0.12 cmanterior - posterior \times$ 0.30 cm superior-inferior. The simulated treatment-time projections were generated by transforming and reprojecting the patient's planning CT and then adding Gaussian-distributed noise to each pixel. A total of 180 testing moves, 90 in NST and 90 in CBCT, were combinatorial in both translations and rotations and were randomly picked within -2 to 2 cm and -5 to 5 degrees, respectively. In CBCT, imaging at a variety of arc angles have been tested; the performance with the smallest arc angle with good performance, 5 degrees, is reported. In both NST and CBCT experiments the number of imaging positions was varied to find the minimum number with sub-voxel accuracy.

Zero mean, constant standard deviation Gaussian noise was added to the simulated IGRT projection images. The standard deviation σ of the noise=0.2·(mean bony intensity−mean soft tissue intensity). This noise level is far higher than that produced in either NST or CBCT. Example projection images are shown in Fig. 5.

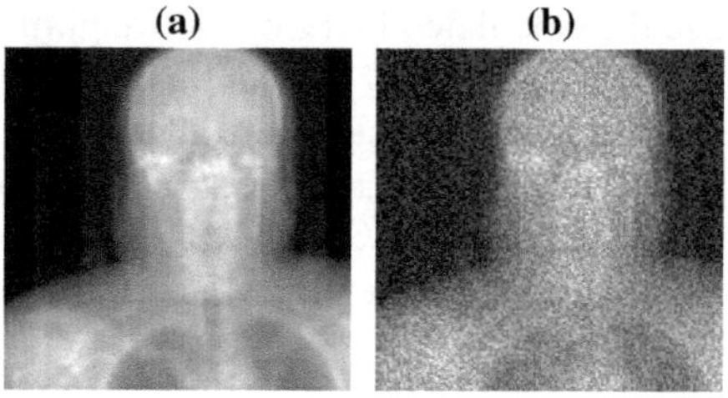

Fig. 5. Sample simulated CBCT projection images: **(a)** before adding noise **(b)** after adding Gaussian noise.

The errors are measured by mean absolute error (MAE) and mean target registration error (mTRE). The MAE in any dimension is the mean, over the test cases, of the error magnitude in the corresponding dimension of **C**. mTRE for a case is the mean displacement error, over all voxels in a $16 \times 16 \times 16\,\text{cm}^3$ bounding box (probable tumor region) centered at the center of the pharynx in the planning image (11).

$$mTRE = \frac{1}{N}\sum_{i=1}^{N} \|T_{true}\cdot \mathbf{x}_i - T_{est}\cdot \mathbf{x}_i\|_2 \tag{11}$$

where N is the number of pixels in the probable tumor region, $\mathbf{x}_i$ is the tuple of the i-th pixel position, and T_{true}, T_{est} are the true and the estimated transformation matrices, respectively.

4.2 Tests on Oblique-Angle NST and Limited-Angle CBCT

We first studied how many projection images are needed to obtain sub-voxel accuracy. The results on patient P5202, displayed in Fig. 6, show that two projection images are enough for both NST and 5° CBCT to have sub-voxel accuracy. Fig. 7 shows the geometries for the two x-ray sources in NST and CBCT, respectively.

Next we studied the accuracy of our method with the geometries shown in Fig. 7. The results, over all three patients are displayed in Tables 1 and 2, are that both the mean absolute errors (MAEs) and their standard deviations (SDs) are less than 0.09

(a) **(b)**

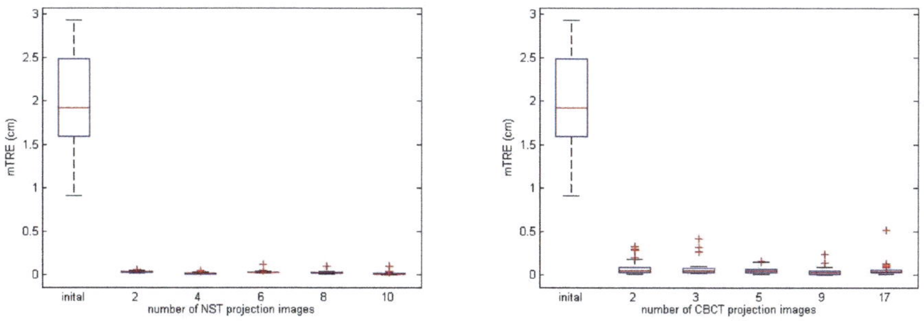

Fig. 6. mTRE statistics on a set of 30 random moves of patient P5202 with **(a)** NST, 4 level of training and the varied numbers of projection images **(b)** CBCT, 4 level of training and the varied numbers of projection images.

(a) **(b)**

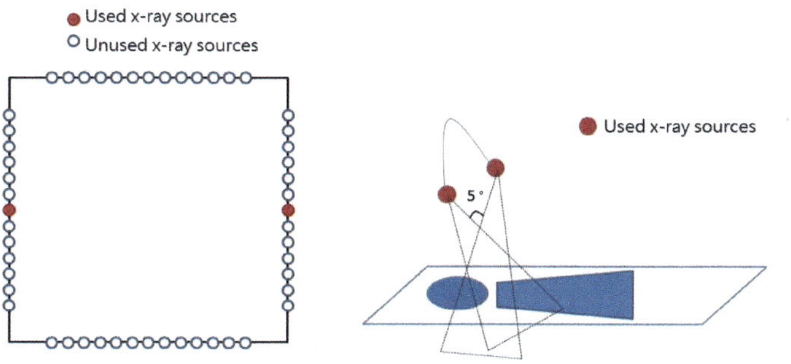

Fig. 7. **(a)** The geometry of the 2 x-ray sources used in generating the simulated NST projections. **(b)** The geometry of the 2 x-ray sources used in generating the simulated 5° CBCT projections.

Table 1. Position and Orientation Errors - NST with 2 projections

	Tx(cm)	Ty(cm)	Tz(cm)	Rx(deg)	Ry(deg)	Rz(deg)	mTRE(cm)
MAE	0.0094	0.0302	0.0262	0.1489	0.0248	0.1540	0.0524
SD	0.0085	0.0211	0.0715	0.1093	0.0174	0.2824	0.0728

Table 2. Position and Orientation Errors - 5° CBCT with 2 projections

	Tx(cm)	Ty(cm)	Tz(cm)	Rx(deg)	Ry(deg)	Rz(deg)	mTRE(cm)
MAE	0.0165	0.0904	0.0353	0.0877	0.0302	0.1334	0.1094
SD	0.0141	0.2007	0.0817	0.0845	0.0217	0.2998	0.2130

deg: degree; axis directions– x: lateral, y: anterior-posterior, z: superior-inferior;
T: translation in the specified axis direction; R: rotation about the specified axis

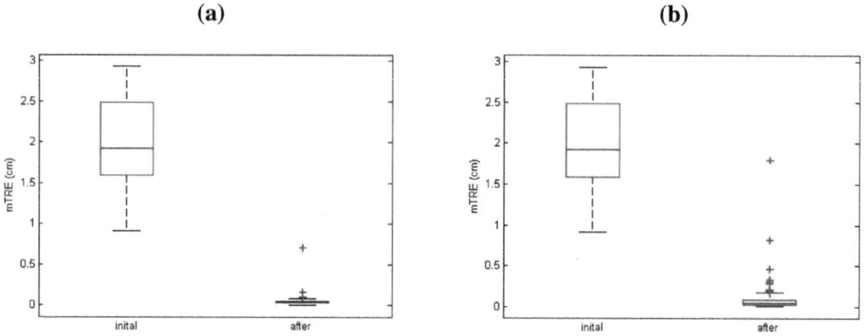

Fig. 8. The mTRE in **(a)** 90 NST and **(b)** 90 CBCT test cases across three patients. The mTREs between the target and the planning images are marked as "initial," and the mTREs between the estimated and the target images are marked as "after." The means and the standard deviations of mTRE after repositioning are 0.0524 ± 0.0728 cm and 0.1094 ± 0.2130 cm for NST and CBCT test cases, respectively.

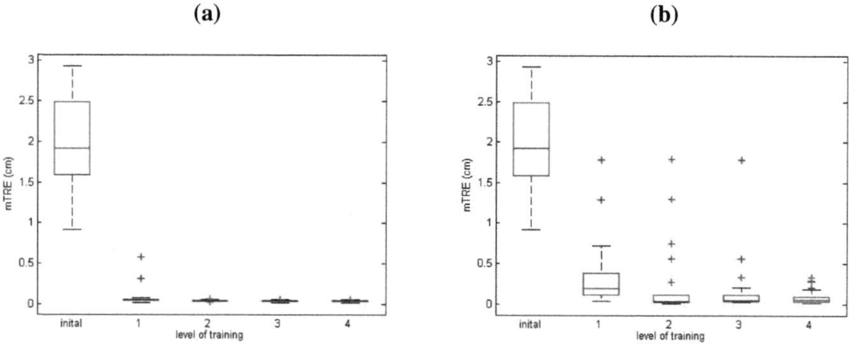

Fig. 9. mTRE statistics on a set of 30 random moves of patient P5202 with **(a)** NST, 2 projection images and the varied levels of training **(b)** CBCT, 2 projection images, 5 degrees and the varied levels of training.

cm in each dimension of translation and less than 0.3 degree in each dimension of rotation. Moreover, in Fig. 8, 97.8% (176 out of 180) of mTREs in our test cases are less than one voxel size, 0.34 cm. Due to the larger tomographic angle of NST (about 20°) rather than CBCT's (2.5°), the NST registration errors are smaller than the registration errors of CBCT. In addition, for both NST and CBCT, the registration errors in Ry are smaller because major projection images lie in the patient's plane which is parallel to the rotation plane of Ry.

Finally we studied the effect of how many levels of training were used. Fig. 9 (a)(b) show that increasing the level of training significantly reduces the registration errors in both NST and CBCT.

5 Discussion and Conclusion

Remarkably, our novel patient repositioning method requires only two projections and as little as 2.5° tomographic angle but yields patient's treatment-time position and orientation for the head and neck with sub-voxel accuracy as well as a fully-volumetric estimated treatment-time 3D image. Our method is a general patient positioning scheme for IGRT across limited-angle imaging modalities. The imaging time required is about 400 ms for NST and about 1 second for CBCT with 2.5° tomographic angle. After image acquisition, the average computation time of our method is 3 seconds (can be extended to half second with multithreading). Thus the full method is fast enough for clinical use.

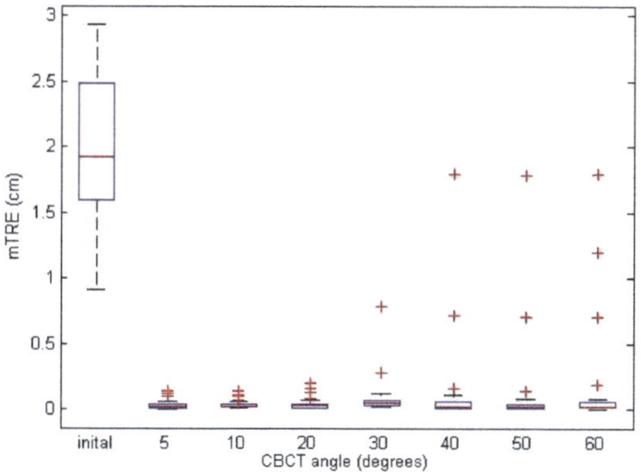

Fig. 10. mTRE statistics on a set of 30 random moves of patient P5202 with 2 CBCT projection images, 4 level of training and the varied angles.

Fig. 10 studies the repositioning errors as a function of CBCT projection angle. In each case two projections are used. Surprisingly, the number of repositioning values with more than sub-voxel errors increases with the CBCT projection angle. We must study this behavior further but suggest it shows a sensitivity of our method to the initial values of the six rigid-motion parameters or it implies that the confidence interval of our 4-level training doesn't completely cover the training domain.

Our method has been tested only with head and neck images simulated from patient CTs. It needs to be tested on real IGRT images. Also, it needs to be tested on CT images with lower contrast, such as those of soft tissue. Initial testing on simulated rigid transformations of soft tissue images, namely those of the breast, have shown good accuracy.

A future extension of our method is to capture not only the rigid transformations but also those in combination with treatment-time organ/tissue warps. The approach will use a shape space of credible warps.

Acknowledgements. This research is sponsored by Siemens Medical Solutions. The authors would also like to thank Gregg Tracton, Michael Lawrence, and David Lalush for their advice on this work.

References

1. Maltz, J.S., Sprenger, F., Fuerst, J., Paidi, A., Fadler, F., Bani-Hashemi, A.R.: Fixed gantry tomosynthesis system for radiation therapy image guidance based on a multiple source x-ray tube with carbon nanotube cathodes. Medical Physics 36(5), 1624–1636 (2009)
2. Andersen, A.H., Kak, A.C.: Simultaneous algebraic reconstruction technique (SART): a superior implementation of the ART algorithm. Ultrasonic Imaging 6(1), 81–94 (1984)
3. Sadowsky, O., Ramamurthi, K., Ellingsen, L.M., Chintalapani, G., Prince, J.L., Taylor, R.H.: Atlas-assisted tomography: registration of a deformable atlas to compensate for limited-angle cone-beam trajectory. In: 3rd IEEE International Symposium on Biomedical Imaging: Nano to Macro, pp. 1244–1247 (2006)
4. Thilmann, C., Nill, S., Tucking, T., Hoss, A., Hesse, B., Dietrich, L., Bendl, R., Rhein, B., Haring, P., Thieke, C.: Correction of patient positioning errors based on in-line cone beam CTs: clinical implementation and first experiences. Radiation Oncology 1(1), 16 (2006)
5. Wu, Q.J., Godfrey, D.J., Wang, Z., Zhang, J., Zhou, S., Yoo, S., Brizel, D.M., Yin, F.F.: On-board patient positioning for head-and-neck IMRT: comparing digital tomosynthesis to kilovoltage radiography and cone-beam computed tomography. International Journal of Radiation Oncology, Biology, Physics 69(2), 598–606 (2007)
6. Yoo, S., et al.: Clinical evaluation of positioning verification using digital tomosynthesis (DTS) based on bony anatomy and soft tissues for prostate image-guided radiation therapy (IGRT). International Journal of Radiation Oncology, Biology, Physics 73(1), 296 (2009)
7. Zhang, J., Wu, Q.J., Godfrey, D.J., Fatunase, T., Marks, L.B., Yin, F.F.: Comparing digital tomosynthesis to Cone-Beam CT for position verification in patients undergoing partial breast irradiation. International Journal of Radiation Oncology, Biology, Physics 73(3), 952–957 (2009)
8. Russakoff, D., Rohlfing, T., Maurer, C.: Fast intensity-based 2D-3D image registration of clinical data using light. In: Proceedings. Ninth IEEE International Conference on Computer Vision, vol. 1, pp. 416–422 (2003)
9. Cootes, T.F., Edwards, G.J., Taylor, C.J.: Active appearance models. IEEE Transactions on Pattern Analysis and Machine Intelligence 23(6), 681–685 (2001)

Image and Video Region Saliency Based on Space and Motion

Jian Li, Martin Levine, Xiangjing An, Zhenping Sun, and Hangen He

Abstract. This paper proposes a new bottom-up paradigm for detecting visual saliency in images and videos, which is based on scale space analysis of the log amplitude spectrum of natural images and videos. A salient region is assumed to be any region exhibiting a distinct pattern whose intensity, color, texture and motion is different from the rest of the image. Thus patterns which appear frequently as well as uniform regions are suppressed to produce salient region pop-out. We show that the convolution of the image log amplitude spectrum with a low-pass Gaussian kernel (at the proper scale) is equivalent to such an image saliency detector. A saliency map can then be obtained by reconstructing the 2-D signal using the original phase spectrum and an appropriately filtered log amplitude spectrum to produce pop-out. The optimal scale for each image feature channel (intensity, color, motion) is determined by minimizing the entropy of its saliency map. This produces four maps which are then fused by a weighted linear combination. Significantly, the approach does not require the setting of any parameters. We demonstrate experimentally that the proposed model has the ability to highlight small and large salient regions and to inhibit repeating patterns in both images and videos.

1 Introduction

Object detection in images and videos has been and continues to be of great interest in computer vision. One aspect regarding such computational methods deals with the issue of whether to sample images densely or at certain salient points. The latter is not only perceptually plausible, but also would reduce computing costs [1, 2, 3, 4, 5]. Researchers have always realized that locating salient regions in a scene is a

Jian Li · Xiangjing An · Zhenping Sun · Hangen He
Institute of Automation, Natl. Univ. of Defense Technology, P.R. China
e-mail: lijian@nudt.edu.cn

Martin D. Levine
Center for Intelligent Machines, McGill Univ., Canada
e-mail: levine@cim.mcgill.ca

J. Angeles et al. (Eds.): Brain, Body and Machine, AISC 83, pp. 95–109.
springerlink.com

key step to efficient object detection. The primary drawback with this approach has been the lack of robust methods for computing saliency, that do not involve setting a myriad of parameters, an issue which could be called curse of uncertainty.

Visual saliency has received extensive attention by both psychologists and computer vision researchers [6, 1, 2, 7, 8, 9, 3, 4, 10, 11]. It is usually assumed that objects in a scene will influence each other by suppression. Consequently, many regions are inhibited, while those that are not will ultimately predominate in the visual cortex to provide a focus of attention [7, 8].

Many bottom-up computational models, which simulate primate perceptual abilities, have appeared in the literature. For example, in [6, 1] a center-surround mechanism is used to define saliency, which is inspired by the putative neural mechanism. It has also been hypothesized that certain visual inputs are intrinsically salient in certain background contexts and that these are actually task-independent [1]. Furthermore, this saliency is a bottom-up process that produces object pop-out.

Recently, a simple and fast algorithm for image saliency extraction, called the spectrum residual (SR), was proposed based on the Fourier Transform [12]. It was argued that the spectrum residual corresponds to the saliency in an image. This was followed by a paper in which the phase spectrum of the Fourier Transform (PFT) was used to compute saliency and which achieved nearly the same performance as SR [13, 5]. The authors also argued that the key to saliency detection is image phase information, but not the amplitude [13]. However, it can be shown that both SR and PFT are, to large extent, equivalent to a gradient operator combined with a low-pass Gaussian filter. Hence, they will mainly respond to the boundaries of objects as well as richly textured regions in natural images. Considering PFT as an example, an image I is transformed into the frequency domain, and the amplitude $A(u,v)$ and phase $p(u,v)$ spectra are calculated. Then, the saliency map $sM(x,y)$ is obtained by computing $sM(x,y) = g(x,y) * \parallel F^{-1}[e^{(0 \cdot log(A(u,v)))+i \cdot p(u,v)}] \parallel^2$, where $g(x,y)$ is a 2-D low-pass Gaussian filter. Obviously, the amplitude information has been totally abandoned, since $A(u,v)$ is replaced by $e^{(0.log(A(u,v)))} = 1(u,v)$ which is a horizontal plane in the frequency domain. However, according to the statistical characteristics of natural images, the low-frequency energy is more significant than the high-frequency energy. Therefore, replacing the amplitude spectrum by a horizontal plane implies suppressing the low-frequency components or, equivalently, enhancing high-frequency components (gradient enhancement). Objectively, a saliency detector should highlight uniform salient regions [14] and suppress the cluttered background. Although the smoothing filter $g(x,y)$ has been adopted to alleviate this problem in both SR and PFT, they are still unable to highlight the uniform salient regions when the size of object is not sufficiently small.

In this paper, we assume that an image or video consists of salient and uniform regions. A salient region is defined as one in which the intensity, color, texture and motion (for videos) are different from the rest of the image, while any existing uniform patterns should be suppressed to promote the pop-out of salient regions. We propose a new saliency detection framework inspired by [7, 1, 12, 13], in which the salient regions are detected by analyzing the spectrum scale-space (SSS). For natural images, the amplitude and phase spectra contain different information. As we know, the convolution theorem states that convolution in one domain equals point-wise multiplication in the other, which is not the approach that we have used. We map an image under consideration from the spatial domain to the frequency domain and obtain a complex matrix (amplitude and phase). Next we convolve the log amplitude spectrum with a Gaussian kernel yielding a low-pass filtered version of the spectrum. The latter, using the original phase spectrum, is transformed into the spatial domain where its absolute value is processed by a low-pass filter. Surprisingly, it is this result that produces a saliency detector.

Additionally, using Gaussian kernels at multiple scales in the frequency domain will produce a range of salient region sizes. In the proposed model, a set of saliency maps is computed at different scales for each feature and entropy is used to determine the best saliency map for each feature channel (for example, the intensity channel, color channels, and motion channel). The intensity and color channels are represented using the opponent color space; the motion channel is obtained by estimating the derivative at each pixel location in the current frame using data from five adjacent frames. Finally, the best saliency maps are fused by calculating a weighted linear combination of the selected individual saliency maps.

The contributions of this paper are as follows: 1) We argue that convolution of the log amplitude spectrum with a low-pass filter is equivalent to an saliency detector; 2) We propose a new bottom-up saliency model, which can highlight both small and large salient regions and inhibit repeating patterns; 3) We show that SR, PFT and the frequency-tuned approach [14] are, to some extent, special cases of the proposed model; and 4) We extend the framework to include video saliency detection.

2 The Methodology

Many computational models of attention search for anomalous patterns in an image [11]. In this paper, a salient region is considered to be a distinctive pattern which possesses a distinct feature distribution when compared with the rest of the image. (See Fig. 1). In this section, we first present the theoretical background behind the proposed framework. Then we introduce a computational approach for producing an image saliency map. Finally, we extend the framework from a 2D image to a 3D video and describe how to compute video saliency.

2.1 Suppressing Repeating Patterns for Saliency Pop-Out

In the proposed model, it is assumed that an image consists of several salient regions and many uniform regions[1]. All of these entities (whether distinct or not) may be considered as visual stimuli that compete for attention in the visual cortex [7].

The primate visual system is more sensitive to distinctive rather than repeating patterns in the image. In a natural image, there exist many kinds of repeating

[1] Background patches are considered to be either uniform regions or textured patterns in this paper.

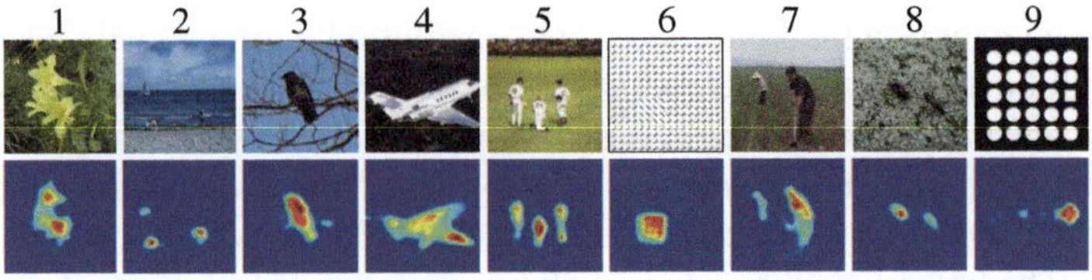

Fig. 1. Saliency detection using the proposed method. Row 1 shows the original images, and row 2 shows the saliency detection results. Col. 2 shows distant objects in the scene (small salient regions); col. 1 and 4 show near objects (broad salient regions); col. 1, 3, 7 and 8 show objects with cluttered backgrounds; col. 6 and 9 show a distinct pattern among many common patterns.

patterns, e.g., grassy patches (see col. 1, 5, 7, 8 in Fig. 1), branches of a tree (see col. 3 in Fig. 1) and so on. We will illustrate that near zero-frequency components in an image are also a kind of special repeating pattern. These components always correspond to, for example, sky, roads or regions with the same intensity or color.

2.1.1 Spikes in the Amplitude Spectrum Correspond to Repeating Patterns

It has been argued [12] that the *spectrum residual* corresponds to image saliency, while contradictorily in [13, 5], the amplitude information was totally abandoned. However, in this paper, we will illustrate that spikes in the amplitude spectrum correspond to repeating patterns, which should be suppressed for saliency detection.

For natural images, repeating patterns always lead to spikes in the amplitude spectrum. Considering a 1-D periodic signal $f(t)$ as an example, suppose that it can be represented by $f(t) = \sum_{n=-\infty}^{\infty} F_n e^{jn\omega_1 t}$, where $F_n = \frac{1}{T}\int_{-T/2}^{T/2} f(t)e^{-jn\omega_1 t}dt$. Then its Fourier transform is given by: $F(\omega) = 2\pi \sum_{n=-\infty}^{\infty} F_n \delta(\omega - n\omega_1)$. From the latter, we can conclude that the spectrum of a periodic signal is a set of impulse functions (spikes). We note that this is based on the assumption that the periodic signal is of infinite length. Therefore, given a more realistic finite length periodic signal, the shape of the spectrum will be obviously different but not degraded greatly.

Suppose there is one salient part of a signal embedded in a finite length periodic signal (see the original signals in Fig. 2). We will illustrate that this salient part will not have a large influence on the spikes in the spectrum. That is to say, 1) The spikes will remain, even though the salient part is embedded in the signal; 2) The embedded salient part will not lead to very sharp spikes in the amplitude spectrum of the signal. As shown in Row 2 of Fig.2, we observe that all of the sharp spikes labeled by solid boxes are not due to the salient part of the signal, but rather the background. The signal to be analyzed is defined by $f(t) = g(t) + g_{\sigma}(t) + s(t)$, where $g(t)$ is a finite length periodic signal, which can be considered as the background signal in $f(t)$. As shown in row 1 of Fig. 2, $g(t)$ is the lower frequency one, which more or less appears throughout the whole length of $f(t)$. It is defined by:

$$g(t) = \begin{cases} p(t) & \text{if } t \in (0,L) \\ 0 & \text{otherwise.} \end{cases} \tag{1}$$

Both $g_\sigma(t) = -p(t) \cdot W(t)$, and $s(t) = -p_s(t) \cdot W(t)$, are included in Eq. 1 to account for the salient parts of the signal. Thus $s(t)$ is the salient part of $f(t)$, which for convenience is also defined as a portion of a periodic function. Both $p(t)$ and $-p_s(t)$ are periodic functions with frequencies f and f_s respectively; $W(t)$ is a rectangular window function that equals 1 inside the interval $(t_0, t_0 + \sigma)$ and 0 elsewhere; we also assume that $(t_0, t_0 + \sigma) \in (0,L)$ and $\sigma \ll L$ (refer to Row 1 of Fig. 2 where the length of salient part is much less than the length of the whole signal). Thus the Fourier Transform of $f(t)$ can be represented as follows:

$$F(f)(\omega) = \int_0^L g(t)e^{-j\omega t}dt + \int_{t_0}^{t_0+\sigma} g_\sigma(t)e^{-j\omega t}dt + \int_{t_0}^{t_0+\sigma} s(t)e^{-j\omega t}dt. \tag{2}$$

As mentioned above, we assume that $\sigma \ll L$. This implies that the first term has very sharp spikes in the amplitude spectrum, while this is not true of the second and third terms. In order to illustrate this, we define a notion of sharpness of an log amplitude spectrum X. Suppose that we smooth X, containing several spikes, using a low-pass filter. Obviously, the sharper the original spike, the more the height of its peak will be reduced. Therefore, we describe the sharpness of signal X by $S(X) = \| X - X * h \|_\infty$, where h is a low-pass Gaussian kernel with scale σ. Taking $g_\sigma(t)$ as an example, we compute the point-wise product of a periodic signal $-p(t)$ and a rectangular window function $W(t)$. According to the convolution theorem, $F(g_\sigma)(\omega)$ equals the convolution of $-F(p)(\omega)$ with $F(W)(\omega)$. Since $F(W)(\omega) = \frac{2sin(\sigma/2)}{\omega}e^{j\omega(t_0+\sigma/2)}$ is a low-pass filter, the spikes in the amplitude spectrum of $-F(p)(\omega)$ will be greatly suppressed. This also occurs for the third term.

As discussed above, the sharpness of $F(f)(\omega)$ is mainly determined by $g(t)$, while the latter two terms in Eq. (2) do not contribute greatly to the spikes in the spectrum. In other words, since the first term corresponds to repeated patterns which lead to spikes in the spectrum, they can be suppressed by smoothing the spikes in the amplitude spectrum of $F(f)(\omega)$.

2.1.2 Suppressing Repeated Patterns Using Spectrum Filtering

Low-pass filters can be employed to suppress spikes in the amplitude spectrum. Consider the very simple example shown in Fig. 2 The input signal (row 1) is periodic, but there is a short segment for which a different frequency signal is apparent. Note that the whole signal is superimposed on a constant value. The short segment is quite distinct from the background for human vision, so a saliency detector should be able to highlight it. Row 2 shows the amplitude spectrum: there are three very sharp spikes (Labeled by solid boxes) which correspond to the constant at zero frequency plus two which correspond to the periodic background. In addition, there are two rounded maxima (labeled by a dashed box) corresponding to the salient parts. The amplitude spectrum is then filtered by a low-pass filter (row 3), and the signal

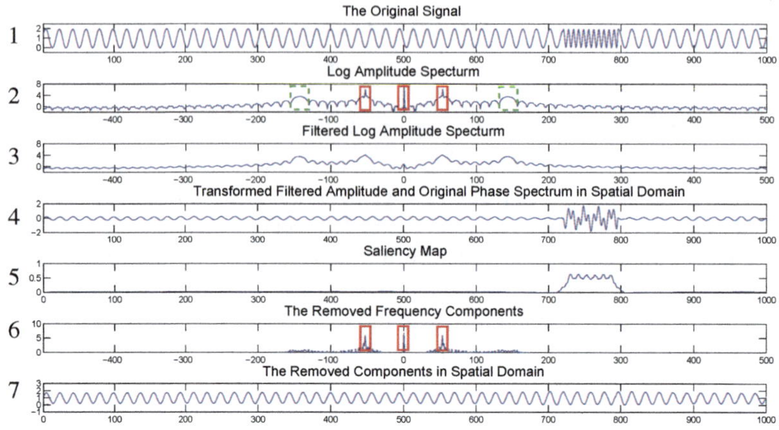

Fig. 2. Suppression of repeated patterns by using spectrum filtering. Analyzing this process, it is clear that the larger the background, the sharper the spikes, leading to the suppression of the amplitude spectrum via filtering. Note that the phase is not shown in this figure.

is reconstructed using the filtered amplitude and original phase spectrum (row 4). It is clear that both the periodic background and the near zero-frequency components are largely suppressed while the salient segment is well preserved. Row 5 shows the saliency map after enhancing the signal shown in row 4 (see step 5 of Algorithm. 1). We can further analyze this in the frequency domain, as shown in row 6, which illustrates the components removed by the previous operations. Here the removed frequency components are mainly the low frequencies near zero-frequency as well as the periodic background. Row 7 presents these removed components in the spatial domain (by measuring the difference between the original and reconstructed signals). Row 6 indicates the frequency spectrum of the signal shown in row 7.

2.1.3 Spectrum Scale-Space Analysis

Repeating patterns can be suppressed by filtering the amplitude spectrum at the appropriate scale. However, which scale is appropriate or the best? As shown in Fig. 3, if the filter scale is too small, the repeating patterns cannot be suppressed sufficiently (see the saliency map in row 2), while if the filter scale is too large, only the boundaries of the salient region are highlighted (see the saliency map in rows 4 and 5). Therefore it is important to select a proper scale for the filter. In fact, we will illustrate that different filter scales are required for different saliency regions. For example, a large-scale filter is needed to detect uniform regions without any rich texture, while a small-scale kernel could be used to detect saliency in regions which are texture-rich or in distant locations in the scene.

In this paper, we propose a Spectrum Scale-Space (SSS) for handling amplitude spectra at different scales, yielding a one-parameter family of smoothed spectra. The SSS representation is parameterized by the scale of the low-pass Gaussian kernel used for suppressing fine-scale structures in the amplitude spectra.

Given an amplitude spectrum, $A(u,v)$, of an image, the SSS is a family of derived signals $L(u,v;k)$ defined by the convolution of $A(u,v)$ with the Gaussian kernel $g(u,v;k) = \frac{1}{2^k t_0} e^{-(u^2+v^2)/(2^{k+1} t_0)}$, where k is the scale parameter, $k = 1,...,K$. K is determined by the input image resolution: $K = \lceil log_2 min\{X,Y\} \rceil$, where X, Y indicate the height and width of the image. Thus SSS is defined as follows[2]:

$$L(u,v;k) = (g(.,.;k) * A)(u,v). \quad (3)$$

As an example, assume a 1-D input signal. We first compute a series of filtered spectra according to the SSS framework; then the saliency map is computed for each scale, as shown in Fig. 3. We can see clearly the significance of scale for saliency detection. In this example, smoothed spectrum 2 gives the best result. As the kernel scale goes to infinity, the spectrum tends to be a constant (horizontal plane in 2D), as shown in the last row of Fig. 3. This is exactly the case proposed in [12, 13, 5].

Fig. 4 shows saliency results obtained using the SSS. The best saliency map is labeled by a red square. The scale of each filter increases from left to right. We observe that broad regions pop out when a smaller scale filter is used while distant objects in the scene or those with rich texture or strong edges pop out when larger scale filters are used. Given a natural image, a set of saliency maps is obtained from which one is selected according to the criterion to be discussed in section 2.2.

Here, we suggest that the frequency-tuned (FT) method [14] is, to some extent, also a special case of the proposed model. In FT, the saliency map is obtained by computing $S(x,y) = \parallel I_\mu - I_\omega hc(x,y) \parallel$, where I_μ is the average *Lab* vector of the input image and $I_\omega hc(x,y)$ is a specific *Lab* pixel vector from the Gaussian filtered version of the original image. FT computes a saliency map by removing the frequencies in the region of the mean. We have illustrated that there is always a very sharp spike in the region of zero frequency, which corresponds to this mean. Hence, if we use a very small scale Gaussian filter to smooth the spectrum (k tends to zero), only that very sharp spike (corresponding to the mean) will be suppressed significantly.

2.2 The Saliency Map

Algorithm 1 below illustrates the steps required to compute saliency and consists of three main components: 1) Preprocessing: Transform the input color image into the opponent color space [13]. This color space includes two color channels (Red-Green and Blue-Yellow) and one intensity channel; 2) Saliency Map: Compute a set of saliency maps by filtering the image amplitude spectrum in each channel with different filter scales; 3) Fusion: The fused saliency map is obtained by linearly combining the selected saliency maps in each channel.

In order to select just one saliency map for each color channel, that is, to select the proper scale of the filter, we compute the entropy of each saliency map in each color channel. The reasoning is that entropy is related to the efficiency of sparse

[2] Note that in this equation and all the follow, the symbol * refers to convolution in the spatial or frequency domain.

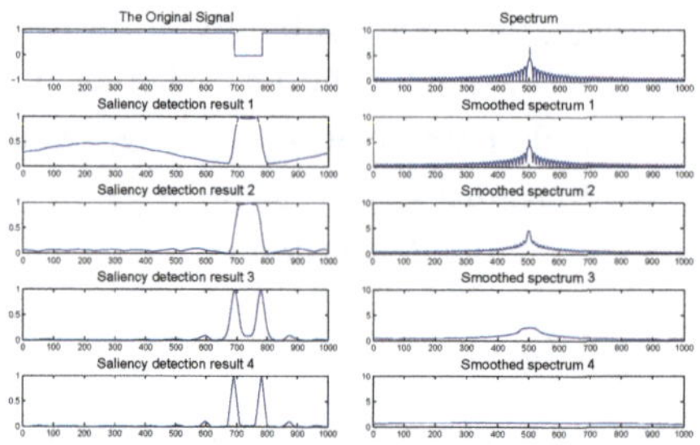

Fig. 3. The original 1D signal is shown in the first row of col. 1 with the 1D saliency maps below. The spectrum of the original signal is shown in the first row of col. 2, followed by the smoothed spectra associated with the paired saliency map in col. 1.

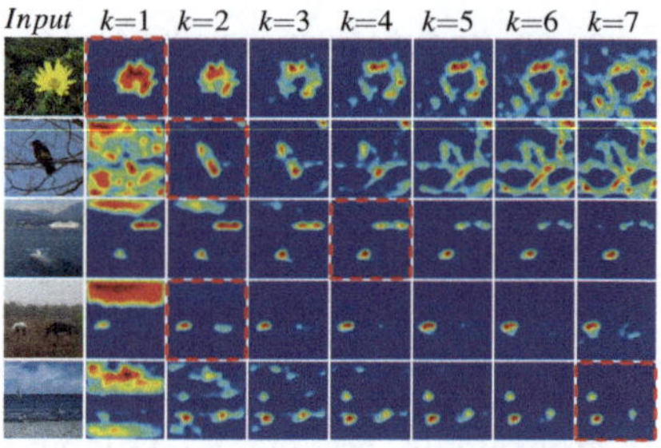

Fig. 4. Five 2-D examples are shown. The first column shows the original 2-D signals (images); The latter part followed the input image in each row presents the set of saliency maps computed by filtering the original image amplitude spectra using different scale kernels.

coding, which is often used to simulate natural scenes in the visual cortex. For each channel, the appropriate scale k_p is defined as follows:

$$k_p = \text{argmin}\left(\text{entropy}(saliencymap(k))\right), \tag{4}$$

where entropy is given by $H(x) = -\sum_{i=1}^{n} p_i log p_i$.

Algorithm 1. Spectrum-Scale-Space Image Saliency Detection.

Input:

The resized color image I with resolution 128×128

Output:

Saliency map sM of the input image I

1: Compute the features lightness L and color components $R-G$, $B-Y$ that form the image I;
2: Calculate weight ω_i for each feature channel according to Eq. 5;
3: For the feature image P_i in each channel, $i \in \{R-G, B-Y, L\}$:
A. Transform the image into the frequency domain using the Fourier Transform, and then calculate the amplitude spectrum $A_i(u,v)$ and phase spectrum $p_i(u,v)$;
B. Compute the spectrum scale-space $\{A_k^i\} = L_i(u,v;k)$ according to Eq. 3;
C. For k=1:K, compute the set of saliency maps $\{sM_k^i\}$ by reconstructing the signal using p_i and A_k^i;
D. Select the best saliency map sM_i from $\{sM_k^i\}$ according to Eq. 4;
4: Compute the final saliency map by fusing the selected saliency maps: $sM = \sum_{i \in \{R-G, B-Y, L\}} \omega_i \cdot sM_i$.
5: In order improve the visual effect, the saliency map is filtered with a low-pass gaussian filter: sM $= \| sM \|^2 * g(u,v)$.
6: **return** sM.

According to the Feature Integration Theory [15] in human visual processing, several primary visual features are processed and represented by separate feature maps that are later integrated into a saliency map. The latter can then be accessed in order to direct attention to the most conspicuous areas in an image. Similarly in the proposed model, the best saliency map from each channel is combined linearly in accordance with their computed weights. In the experiments described below, the weight for each color channel is determined by two factors: (1) The range of the "gray" levels in the image histogram; (2) The entropy value of the best saliency map. The weight is calculated according to Eq. 5:

$$weight_i = width_i \cdot entropy_i, \quad i \in \{R-G, B-Y, L\}, \tag{5}$$

Then, he weight in each channel is normalized to produce the final saliency map.

2.3 Computing the Video Saliency Map

In this section, we extend the above approach to finding saliency in videos containing moving objects. Many frameworks have been proposed based on both spatial and motion cues [1, 13, 5]. In this paper, we simply take the motion cue as another feature channel in addition to the three channels already described in 2.2.

For the nth frame $f(x,y;n)$ of a video, we capture the motion information by estimating the derivative at each location (x,y):

$$M(x,y;n) = \frac{-f(x,y;n+2) + 8f(x,y;n+1) - 8f(x,y;n-1) + f(x,y;n-2)}{12}. \tag{6}$$

Eq. 6 is a robust method for estimating the derivative. In fact, there is considerable noise in the temporal domain due to changing illumination conditions or fixed camera noise. Hence such a robust estimate method is necessary to alleviate this problem. The computed motion map is included in the overall framework (as described in step 3 of Alg. 1). Why it is reasonable to compute the motion saliency map using SSS? As we know, we can capture moving objects by using background differences when the camera is stationary. However, this is a very unstable method, because of noise in the background due to illumination variations. This exacerbated even more when the camera is non-stationary. Nevertheless the background difference method can be successfully applied to the proposed framework. The is because the noise caused by illumination variations or non-stationary backgrounds is distributed uniformly with very high probability along the frequency axis. These background effects behave as uniform or repeating patterns and will be suppressed to permit the pop-out of moving objects in the scene.

Suppression using the SSS framework is illustrated in Fig. 5. It is clear that, the SSS framework can suppress background noise caused by both changing illumination conditions and moving cameras. In row 1, the moving pedestrian definitely produces a high-value region in the 5-point difference image. While changes in illumination will also cause the same consequences, these will not generally cluster in

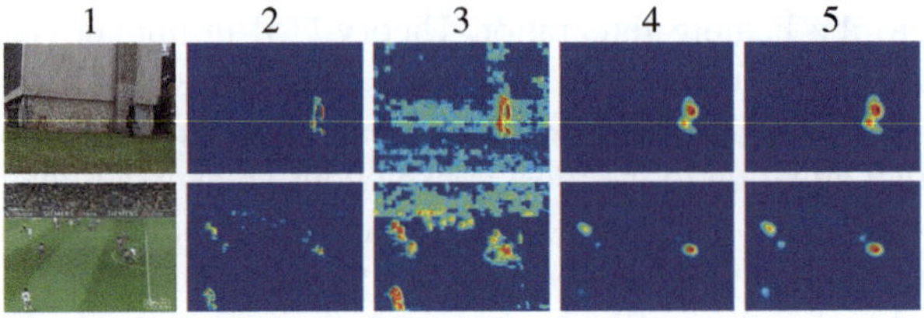

Fig. 5. Background noise suppression in videos using SSS. Col. 1 shows the n-th frame of a video. Col. 2 indicates the 5-point derivative image computed by Eq.6. Note that in order to clearly highlight the noise, we display the log of the derivative image in Col.3 and 5. Col. 4 presents the saliency maps for the images in col. 2. Col.5 shows the log image of the saliency map in col. 4.

specific locations. In fact, these are usually broadly distributed. Therefore, a salient region will not occur, as illustrated in Col 2 and 3. We observe, in Col. 4 and 5 that the saliency maps are very clean, with the moving objects well highlighted. A similar situation prevails for the image in Row 2. The two cases differ due to background noise created by the moving camera. Nevertheless, we observe that the background noise in the saliency map is largely suppressed.

3 Experiments

We have used four types of signals to test the proposed framework: 1-D signals, 2-D signals (psychophysical patterns), natural images and videos.

3.1 Saliency Detection in 1-D Signals and 2-D Patterns

Fig. 6 shows saliency detection results for 1-D signals and comparisons with SR. The first row of each sub-figure shows the original 1-D signal; the second row shows the amplitude spectrum and the third row shows the filtered amplitude spectrum; rows 4 and 5 indicate the results produced by our approach and SR, respectively. In Fig. 6 (a), we find that SR will greatly enhance the boundary or edges of the salient region; however, the central part of the salient region is not highlighted uniformly. Sometimes, SR is unable to respond to certain salient regions when no strong edges are evident in the signal, as shown in fig. 6 (b).

Specific types of patterns are often used to test saliency or attention models. Fig. 7 shows saliency detection results for these 2-D patterns. The first column indicates the 2-D input signals; Col. 2-6 show the saliency detection results by using SSS, SR, PFT, frequency tuned approach [14] and Ittis approach, respectively. The relative efficacy of the method described in this paper is clearly evident.

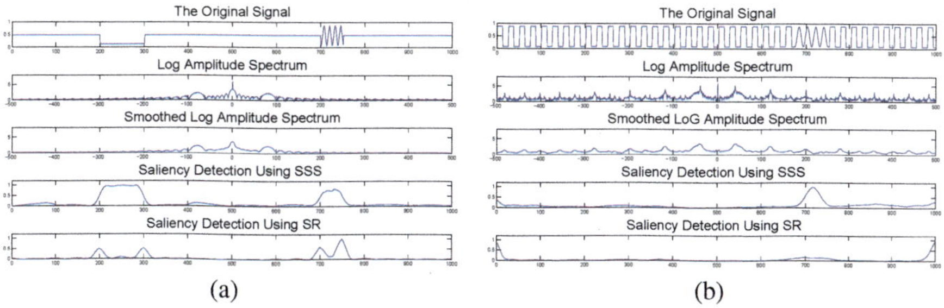

Fig. 6. Saliency detection in 1-D signals.

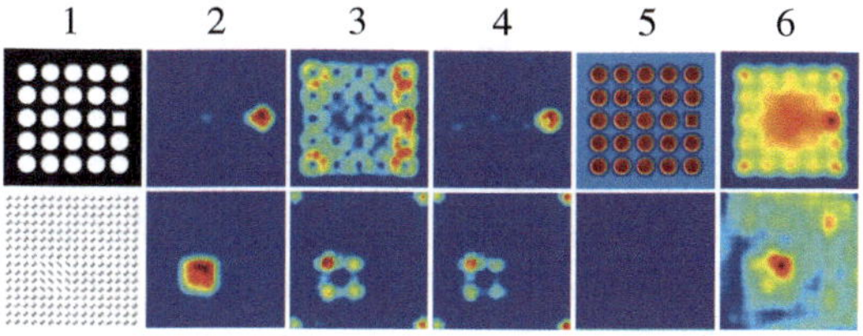

Fig. 7. Saliency detection in 2-D signals.

3.2 Saliency Detection in Natural Images

We have computed the saliency in natural images using SSS, and compared the results with those given by SR [12], PFT [13], FT [14] and Ittis approach [6]. Among these images are scenes containing distant objects with distractors (see col. 1 and 2). The beach and skyline in col.1 of Fig. 8 can be considered as such a distractor; some images in the figure contain near objects against a cluttered background (col. 2, 3, 4, 5, 6, 10, 13 of Fig. 8); some contain a distinct object among several other types of objects (see col. 8, 9 of Fig. 8).

As shown in Fig. 8, SSS highlights both small (see col. 1, 2, 3, 7 and 12) and large salient regions (see col. 2, 4, 5, 6, 10 and 13), and has the ability to inhibit repeated patterns (see col. 2, 4, 5, 6, 8, 9, 11, 12 and 10). As discussed in section 1, we observe that both SR and PFT will enhance object boundaries and cluttered regions heavily but cannot highlight salient regions uniformly. As shown in col. 2, 3, 4, 5, 6 and 10, only the boundaries of larger objects are highlighted. In some cases, the cluttered background is emphasized, as shown in col. 4, 5, 6, 11, 12 and 13. The frequency-tuned approach highlights the salient regions uniformly and preserves boundaries very well. However it suffers from three issues: 1) It cannot highlight those objects which have the same intensity (or color) as the average value of its surrounding patch or the whole image (see col. 1, 2, and 3), cannot suppress certain non-salient distracters, as in col. 1(beach and skyline), col. 2 (sky), col. 4 (branches) and col. 3 (sky), 3). Neither can it inhibit the repeated patterns, as shown in col. 9. Note that t distracters such as sky, grass, ground or skyline are considered as repeated patterns

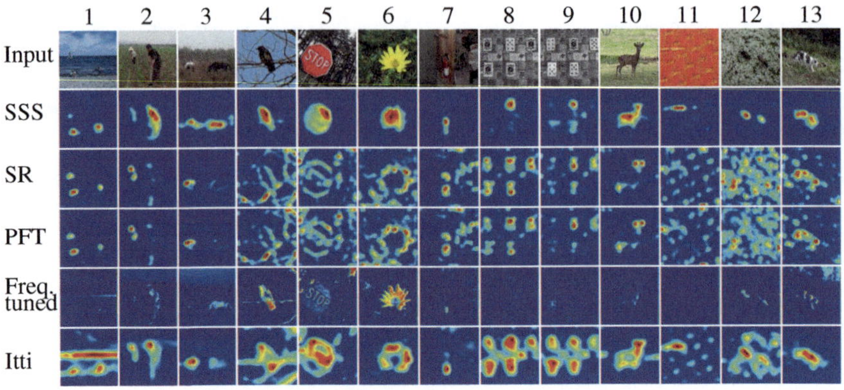

Fig. 8. Saliency detection in natural images.

in our model. The last row of Fig.8 presents the results obtained by Ittis method. In most cases this approach can highlight the salient region uniformly (see col. 2, 3, 4, 8, 9 and 13). However, it sometimes highlights the distractors as well, as observed in col. 1 (skyline), col. 4(branches), col. 5 (trees) and col. 12 (grass). In addition, it cannot inhibit repeated patterns, as seen in col. 8 and 9.

3.3 *Computing Video Saliency Maps*

We tested the performance of the proposed framework to achieve video saliency detection by using three types of videos: 1) Surveillance video: There is no movement in the background because the camera is fixed. Thus all motion information

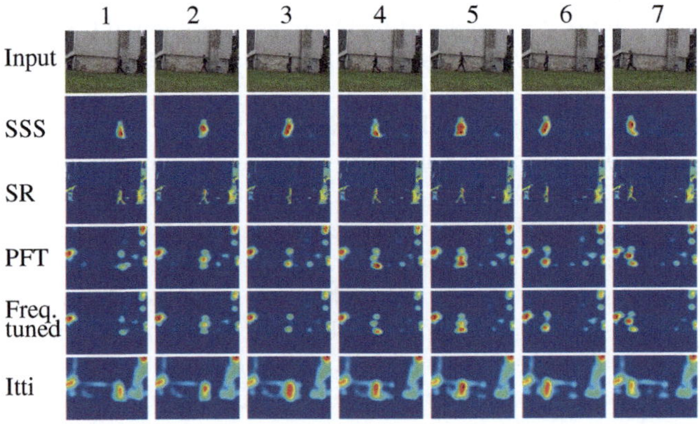

Fig. 9. Saliency detection in a surveillance video showing a person moving across the field of view at a relatively close distance. Col. 1 shows frames from the original video; only a sampled subset of all of the consecutive frames from the captured video is shown. Col. 2 -6 present the results obtained using SSS, FT, SR, PFT and Ittis approach, respectively.

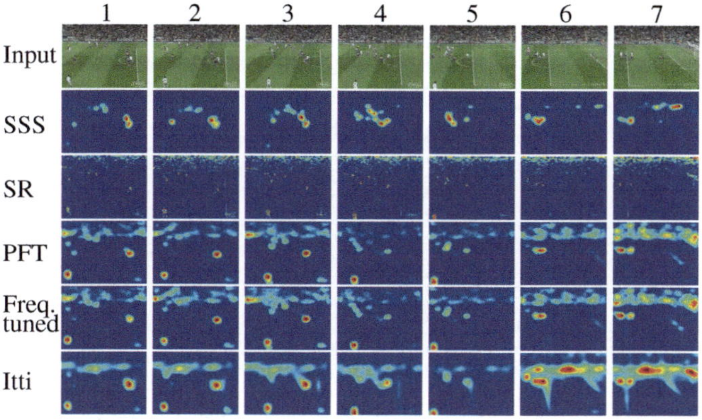

Fig. 10. Saliency detection in a scene using a dynamic camera. The video shows football players moving across the field of view at a relatively far distance. Col. 1 shows frames from the original video; only a sampled subset of all of the consecutive frames from the captured video is shown. Col. 2 -6 present the results obtained using SSS, FT, SR, PFT and Ittis approach, respectively.

can be captured by adjacent frame differentiation, while the static objects can be located by using spatial information. The main issue is how to deal with changing illumination conditions. 2) Dynamic camera with moving objects relatively remote from the camera: This case is more difficult to deal with compared to surveillance videos because both the mobile objects and background are moving. In addition the moving objects are remote from the camera and therefore are small. 3) Dynamic camera with moving objects relatively close to the camera: This kind of video was employed to test viewing close objects, and therefore, of comparatively large size.

Fig. 9 shows saliency detection in a surveillance video for the following: SSS, FT, SR, PFT and Ittis approach. Compared with the results obtained by the other frameworks, the outcomes using SSS clearly highlight the salient object while suppressing the background noise to a greater degree. This will prove to be extremely useful if the objective of determining the saliency map is to ultimately segment the image into salient and non-salient objects.

Fig. 10 shows the results of a scene taken at a distance using a non-stationary camera. Because of the moving background, there is considerable noise. The results achieved by SSS clearly indicate that the background can strongly be suppressed. Inspecting the other results, we observe that the white line painted on the ground and the audience in the stands are highlighted significantly, even though this is not the most salient part of the scene for human observers.

As discussed in section 3.2, the proposed SSS has the ability to highlight equally both remote objects (of small size) and near objects (of large size) in the scene. In Fig. 11, both SSS and Ittis method can highlight near objects equally well. However, SSS suppresses both the background noise and non-salient objects properly. The other methods have difficulty with this video and are able to only highlight some parts of near objects.

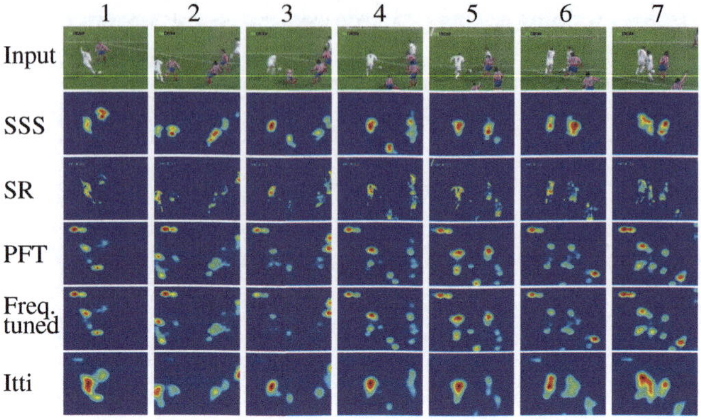

Fig. 11. Saliency detection in a scene using a dynamic camera where the objects are relatively close to the camera. Col. 1 shows frames from the original video; only a sampled subset of all of the consecutive frames from the captured video is shown. Col. 2 -6 present the results obtained using SSS, FT, SR, PFT and Ittis approach, respectively.

To some extent, this is not a fair comparison since the experiments in 3.3 are designed to show the performance of video saliency detection and employ motion information. However, this is not the case with the other frameworks.

4 Discussion

This paper proposes a new saliency detection framework for videos based on analyzing the spectrum scale-space. We assume that a natural image consists of several salient regions and many background regions. A salient region is defined as a distinct pattern where the intensity, color, and motion properties are different from the rest of the image. The objective of a saliency detector is to create a pop-out of visually significant objects in a scene while ignoring so-called background clutter. The proposed approach is able to highlight both small and large salient regions and to inhibit repeated patterns. In this paper, we also illustrate that both SR and PFT are special cases of the proposed model when the scale parameter goes to infinity, while FT is a special case when the scale parameter goes to zero. In addition, we also extend the proposed framework to compute video saliency by combining spatial and temporal channels in a single feature vector.

To validate the proposed approach, we have performed saliency computations on both synthetic data, natural images and videos and then compared the results with those given by SR, PFT, FT and Ittis method. Experimental results show that our method has three advantages: 1) It can highlight both small and large salient regions; 2) it can inhibit common distracters; and 3) it can inhibit repeated patterns. In future work, we will focus on a hierarchical saliency model. As shown in col. 9 and 10 of Fig. Fig. 8, there are five cards in each of these two images. For human vision, these five cards are distinctive and draw our attention. However, among these five cards, there is one which is more distinct than the others. A hierarchical model is needed to deal with this issue.

References

1. Laurent, I., Christof, K.: Computational modelling of visual attention. Nature Reviews Neuroscience 2(3), 194–203 (2001)
2. Yantis, S.: How visual salience wins the battle for awareness. Nature Neuroscience 8(8), 975–977 (2005)
3. Bernhard, S., et al.: A Nonparametric Approach to Bottom-Up Visual Saliency. Advances in Neural Information Processing Systems 19, 689–696 (2007)
4. Vijay, M., Dashan, G.L., et al.: The discriminant center-surround hypothesis for bottom-up saliency. In: Advances in Neural Information Processing Systems, vol. 20, pp. 497–504 (2008)
5. Chenlei, G., Liming, Z.: A novel multiresolution spatiotemporal saliency detection model and its applications in image and video compression. IEEE transactions on image processing 19(1), 185–198 (2010)
6. Christof, K., Laurent, I., et al.: A model of saliency-based visual attention for rapid scene analysis. IEEE Transactions on Pattern Analysis and Machine Intelligence 20(11), 1254–1259 (1998)
7. Beck, D.M., Kastner, S.: Stimulus context modulates competition in human extrastriate cortex. Nature Neuroscience 8(8), 1110–1116 (2005)
8. Neil, B., John, T.: Saliency Based on Information Maximization. In: Advances in Neural Information Processing Systems, vol. 18, pp. 155–162 (2006)
9. Christof, K., Jonathan, H., et al.: Graph-Based Visual Saliency. In: Advances in Neural Information Processing Systems, vol. 19, pp. 545–552 (2007)
10. Wolfgang, E., Christof, K., et al.: Predicting human gaze using low-level saliency combined with face detection. In: Advances in Neural Information Processing Systems, vol. 20, pp. 241–248 (2008)
11. Xiaodi, H., Liqing, Z.: Dynamic visual attention: searching for coding length increments. In: Advances in Neural Information Processing Systems, vol. 21, pp. 681–688 (2009)
12. Xiaodi, H., Liqing, Z.: Saliency detection: A spectral residual approach. In: IEEE Conference on Computer Vision and Pattern Recognition CVPR 2007, pp. 1–8 (2007)
13. Chenlei, G., Qi, M., et al.: Spatio-temporal saliency detection using phase spectrum of quaternion fourier transform. In: IEEE Conference on Computer Vision and Pattern Recognition CVPR 2008, pp. 1–8 (2008)
14. Hemami, S., Estrada, F., et al.: Frequency-tuned Salient Region Detection. In: IEEE Conference on Computer Vision and Pattern Recognition CVPR 2009, pp. 1597–1604 (2009)
15. Treisman, A.M., Gelade, G.: A feature-integration theory of attention. Cognitive Psychology 12(1), 97–136 (1980)

Generalized PCA via the Backward Stepwise Approach in Image Analysis

Sungkyu Jung, Xiaoxiao Liu, J.S. Marron, and Stephen M. Pizer

Abstract. Principal component analysis (PCA) for various types of image data is analyzed in terms of the forward and backward stepwise viewpoints. In the traditional forward view, PCA and approximating subspaces are constructed from lower dimension to higher dimension. The backward approach builds PCA in the reverse order from higher dimension to lower dimension. We see that for manifold data the backward view gives much more natural and accessible generalizations of PCA. As a backward stepwise approach, composite Principal Nested Spheres, which generalizes PCA, is proposed. In an example describing the motion of the lung based on CT images, we show that composite Principal Nested Spheres captures landmark data more succinctly than forward PCA methods.

1 Introduction

Principal component analysis (PCA) is a widely used data exploration method in a variety of fields, for many purposes including dimension reduction and visualization of important data structures. In image analysis, the dimensionality of objects under investigation is usually very high, so dimension reduction through PCA is essential in some analysis; see for example, [1].

The classical PCA is based on the Euclidean properties of vector space, especially inner products and orthogonality. PCA is easily applicable for the many data types with these properties, an example of which is Functional PCA ([2, 3]), where the data set consists of smooth curves and the goal is to understand the variation in

Sungkyu Jung · J.S. Marron
Department of Statistics and Operations Research, University of North Carolina at Chapel Hill
e-mail: {sungkyu,marron}@email.unc.edu

Xiaoxiao Liu · Stephen M. Pizer
Department of Computer Science, University of North Carolina at Chapel Hill
e-mail: {sharonxx,pizer}@cs.unc.edu

J. Angeles et al. (Eds.): Brain, Body and Machine, AISC 83, pp. 111–123.
springerlink.com

a set of curves. By a basis expansion of curves, the Euclidean properties are still well-defined, and the Functional PCA is a complete analog of the classical PCA.

Two useful viewpoints on PCA are the forward and backward stepwise approaches. In the traditional forward view, PCA is constructed from lower dimension to higher dimension. In the backward point of view, PCA is constructed in reverse order from higher to lower dimensions. These two approaches are equivalent in Euclidean space but lead to different methodologies in non-Euclidean data discussed next.

A growing number of data types are non-Euclidean, so the classical PCA idea does not apply. This paper focuses on the *mildly non-Euclidean* data, which are also referred to as manifold data, as in that context, the data objects are on the surface of a curved manifold forming a feature space. Data on curved manifolds have long been investigated. Among those the following are best studied:

Directional data. Angles or directions lie on the unit circle or the unit sphere (or a hemisphere), which include wind or ocean current directions, orientation of cracks on rocks, and directions from the earth to celestial objects. A substantial amount of literature can be found in the area of circular, angular or directional statistics, see [4], [5].

Statistical shape space. Landmark-based shape analysis analyzes data lying on special manifolds. A shape is defined as an equivalence class under translation and rotation, scaling in many cases and sometimes reflection. Thus, shape spaces are constructed by removing the translation, scale, and rotation from the set of landmarks, as proposed and investigated by both Kendall [6] and Bookstein [7] and described well in [8].

Due to advances in technology, a demand to analyze different types of manifold data is growing. These modern data are mostly from medical imaging and include

Medial shape representations. Shapes of 2-*d* or 3-*d* objects are represented in a parametric model, called *m-reps* in short, including directions and angles as parameters. The data space here is a manifold that is a direct product of Euclidean space and unit spheres. See [9].

Diffusion Tensor Imaging. DTI [10] is a recently developed and widely studied MRI technique that measures the diffusion of water molecules in a biological object. Random motion of water molecules in each voxel of an image is represented by a 3-*d* tensor, i.e. a non-negative definite 3×3 matrix. Each tensor lies in a lower dimensional sub-manifold of $\mathbf{R}^9$ since it has to be non-negative definite. DTI data, consisting of multiple tensors, thus naturally lie in a manifold.

Diffeomorphisms. A common methodology for comparing shapes in image analysis is to use diffeomorphisms ([11], [12]), i.e. smooth space warping functions. This method delivers a new approach to shape analysis. A shape is considered as a distortion (i.e. diffeomorphism) of some template. Thus a set of shapes is represented as a set of diffeomorphisms and the variation in the population of diffeomorphisms can be studied to understand variation in shapes. The set of diffeomorphisms forms a very high dimensional manifold.

Conventional statistical analysis, including PCA, is not directly applicable to these manifold data. On the other hand, there is a growing need of PCA-like methods, because the dimensionality of the data space is often very high. Previous approaches for generalized PCA to manifold data are listed and discussed in Section 2. Many commonly used methods can be viewed as the forward approach. However, [13] points out that the backward viewpoint is seen to provide much more natural and accessible analogues of PCA than the standard view. This is discussed further in Section 2.2.

Section 3 is devoted to proposing a methodology of generalized PCA to the surface point distribution model (PDM). The method, *composite PNS*, can be viewed as an extension of Principal Nested Spheres, proposed by [14] and discussed in Section 3.1, which also can be viewed as a backward generalization of PCA to manifold data. The procedure of the proposed method is illustrated in Section 3.2.

Advantages of the proposed method are presented by some experimental results in Section 3.3. We use this approach to describe the motion of the lung using landmark data extracted from CT images. We show that composite Principal Nested Spheres captures more variation of this landmark data in fewer dimensions than the standard PCA.

This article argues that the backward approach leads to a natural extension of PCA for dimension reduction on non-linear manifolds and that in a variety of application areas this extension appears superior to alternative methods in the literature. In particular, composite PNS is proposed, and in real and simulated examples it is shown to outperform classical and kernel-based PCA.

2 Forward and Backward Stepwise View of PCA

The forward and backward stepwise views of PCA, either in Euclidean space or for manifolds, are discussed in this section.

2.1 Mathematical Development for Euclidean PCA

Let $X_1,\dots,X_n$ be d-dimensional column vectors that are inputs for Euclidean PCA. The data matrix is formed by aggregating the data vectors: $\mathbf{X} = [X_1,\dots,X_n]$. A *forward stepwise* view to Euclidean PCA is understood by increasing the dimension of the best approximating (affine) subspace, as described in the following steps:

1. Find a center point that best represents the data, by minimizing the sum of squared distances: the empirical mean $\bar{X} = \frac{1}{n}\sum_{i=1}^{n} X_i$.
2. Find a line, an affine 1-d subspace, that best approximates the data, again by minimizing the sum of squared distances from the data to the line. Pythagorean theorem shows that this line must pass through the sample mean $\bar{X}$. This affine one dimensional subspace can be written with a direction vector $\mathbf{u}_1$ so that

$$AS_1^1 = \{\bar{X} + cu_1 : c \in \mathbf{R}\}.$$

The direction $\mathbf{u}_1$ is the first principal component (PC) direction. The orthogonal projections of the data X_i onto AS_1^1 are then in the form $\bar{X} + c_i\mathbf{u}_1$, which are the best rank 1 approximation of the data. The amount of deviation c_i from the center is called PC scores.

3. Next find a line in the affine subspace orthogonal to $\mathbf{u}_1$, that best represents the data. The line is denoted with the second PC direction vector $\mathbf{u}_2$ by $AS_2^1 = \{\bar{X} + c\mathbf{u}_2 : c \in \mathbf{R}\}$. Since $\mathbf{u}_1$ and $\mathbf{u}_2$ are orthogonal, the best two dimensional approximation of the data is contained in the affine 2-d subspace

$$AS^2 = AS_1^1 \otimes AS_2^1 = \{\bar{X} + c_1\mathbf{u}_1 + c_2\mathbf{u}_2 : c_1, c_2 \in \mathbf{R}\},$$

   where $\otimes$ represents the direct product. PC scores for the second PC are found again through the projections of the data onto AS_2^1 (or AS^2), similar to the 1-d case.

4. Higher order components can be found iteratively for $k = 3, 4, \ldots, d$, that results in k-dimensional affine subspaces

$$AS^k = \otimes_{j=1}^k AS_j^1 = \{\bar{X} + \Sigma_{j=1}^k c_j\mathbf{u}_j : c_1, \ldots, c_k \in \mathbf{R}\}.$$

In this forward formulation of PCA the best approximating affine subspaces are constructed from the lowest dimension to higher dimension, i.e.

$$\{\bar{X}\} \subset AS_1^1 \subset AS^2 \subset \cdots \subset AS^d.$$

This formulation is most useful for heuristic understanding of the method. A practical formulation uses the fact that the PC direction vectors $\mathbf{u}_j$ are eigenvectors of the sample covariance matrix $S = \frac{1}{n-1}(\mathbf{X} - \bar{X})(\mathbf{X} - \bar{X})^T$ or the left singular vectors of the centered data matrix $(\mathbf{X} - \bar{X})$.

The viewpoint that seems most useful for generalization of PCA to manifold data is the *backward stepwise* view. In backward PCA, principal components are found in reverse order, i.e. AS_ks are fitted from the largest dimension, which leads to

$$\mathbf{R}^d = AS^d \supset AS^{d-1} \supset \cdots \supset AS_1^1 \supset \{\bar{X}\}.$$

In particular, AS^{d-1} is found from AS^d by removing the PC direction $\mathbf{u}_d \in AS^d$, which is orthogonal to AS^{d-1}. Deriving AS^{d-1} from AS^d is equivalent to the finding of the $(d-1)$-dimensional linear subspace of AS^d that minimizes the sum of squared distances. The projections X_i^P of X_i to AS^{d-1} are then the best $(d-1)$-dimensional approximation of the data, and the signed length of projections (from X_is) become the PC scores. An application of the Pythagorean theorem yields that AS^{d-2} can be found in the same manner from the projections X_i^P.

In Euclidean space, the forward and backward approaches are equivalent. However, in non-Euclidean spaces, this is usually not the case, and the choice of viewpoint affects the generalizations of PCA.

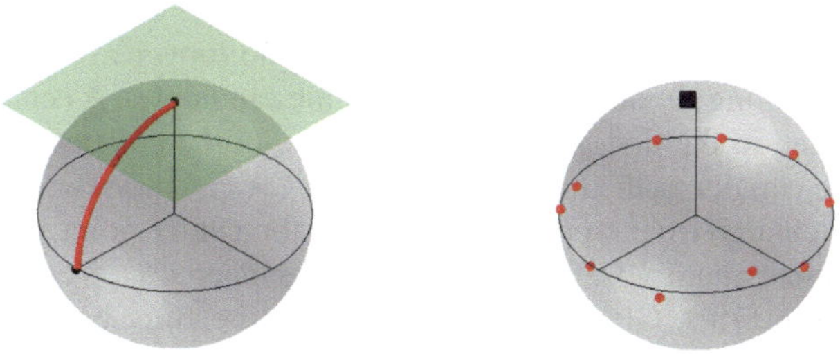

Fig. 1. (Left) The usual unit sphere S^2 with a geodesic segment (great circle segment) joining the north pole and a point in the equator. The tangent plane at the north pole is also depicted. (Right) Plot of 10 points along the equator with random perturbation and the geodesic mean (black square) near the north pole illustrates the case where the geodesic means on S^2 does not represent the data well.

2.2 PCA Approaches for Manifold Data

A widely used approach to manifold PCA, called Principal Geodesic Analysis (PGA, [15]), generalizes PCA in a forward stepwise manner. The first step in PGA is to find a center point for the manifold data. While the sample mean (i.e. the average) is not defined, a useful notion for generalization of mean is the Fréchet mean, defined as a minimizer of $\min_{x \in \mathscr{M}} \sum_{i=1}^{n} \rho^2(x, x_i)$, where $\mathscr{M}$ is the data space and ρ is a metric defined on $\mathscr{M}$. The Fréchet mean is widely applicable, since it only requires a metric on the manifold. In Euclidean space, the sample mean is the Fréchet mean with the usual metric $\rho(x,y) = \|x - y\|$. On curved manifolds, distances are commonly measured along geodesics. A geodesic is an analog of straight lines in Euclidean space; it is roughly defined as the shortest path between two points (see Fig. 1). The geodesic distance function measures the shortest arc length between two points. With the geodesic distance as its metric, the Fréchet mean is often called geodesic mean.

Having the geodesic mean as the center point in PGA, the second step is to find a geodesic (instead of a line) that best represents the data, among all geodesics that pass through the geodesic mean. The higher order components are again geodesics that are orthogonal (in a sense) to the lower order geodesics. In practice, these geodesic components are computed through the projection of the data onto the tangent space at the geodesic mean. The PGA and similarly defined forward approaches are developed for various types of data; see e.g. [15] for m-reps data, [16] for DTI data, and [8] for shape data.

However, there has been a concern that the geodesic mean and tangent space approximation can be very poor. As a simple example, consider the usual unit sphere S^2 in $\mathbf{R}^3$ and the data distributed uniformly along the equator of the sphere as illustrated in Fig. 1. In this case, the equator itself is the geodesic that best represents the data. However, the geodesic mean is located at near the north or the south pole, far from any data. PGA finds principal geodesics through this geodesic mean, which fail to effectively describe the variation in the data.

This observation motivated [17] to propose Geodesic PCA (GPCA). In GPCA, the geodesic mean or any pre-determined mean is no longer used; instead it finds the best approximating geodesic among all possible candidates. A center point of the data is then found in the first geodesic component. In the equator example above, GPCA finds the equator as the first component. GPCA can be viewed as a backward approach, particularly when applied to S^2, since the center point is found last. In higher dimensional manifolds, for example in hyperspheres $S^p, p > 2$ and Kendall's shape spaces (see [18]), GPCA does not appear to be fully backward, since the method is built by considering lower dimensional components first, only with an exception for center point. Nevertheless, the advantage of the method indeed comes from the backward viewpoint, i.e. from reversing the order of the first two steps.

Another method that can be viewed as the backward stepwise approach is Principal Arc Analysis (PAA), proposed by [19], which is a non-geodesic generalization of PCA. PAA is motivated by data distributed along a *small circle* on S^2. Since the major variation is no longer along a geodesic, no geodesic based methods including PGA and GPCA capture the variation effectively. PAA begins with the full sphere S^2 and finds the small circle as the best fitting 1-d approximation of the data, followed by a center point contained in the small circle. PAA was shown to provide just this type of effective data approximation in S^2 and also in m-reps data in [19].

In generalizations of PCA for higher dimensional manifolds, including hyperspheres S^p and Kendall's shape spaces, the backward stepwise principle led to a new fully backward generalization of PCA: Principal Nested Spheres (PNS), proposed by [14]. In taking the backward approach, it inherits the advantages of GPCA. In using non-geodesic approximation, it inherits advantages of PAA. A detailed description of the method can be found in Section 3.1. PNS has been shown to provide more representative description of the data (compared to other forward stepwise approaches) in a number of standard examples in [14]. A discussion of application of PNS to Euclidean data, in Section 3, shows how beneficial a backward generalization of PCA could be even for Euclidean data.

Many kernel PCA ([20]) methods can also be used for dimension reduction and feature extraction, but the nature of these methodologies is quite different from PCA extensions for manifold data. Kernel PCA maps the data implicitly into a Hilbert space, then PCA is applied to the mapped data. While this approach gives a wide possibility of non-linear feature extraction, the interpretation of the result is difficult and reconstruction of components in the original space is impossible.

3 Method

In this section, a method for Euclidean data that possesses the advantage of backward generalization of PCA is discussed. In particular, when the dataset is a set of the surface point distribution models (PDM) representing the shape of an object, the backward generalization of PCA to shape space, Principal Nested Spheres (PNS), fits well. We summarize PNS in more detail and discuss the composite PNS,

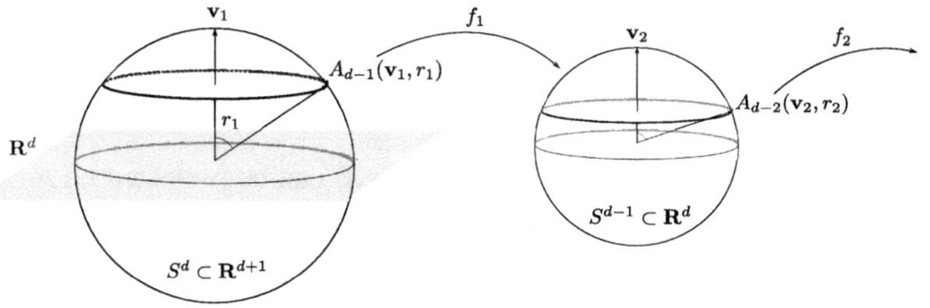

Fig. 2. The nested sphere $A_{d-1}(v, r_1)$ in S^d and its relation to S^{d-1}, through some isomorphism f_1. Recursively, A_{d-2} is found in S^{d-1}.

followed by experimental results which shows that the composite PNS gives more effective description of the PDMs in lower dimension than Euclidean PCA.

3.1 Principal Nested Spheres

The analysis of PNS is summarized in this section. PNS is essentially a decomposition method for hyperspheres and Kendall's shape space, which generalizes PCA in a non-geodesic way. Detailed geometric properties and statistical discussions of PNS can be found at [14]. As mentioned in Section 2.2, the first step in PNS is to reduce the dimension d of S^d to $d-1$. Specifically, we wish to find the best approximating sub-manifold of dimension $d-1$. PNS solves this problem with a flexible class of sub-manifolds in the form of nested spheres.

A k-dimensional nested sphere A_k of S^d is nested within (i.e. sub-manifold of) higher dimensional nested spheres; and A_k itself can be thought of as a k-dimensional sphere. As an example, A_{d-1} of S^d is defined with an axis $v_1 \in S^d$ and distance $r_1 \in (0, \pi/2]$ as follows,

$$A_{d-1}(v_1, r_1) = \{x \in S^d : \rho_d(v_1, x) = r_1\},$$

where ρ_d is the geodesic distance function defined on S^d. The parameter v_1 drives the 'direction' that is not contained in A_{d-1}. In relation to the backward view of Euclidean PCA in Section 2.1, the direction coincides to $\mathbf{u}_d$, which is orthogonal to AS^{d-1}. The distance from v_1 to any point in A_{d-1} is r_1, which is responsible for the curvature of A_{d-1}. This flexibility of curvature in A_{d-1} allows PNS to capture a certain form of non-geodesic variation.

Lower dimensional nested spheres are defined similarly. Since A_{d-1} is essentially a sphere, A_{d-2} can be defined again with a pair (v_2, r_2) and in a way that A_{d-2} is nested within A_{d-1}. Iteratively, one can continue to build a sequence of nested spheres $S^d \supset A_{d-1} \supset \cdots \supset A_1$. Fig. 2 shows a geometric structure of nested spheres that are recursively defined and fitted.

In PNS with a data set $X_1, \ldots, X_n \in S^d$, the pair (v, r) of nested spheres is fitted to the data iteratively so that the fitted nested spheres represent the data. [14] proposed minimizing the sum of squared distances to the data, i.e. the $d-1$ dimensional PNS is

$$\hat{A}_{d-1} = \operatorname{argmin} \sum_{i=1}^{n} \rho_d(A_{d-1}, X_i)^2, \tag{1}$$

where $\rho_d(A_{d-1}, X_i)$ is defined as follows. Each X_i can be projected on A_{d-1} along the minimal geodesic that joins X_i to A_{d-1}. Denote X_i^P for the projection. The length of the minimal geodesic is the distance, that is $\rho_d(A_{d-1}, X_i) = \rho_d(X_i^P, X_i)$. Note that each observation gets a signed residual $z_{d,i}$.

The second (or the $d-2$ dimensional) PNS is found with the projections X_i^P. Since X_i^P's are on $\hat{A}_{d-1}$, one can use the method (1) by treating $\hat{A}_{d-1}$ and $\{X_i^P\}$ as S^{d-1} and $\{X_i\}$, respectively. Simply put, $\hat{A}_{d-2}$ is fitted to X_i^P's by minimizing the sum of squared distances. In general, we recursively find the sequence of PNS from the (iteratively) projected data.

The lowest level principal nested sphere $\hat{A}_1$ is then a small circle, with intrinsic dimension 1. The Fréchet mean of $X_1^P, \ldots, X_n^P \in \hat{A}_1$ is used as the best 0-dimensional representation of the data in the framework of PNS. Denote the Fréchet mean as $\hat{A}_0$, and keep the signed deviations $z_{1,i}$ of X_i^P for later use.

As a result, PNS constructs the sequence of the best approximating sub-manifolds

$$S^d \supset \hat{A}_{d-1} \supset \cdots \supset \hat{A}_1 \supset \{\hat{A}_0\},$$

for every dimension. The backward principle is essential to PNS, since the forward stepwise generalizations of PCA are not be equivalent to PNS (see Section 2.2) and are even not clearly defined for non-geodesic variation.

Furthermore, we wish to represent the data in an Euclidean space for further analysis (e.g. the method of composite PNS, discussed later in Section 3.2). Recall that in the procedure above, we have collected the signed residuals $z_{k,i}$. The *Euclidean-type representation* of the data by PNS is obtained by combining those residuals into a $d \times n$ data matrix

$$\mathscr{X} = \begin{pmatrix} z_{1,1} & \cdots & z_{1,n} \\ \vdots & \ddots & \vdots \\ z_{d,1} & \cdots & z_{d,n} \end{pmatrix},$$

where each column is the corresponding sample's coordinates in terms of the PNS. Each entry in row k works like the kth principal component score.

The procedure is computationally fast in a moderate size of dimension and samples, when using the computational algorithm proposed in [14] for the optimization task (1). However in the high dimension low sample size situation where for example $d > 1000$ and $n < 100$, strict application of the iterative procedure results in a very slow computation. [14] have shown that the intrinsic dimensionality of the data can be reduced to $n-1$ without losing any information and that the first $d-n$ PNS can be found trivially by an application of singular value decomposition. This fact is used when it applies, including the experiments in Section 3.3.

3.2 Application of PNS to Scaled Point Distribution Models

The surface point distribution model (PDM) representing the shape of a human organ (or other solid object) is denoted by $X_i = [\mathbf{x}_1(i), \dots, \mathbf{x}_p(i)]^T$ where $\mathbf{x}_j(i) = (x_{ij}, y_{ij}, z_{ij})$ is the jth point on the surface and p is the number of points on the surface. The subscript i denotes the ith sample or time point, and let n be the total number of time points. A scaled PDM (SPDM) is derived from a PDM by moving each point towards some designated center point by some fixed factor such that the sum of squared distances from the center point is unity. Thus an SPDM is a PDM that lies on a unit hypersphere, which reflects the shape. The PCA-like analysis of such data should reflect not only variability on the hypersphere but also the correlation between the scale factor, which reflects the size, and the shape.

Wanting to apply a backward generalization of PCA, we might think to use PNS, but it applies only to the variability on the hypersphere. In the composite PNS we propose, the variables are separated into the size and the shape variables. The dimension of the shape space is reduced by PNS. Then the size variable is post-combined with the result of PNS, to incorporate the correlation between size and shape.

A procedure for the composite PNS is as follows:

1. (Centering) Let $\widetilde{X}_i = X_i - \frac{1}{np}\sum_{ij}\mathbf{x}_j(i)$ be the ith uniformly translated PDM, so that $\frac{1}{np}\sum_{ij}\widetilde{\mathbf{x}}_j(i) = (0,0,0)$.
2. (Scaling) Let $S_i = (\sum_{j=1}^{p} \|\widetilde{\mathbf{x}}_j(i)\|^2)^{\frac{1}{2}}$ be the size of the ith PDM, measured by the sum of squared distances to the center. The scaled PDM is $\widetilde{X}_i^* = \widetilde{X}_i / S_i$, so that the size of $\widetilde{X}_i^*$ is 1 for all i. Then the pair ($\widetilde{X}_i^*$, S_i) represents the shape and size of X_i, respectively.
3. (Shape analysis by PNS) Find principal nested spheres and PNSmean, as described in the previous subsection with inputs $\{\widetilde{X}_i^*\}$, and denote the resulting Euclidean-type representation as an $m \times n$ matrix $\mathscr{Z} = (z_{ki})$, where z_{ki} is the ith sample's deviation from the PNSmean along the kth principal arc, and $m \leq n-1$ is the number of components, which may be chosen by practitioners.
4. (Size analysis in log scale) Since the size S_i is strictly positive, it makes most sense to compare variability in a log scale. Let $\bar{S}_n = (\prod_{i=1}^{n} S_i)^{\frac{1}{n}}$ be the geometric mean of the size, which is the exponential of the arithmetic mean of $\log(S_i)$. Define $S_i^* = \log(S_i/\bar{S}_n)$.
5. (Composite space for shape-and-size of PDM) In order to incorporate the correlation between the size variables S_i^* and the shape variables $\mathscr{Z}$, define a composite data matrix
$$\mathscr{Z}_s = \begin{pmatrix} \mathscr{S} \\ \mathscr{Z} \end{pmatrix},$$
where $\mathscr{S} = (S_1^*, \dots, S_n^*)$ and each column contains the size (S_i) and shape $(z_{1i}, \dots, z_{mi})$ information of each sample.
6. (Principal arcs and scores) Let the spectral decomposition of the $(m+1)$-dimensional square matrix $\frac{1}{n-1}\mathscr{Z}_s\mathscr{Z}_s^T$ be $U\Lambda U^T$, where $U = [\mathbf{u}_1, \dots, \mathbf{u}_{m+1}]$ is the orthogonal matrix of eigenvectors and Λ is the diagonal matrix of eigenvalues

$\lambda_1, \ldots, \lambda_{m+1}$. Similar to the conventional PCA, the eigenvectors $\mathbf{u}_k$ represent the direction of important variation in the space of $\mathscr{Z}_s$ which leads to the *principal arc* when converted back to the original space of PDMs. Likewise, the eigenvalues λ_k represent the variation contained in each component. *Principal Arc scores* for each component are computed by $\mathbf{u}_k^T \mathscr{Z}_s$, which is the vector of the kth scores of all n samples.

The analysis of composite PNS can be used in a same fashion as Euclidean PCA is used. Both provides a nested sequence of subspaces (or sub-manifolds) for dimension reduction, and PC scores (or PA scores) that are important for visualization of important data structure, and for further analysis such as PC regression.

The advantage of composite PNS comes from the flexible class of sub-manifolds instead of subspaces. As shown in Section 3.3, the proposed method gives a more effective decomposition of the space compared to Euclidean PCA and PGA.

3.3 Experimental Results

Respiratory motion analysis in the lung is important for understanding the motilities of tumors in the lung of an individual patient for radiation therapy applications. The PDM of the lung boundary is used as the surrogate signal for characterizing the respiratory motion [21]. The usual PCA has been used for extracting shape deformation statistics of a patient's lungs from a sequence of 3-d CT images collected at ten time points within one breathing cycle. In preparation for PCA (also for composite PNS) on this data set, the geometric correspondence over the the training samples is optimized via an entropy-based correspondence algorithm [22].

We consider two examples. The first data set is from 4D Nurbs-based Cardiac-Torso (NCAT) phantom thorax CTs, which were produced at ten phases sampled in one breathing cycle. The second data set is from Respiration-correlated CT of a real patient. The CT data sets are provided by a 4-slice scanner (lightSpeed GX/i, GE Medical System), acquiring repeat CT images for a complete respiratory cycle at each couch position while recording patient respiration (Real-time Position Management System, Varian Medical Systems). The CT images are retrospectively sorted (GE Advantage 4D) to produce a series of 3D images at different respiratory time points.

The difficulty of the problem is two-fold; the dimension is very high ($d = 10550$, which could be much higher depending on the number of points on the surface) while the sample size is small ($n = 10$) and the major variation is non-linear, as shown in Fig. 3 for the NCAT data sets.

Fig. 3 shows scatter plots of NCAT lung data by the usual PCA (in the left panel) and by composite PNS (in the right panel). The dimension of the data space is reduced to 3 to visualize the structure of major variation. The non-linear variation apparent in the PCA subspace is represented as a linear motion in the sub-manifold of composite PNS. In particular, the quadratic motion in the PC 1–2 plane is efficiently captured by the 1-dimensional principal arc. Observe that the sum of variances contained PC 1–2 is roughly the amount of variation in the first principal arc.

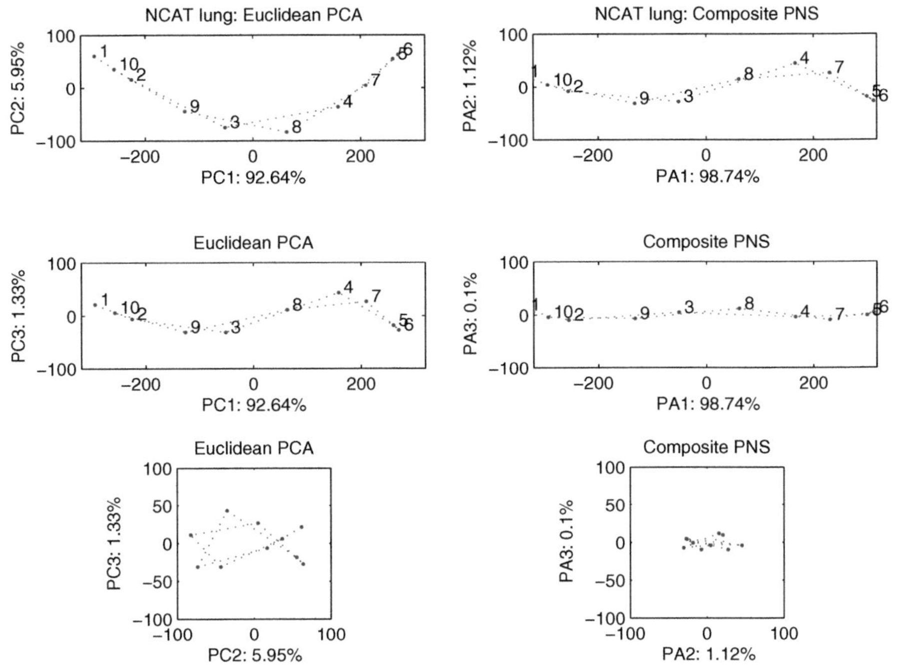

Fig. 3. (Left) Different views of scatterplot of NCAT lung data by PC scores in the first three components of Euclidean PCA. Time points are labeled as 1-10 in the scatterplot and the proportion of variance contained in each component appears in the labels of axes. Major variation in the data is non-linear and exists through all three components. (Right) Different views of scatterplot of the NCAT lung data by PA scores of composite PNS. The major quadratic variation in the first two PCs are captured in the first principal arc, and thus the variation appears more linear. The first component in composite PNS contains more variation (98.74% of the total variation) than 92.64% of PCA.

We also have compared the method with PGA (in place of PNS) and kernel PCA ([20]). The resulting proportions of variations in PGA are similar to PCA: 92.72%, 5.88% and 1.32% in the first three components of PGA. Kernel PCA with kernel $k(\mathbf{x},\mathbf{y}) = \exp(-10^6 \|\mathbf{x}-\mathbf{y}\|^2)$ gives better results than PGA and PCA in the first three proportions of eigenvalues (97.94%, 1.89% and 0.09%), but inferior to composite PNS. Discussion related to kernel PCA is deferred to the end of this section.

The data set from the real patient gives a similar result, where the cumulative proportions of variances in the first three sub-manifolds (96.38%, 97.79%, and 98.63%, respectively) are higher than those of PCA (93.52%, 96.25% and 97.74%).

The major lung motions contained in the first principal arc is illustrated in Fig. 4. We show the coronal and axial slices of lungs corresponding to the PNSmean and ± 1.5 standard deviations.

We also measure the discrepancy between the PDM at each time point and its 1-d approximation by PCA or composite PNS. The discrepancy here is computed by the square root of sum of squared distances between corresponding points. In the patient lung data, the discrepancy of 1-d approximations by composite PNS is uniformly smaller than that by PCA, as summarized in Table 1.

Table 1. Discrepancy of 1-d approximations at each time point of the real patient lung motion.

time point	1	2	3	4	5	6	7	8	9	10
PCA	65.2	69.9	88.7	77.7	38.9	74.4	44.1	69.8	74.6	57.6
composite PNS	38.2	66.9	66.1	55.6	37.8	36.7	30.4	63.0	60.2	44.6

Note that kernel PCA was developed for extracting non-linear features from the data set. It is possible for the kernel PCA to perform similar to our method, with suitable choice of kernel. However, exact reconstruction of principal components in the original space is sometimes impossible and thus comparisons such as Table. 1 and images such as Fig. 4 can not be obtained.

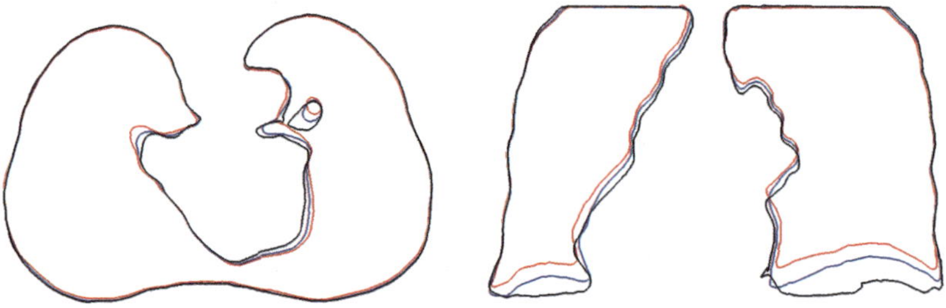

Fig. 4. Axial view (left) and coronal view (right) of boundaries of lungs. Illustrated is the variation of shapes captured in the first principal arc.

4 Conclusion

The backward PCA approaches have proven useful for dimension reduction of non-linear manifolds. In particular, PNS enjoys the advantages of the fully backward approach that enable it to yield more succinct description of the data, as shown in the example of size and SPDM shape changes with application to the lung motion. Image analysis benefits from taking attention to analysis of shapes, and thus statistical analysis in that domain might be beneficial. Particularly, the idea of PNS can be generalized to a variety of applications over both computer vision and medical imaging.

References

1. Rajamani, K.T., Styner, M.A., Talib, H., Zheng, G., Nolte, L.P., Ballester, M.A.G.: Statistical deformable bone models for robust 3d surface extrapolation from sparse data. Medical Image Analysis 11, 99–109 (2007)
2. Ramsay, J.O., Silverman, B.W.: Applied Functional Data Analysis: Methods and Case Studies. Springer, New York (2002)
3. Ramsay, J.O., Silverman, B.W.: Functional Data Analysis, 2nd edn. Springer, New York (2005)
4. Fisher, N.I.: Statistical analysis of circular data. Cambridge University Press, Cambridge (1993)
5. Mardia, K.V., Jupp, P.E.: Directional statistics. John Wiley & Sons Ltd., Chichester (2000)
6. Kendall, D.G., Barden, D., Carne, T.K., Le, H.: Shape and shape theory. Wiley Series in Probability and Statistics. John Wiley & Sons Ltd., Chichester (1999)
7. Bookstein, F.L.: Morphometric tools for landmark data. Cambridge University Press, Cambridge (1991), Geometry and biology, Reprint of the 1991 original
8. Dryden, I.L., Mardia, K.V.: Statistical shape analysis. Wiley Series in Probability and Statistics. John Wiley & Sons Ltd., Chichester (1998)
9. Siddiqi, K., Pizer, S.M.: Medial Representations: Mathematics, Algorithms and Applications. Springer, Heidelberg (2008)
10. Pennec, X., Fillard, P., Ayache, N.: A riemannian framework for tensor computing. International Journal of Computer Vision 66(1), 41–66 (2006)
11. Joshi, S.C., Miller, M.I.: Landmark matching via large deformation diffeomorphisms. IEEE Trans. Image Process. 9(8), 1357–1370 (2000)
12. Joshi, S., Davis, B., Jomier, M., Gerig, G.: Unbiased diffeomorphic atlas construction for computational anatomy. Neuroimage 23, S151–S160 (2004)
13. Marron, J.S., Jung, S., Dryden, I.L.: Speculation on the generality of the backward stepwise view of pca. In: Proceedings of MIR 2010: 11th ACM SIGMM International Conference on Multimedia Information Retrieval, Association for Computing Machinery, Inc., Danvers, MA, pp. 227–230 (2010)
14. Jung, S., Dryden, I.L., Marron, J.S.: Analysis of Principal Nested Spheres. Submitted in Biometrika (2010)
15. Fletcher, P.T., Lu, C., Pizer, S.M., Joshi, S.: Principal geodesic analysis for the study of nonlinear statistics of shape. IEEE Trans. Medical Imaging 23, 995–1005 (2004)
16. Fletcher, P.T., Joshi, S.: Riemannian geometry for the statistical analysis of diffusion tensor data. Signal Processing 87(2), 250–262 (2007)
17. Huckemann, S., Ziezold, H.: Principal component analysis for Riemannian manifolds, with an application to triangular shape spaces. Adv. in Appl. Probab. 38(2), 299–319 (2006)
18. Huckemann, S., Hotz, T., Munk, A.: Intrinsic shape analysis: Geodesic PCA for Riemannian manifolds modulo isometric lie group actions. Statistica Sinica 20(1), 1–58 (2010)
19. Jung, S., Foskey, M., Marron, J.S.: Principal arc analysis on direct productmanifolds. To appear in Annals of Applied Statistics (2010)
20. Schoölkopf, B., Smola, A.J., Mller, K.R.: Nonlinear component analysis as a kernel eigenvalue problem. Neural Computation 10, 1299–1319 (1998)
21. Liu, X., Oguz, I., Pizer, S.M., Mageras, G.S.: Shape-correlated deformation statistics for respiratory motion prediction in 4D lung. SPIE Medical Imaging 7625 (2010)
22. Oguz, I., Cates, J., Fletcher, T., Whitaker, R., Cool, D., Aylward, S., Styner, M.: Cortical correspondence using entropy-based particle systems and local features. In: 5th IEEE International Symposium on Biomedical Imaging, ISBI 2008, pp. 1637–1640 (2008)

Performance of MRF-Based Stereo Algorithms for Cluttered Scenes

Fahim Mannan and Michael Langer

Abstract. This paper evaluates the performance of different Markov Random Field (MRF) based stereo algorithms for cluttered scenes. These scenes are generated by randomly placing objects within a 3D volume. The scenes, which model natural cluttered scenes such as forests or bushes, contain many depth discontinuities and monocularly visible pixels. Widely used benchmark datasets do not contain stereo pairs with dense clutter, so we address how well existing stereo algorithms perform for such scenes. We use Expansion, Swap, Max Product Belief Propagation (BP-M), Sequential Tree Reweighted Message Passing (TRW-S) and Sequential Belief Propagation (BP-S), all with different forms of data and smoothness terms. The results are compared with the ground truth disparity and energy. We found Expansion, TRW-S, and BP-M with the Potts model to work well for most scenes, in that the correct binocular correspondence is found for most points that are binocularly visible. We also found that the energy for the ground truth is much larger than what is found by the optimizers. This shows that there is room for improving the model for cluttered scenes.

1 Introduction

Over the years a number of approaches have been proposed for solving stereo reconstruction problem. Most of these approaches focus either on speed or accuracy of reconstruction. Some of them are designed to handle difficult cases like slanted or non fronto-parallel surfaces, e.g. see [1] for one of the earliest works and many recent works listed in `http://vision.middlebury.edu/stereo/eval/`.

One of the main steps of stereo reconstruction is solving the correspondence problem where a disparity value – indicating the displacement between the corresponding left and right pixels – is assigned to each pixel of one of the images

Fahim Mannan · Michael Langer
School of Computer Science, McGill University
e-mail: `{fmannan, langer}@cim.mcgill.ca`

J. Angeles et al. (Eds.): Brain, Body and Machine, AISC 83, pp. 125–136.
springerlink.com

or a global "cyclopean" image. The correspondence problem is usually formulated within a variational framework where energy functions are defined over the possible set of disparities and the optimal energy corresponds to the optimal disparity assignment. In this paper we restrict our attention to the Markov Random Field (MRF) based formulation of the problem, where the goal is to estimate the Maximum-a-Posteriori (MAP) probability of a given MRF. Such formulations are known as MAP-MRF. In a variational framework and also in MAP-MRF, the scene under consideration is often assumed to be smooth. This implies that, for a given pixel, all its neighboring pixels are likely to have similar disparity. As a result, depth discontinuities are penalized to encourage smoothness. If we look at some of the stereo benchmark datasets – e.g. the most widely used Middlebury dataset [2] – we can see that many of the images are of indoor scenes with fairly good binocular visibility. For such scenes this type of smoothness assumption is very reasonable. However, certain natural scenes like leaves on trees, bushes and hedges, can have significant depth discontinuities and occlusions. In other words, neighboring pixels are less likely to be on the same surface and as a result, they will have very different disparities. We refer to such scenes as cluttered scenes. More formally, we will define a class of cluttered scenes that have a large number of objects with sizes drawn from some distribution and positions uniformly distributed in 3D space. It is easy to see, that for cluttered scenes, the smoothness assumption is not necessarily correct and thus could affect the overall performance of MRF stereo algorithms.

The question we are interested in is, how well do the existing formulations perform for cluttered scenes? To our knowledge no one has specifically addressed this question even though cluttered scenes occur very frequently. The contribution of this paper is in evaluating how a subset of the existing MRF-based models perform for cluttered scenes and also to empirically identify a set of parameters that works well in most of the cases.

To keep the analysis from getting too complex we only consider the basic formulation that consists of a data and a smoothness term. More specifically we are only interested in the form of these two terms that gives good results. Therefore we do not consider additional constraints like visibility or uniqueness. One challenge in doing such analysis is the availability of benchmark dataset. Most benchmark datasets do not contain significant number of images of cluttered scenes. Therefore we generate synthetic scenes that reasonably simulate the characteristics of natural cluttered scenes. The synthetic scenes are generated based on the model presented in [3]. This also allows us to generate scenes with specific characteristics (e.g. density, occlusion) and also compare statistics of the resulting images with the model.

There are several areas that can benefit from cluttered scene stereo. One such application area is forest measurement, where the problem is to determine the growth of a forest by making measurements such as leaf area index (LAI), canopy size, etc. [4, 5, 6]. Existing methods rely on either making measurements using range finders or by estimating the amount of sunlight falling in an area.Cluttered scene stereo can significantly simplify the process of making such measurements.

The paper is organized as follows. In section 2 we give an overview of the form of energy function that is optimized, the algorithms being used, and discuss some of the related performance comparisons that have been done. Section 3 describes the model that we use for cluttered scenes and in section 4 we discuss how stereo pairs are generated using the model. We also discuss some of the challenges of generating these synthetic stereo pairs and ways to overcome these challenges. In section 5 we describe the experiments that were done and discuss the results and observations. Finally section 6 presents conclusions and some possible future directions.

2 Previous Work

In the MRF-based formulation of the stereo problem, each pixel or a group of pixels is considered as a random variable that can take on values from a set of disparity labels [7]. The problem then becomes solving the MAP assignment problem. The solution can be obtained by minimizing an energy function that maps an assignment of labels to real numbers. More specifically, for stereo pair images I_l and I_r with disparity values f_p and f_q for pixels p and q respectively, the energy function that we optimize is of the following form [8]:

$$E = E_p(f_p) + \lambda\, E_{p,q}(f_p, f_q) \quad (1)$$

$$E_p(f_p) = d(I_l, I_r, p, q)^{k_d} \quad (2)$$

$$E_{p,q}(f_p, f_q) = w_{pq}\, min(|f_p - f_q|^{k_s}, V_{max}) \quad (3)$$

Equation 2 is called the *data term* and 3 the *smoothness term*. The function d in the data term is the dissimilarity measure between corresponding pixels (i.e. pixels p and q in I_l and I_r respectively). It can be absolute difference, sum of squared error or the more sophisticated Birchfield-Tomasi measure [9]. The data term can be either the absolute or squared value (for $k_d = 1$ and 2 respectively) of the dissimilarity measure. In the smoothness term, w_{pq} is a scalar weight defined for every neighboring pixel pair p and q, and often depends on the color gradient. The smoothness term is usually a function of label difference. The parameter λ specifies how much weight should be given to the smoothness term.

It should be noted that in our problem we only consider integer disparity values as is usually considered in many stereo problems e.g. [2, 8]. However the choice of datatype does not have any impact on the choice of optimization algorithm. Furthermore our synthetic scenes are chosen to have integer disparities, which makes the choice of having integer disparities inconsequential.

There are several algorithms available for solving this discrete optimization problem. Some of the widely used algorithms are Expansion, Swap [10], and variants of belief propagation [11] such as BP-M, TRW-S, and BP-S [12, 13, 14]. The motivation behind using different optimization algorithms is that most of the algorithms do not have any guarantee on the quality of the solution for the types of energies that we use. Expansion does provide some bound on the approximate solution but only for submodular functions, and our energy formulations are not necessarily submodular. Similarly TRW-S guarantees that the lower bound of the energy will increase but does not provide any bound on the energy. Therefore, to have a more complete analysis we decided to compare these algorithms.

Scharstein et al. in [2] categorizes and compares different techniques for stereo reconstruction. Compared to their work we only consider MRF based stereo formulation and their performance for a special class of scenes. Tappen et al. in [12] compares Graph Cuts and Belief Propagation using the same energy formulation. The authors found that both algorithms yield optimal solutions whose energy is lower that of the true scene. This indicates that the energy formulation was not sufficient to model the problem for cluttered scenes. T Our paper is motivated by their work. We use a wider range of algorithms and energy terms, but restrict the analysis to cluttered 3D scenes. Finally, in [8], the authors study the performance of different minimization algorithms on a set of benchmark problems. While their objective was to find the algorithm with the lowest energy configuration and fastest convergence, our objective is to find the range of parameters that gives the lowest error rate and also to correlate that with the energy value and underlying statistics of the input image. Furthermore we only concentrate on performance of stereo for cluttered scenes.

3 Cluttered Scene Modelling

This paper addresses a particular class of natural scenes, which we are calling "cluttered scenes." An example is the foliage of a tree or shrub, or a forest. In [15], Huang et al. addresses the problem of modeling the statistics of natural scenes. They used a database of range images, including many natural scenes such as forests, and investigate different statistical properties of range images. They found range images exhibit similar scale-invariant statistics as intensity images and showed that a "dead-leaves model" [16] fit some of the statistics very well. The dead leaves model is not a 3D model, however, and so it cannot be used to generate stereo pairs.

In [3], a statistical model of 3D cluttered scenes was presented, and this is the model that we use in this paper. In this section we give a brief overview of that model and discuss how it is used in our work.

The model presented in [3] assumes that objects of fixed size are uniformly distributed within a 3D volume. Under this scenario if the integer disparity value of a pixel p is f_p, and γ is a constant that depends on the the average number of surfaces per unit volume and the area of surfaces, then the probability density of disparity can be modeled using a Poisson distribution as follows:

$$p(f_p) = \frac{\gamma\sigma}{f_p^2} \exp\{-\gamma\sigma \left(\frac{1}{f_p^2} - \frac{1}{f_0^2} \right)\}. \tag{4}$$

Here, f_0 is the largest disparity or disparity for the closest object and σ is the scaling factor which is the product of focal length, baseline and ratio of projection plane and image width.

Binocular visibility of points in space is also modeled in [3] under the assumption that one of the viewer's field-of-view is 180 degrees. In this paper, we extend that formulation by considering bounded view volume. The resulting equation is

$$p\left(f_p^r|f_p^l\right) = p_{180}\left(f_p^r|f_p^l\right)\frac{w_z - T_x}{w_z}. \tag{5}$$

Here, $p_{180}\left(f_p^r|f_p^l\right)$ is the conditional probability of binocular visibility as given in [3], and w_z and T_x are the width of the view volume at a certain depth z and baseline respectively. Therefore, the second term is the fraction of the view volume visible to the other eye. As in [3], we could multiply the above two probabilities (eq. 4 and 5) to derive the disparity probability density of binocular pixels, which are pairs of pixels in the two images that see the same surface points.

We also wish to choose scenes that have different probabilities of smoothness and discontinuity. For this we compute the joint probability density $p(f_p, f_q)$ of disparity values for neighboring pixels *in the same image*. Figure 2 will show negative log probabilities for neighboring pixels having either different $-\log(p(f_p \neq f_q|f_p))$ or same disparity values $-\log(p(f_p = f_q|f_p))$. We consider these probabilities because the smoothness term penalizes a discontinuity between neighboring pixels and encourages continuity by giving a penalty of 0. Because of space constraints, we leave out the model equations and their formal derivations.

4 Synthetic Stereo Pair Generation

We use OpenGL for rendering synthetic stereo pairs. The scenes are generated using the perspective projection model. For each scene, the object sizes are either constant or obey a $1/r^3$ distribution (for scale-invariant scenes [16]). First, the centers are generated uniformly within a predefined region which is determined from the depth range and baseline distance. Texture mapped squares are then rendered at those positions. Texture mapping is necessary because the stereo formulations that we use do not work well for untextured surfaces. For our experiments we randomly generate the (R,G,B) color values for a 2×2 texture and linearly interpolate the intermediate values while mapping to a smaller or larger surface (using OpenGL's `GL_LINEAR` function).

In our setting, the optical axes of the cameras are parallel to each other and the squares are parallel to the image plane. The reason for such configuration is that we only use first order priors, and these priors work best for fronto-parallel surfaces. This allows us to avoid any performance issues caused by the shape or orientation of the objects and to only evaluate how the depth and size of objects affect the performance of a model. In this way, we generate scenes that conform to the cluttered scene model as well as some of the simpler stereo models.

While generating the scenes we choose the size of the objects in such a way that the smallest object projects to at least 4 pixels (half width is 2 pixels). In the non scale-invariant case this is determined by the size of the object on the farthest plane. In the case of a scale-invariant scene the range depends on the size of the smallest object on the farthest plane and largest on the nearest plane. In our setting, the range of object widths in pixels is between 2 to 128. In scene coordinates, the width range is 0.034 to 0.536 for a depth range of 2 to 8 and 0.134 to 2.14 for a depth range of 8 to 32. For generating scale-invariant scenes we uniformly divide the range of widths and generate a certain number of objects of each width such that we have the desired size distribution.

One of the main difficulties in generating synthetic scenes with a very accurate ground truth disparity map is to ensure that the integer pixel size of the projection of an object in both images is the same and the ground truth disparity value is equal to the theoretical disparity. These two conditions do not hold in general because of rounding error and anti-aliasing in OpenGL. To avoid such problems we perform the perspective calculations outside the OpenGL pipeline and render the scene under orthographic projection with appropriate size and displacement. It should be noted that since we are rounding to the nearest integer the actual distance and size of objects will be slightly different from the specified values. This causes some jaggedness in the ensemble statistics but the overall characteristics remain the same. We emphasize that the goal is to generate scenes with particular statistical properties rather than scenes that have all the properties of natural scenes (e.g. non-integer disparities, different orientations and slants, spatial clumping, etc.).

Once the scene is rendered, the ground truth disparity at each pixel of the left image is computed as follows: First the depth value at a pixel in the left image is used to find the corresponding pixel in the right image. The depth values of the two pixels is then used to determine visibility and disparity. If the corresponding pixels are at the same depth, both are visible (binocular visibility) to the left and right camera and have a disparity equal to the difference in their coordinates (in this case only the horizontal displacement). Otherwise if the pixel in the right image is closer to the camera (i.e. smaller depth value), then the pixel in the left image is occluded in the right view. We call this pixel monocular (this has also been referred to as half-occluded pixel in the literature e.g. [17]) and mark it as occluded in the disparity map. A pixel is also considered occluded, if the corresponding pixel in the right image is outside the view volume of the right camera.

5 Experiments

In this section, we describe the parameters that were used for generating the test scenes and for performing the experiments. We also discuss the effect of those parameters.

5.1 Choosing Scene Parameters

To evaluate the performance of stereo algorithms on different classes of cluttered scenes, we will need to understand the data and smoothness parameter requirements and how they interact with the parameters of each cluttered scene. More specifically we want to find out which set of energy parameters and algorithm combination work best for which type of cluttered scenes. The first step in this analysis is to categorize cluttered scenes based on their parameters. These parameters can be chosen based on the model provided in [3] and the extension discussed in this paper. Table 1 shows the categories we use, which have various γ, and range of depths and object size. A few samples of these scenes are shown in Figure 1.

Table 1. Parameters for 12 cluttered scene categories. For each γ and z range, there are three sizes r (large, small, scale invariant).

scene	γ	z range	r
1			0.1
2		2-8	0.025
3	0.54		0.034 - 0.536
4			0.4
5		8-32	0.1
6			0.134 - 2.14
7			0.1
8		2-8	0.025
9	0.1		0.034 - 0.536
10			0.4
11		8-32	0.1
12			0.134 - 2.14

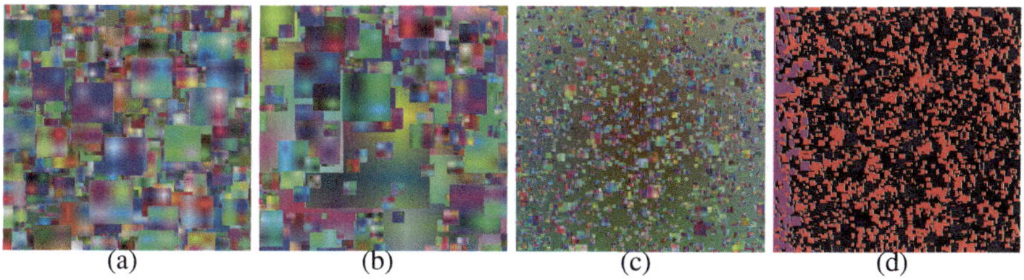

Fig. 1. (a,b,c) are example images of scenes 1, 6, 8 of Table 1. (d) shows the monocular pixels (non-black) for (c). The color indicates whether the monocular pixel is inside (purple) or outside (red) the field of view of the other image.

For each scene category, mean probabilities are calculated using 100 images of size 256×256. See Fig. 2. Pairs of scenes with the same γ have similar column 1's (i.e. similar curves), as expected from Eq. 4. For example, compare the 2nd and 3rd row of Fig. 2. For scenes with different γ, the curves in column 1 are quite different (e.g. 1st and 7th row of the figure). At first glance, column 1 of the 1st and 2nd row might seem different but the 2nd row is showing the left part of the curve (i.e. for smaller disparity) in the 1st row. Also note that the binocular curves in column 1 are, in general, indeed lower than the monocular ones, as expected, and that the difference between monocular and binocular curves tends to more prominent for small disparities i.e. far away.

We next turn to the 2nd and 3rd columns which compare disparities at neighboring points. The 2nd column shows $-\log p()$ of a discontinuity as a function of disparity. While at first glance, it appears the curves are similar, in fact they are quite different: the scales on the axes vary significantly between plots. The 3rd column shows the $-\log p$ of no discontinuity. Here, both the shapes of the curves and the scales vary significantly between plots. Thus, we have illustrated that the scene categories indeed differ significantly in their smoothness and discontinuity probabilities, and that these differences in probabilities depend on disparity. We therefore expect a different pattern of errors in these scene categories when we run stereo algorithms on them.

5.2 *Performance Evaluation*

We use the code from [8] for running the algorithms on the dataset with different forms of data term, smoothness exponent, truncation, and smoothness scaling parameter. While measuring accuracy we only consider binocular pixels, *i.e.* pixels visible to both eyes. This is because the form of the energy functions used in the experiments work best for such pixels, since there are no visibility constraints and the algorithms also do not extrapolate depths using e.g. plane fitting for monocular pixels.

In our analysis, we define the error rate to be the fraction of binocular pixels that differ from the ground truth. Although in many previous work a non-zero error threshold is used for making the error measure more robust, in our case, the synthetic ground truth is exact and so we set the error threshold to zero. The error and energy statistics are computed from 5 images of each category. To make the experiments more realistic we add Gaussian noise with standard deviation 5 to the right view. For such noisy inputs, a gradient threshold (Eq. 3) gives better results. Therefore we only include results where w_{pq} is proportional to the gradient penalty.

The error rates for the outputs are shown in Fig. 3. We show the results for the TRWS algorithm, since it usually produces a solution that either has the lowest error rate or else is within 5% of the lowest error. For the parameters, we use $k_d = 2$, $k_s = 1$ and $V_{max} = 1, 2, 10, 100$. Both $k_d = 1$ and 2 can achieve similar error rates, with the main difference being the range of λ value. For our dataset, with $k_d = 1$, the range of λ is 1–15, and for $k_d = 2$, the range is between 10–120. For the smoothness parameter, $k_s = 1$ gives better results than $k_s = 2$ when $V_{max} = 10, 100$, but for smaller V_{max} the results are the same. From Fig. 3 it can be seen that smaller values of V_{max} perform well in all cases, namely $V_{max} = 1$ (Potts) or 2.

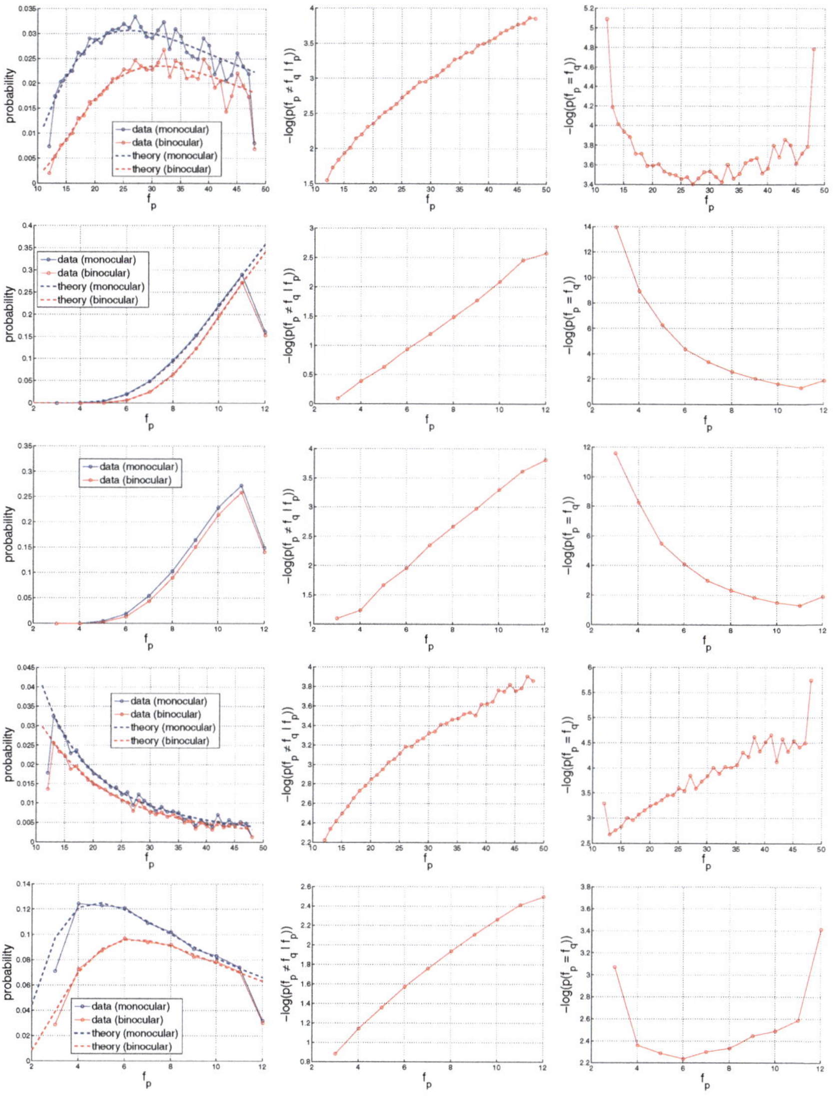

Fig. 2. Probabilities (normalized histograms) for few of the samples from the input dataset (rows 1–5 corresponds to scenes 1,5,6, 7, and 11 respectively). The first column shows the probability of disparity. Discontinuity and continuity probabilities are shown in the second and third column respectively.

Figure 4 compares ground truth energy and energy for computed solutions, for various V_{max} and k_s parameters, and various optimizers. Overall it can be seen that the energies for the ground truth solution are much larger. This gap is mainly due to the large number of monocular or half-occluded pixels. The energy formulations do

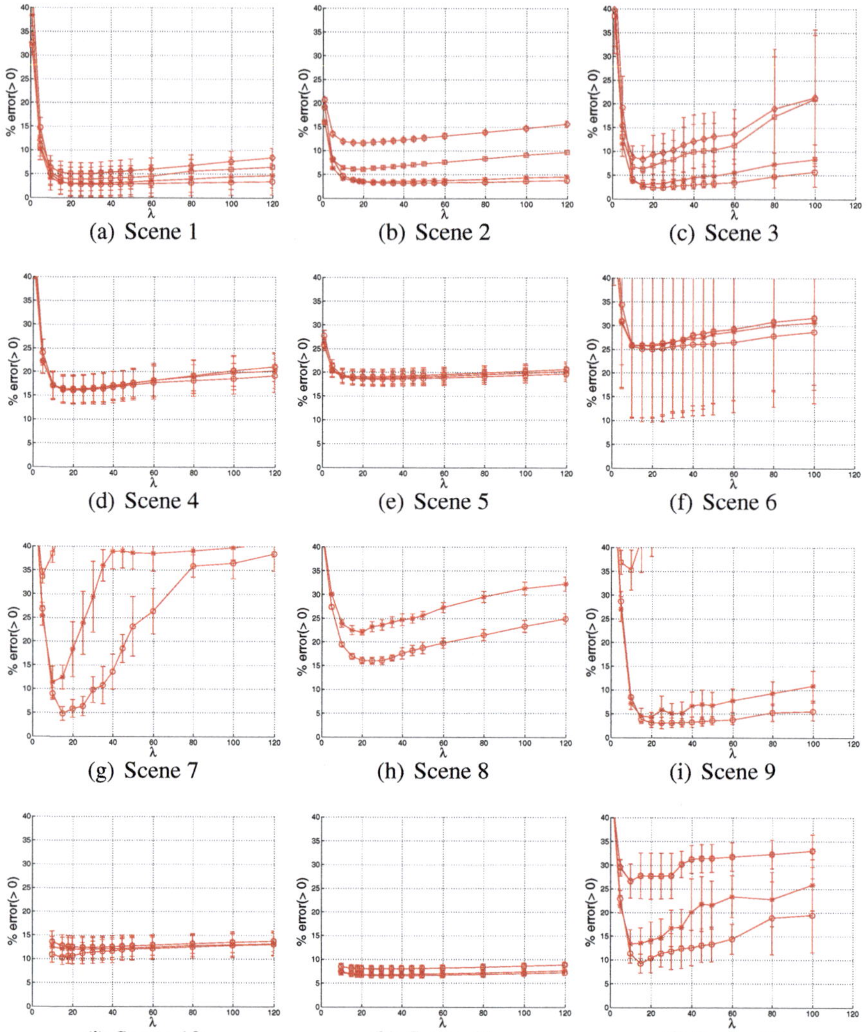

(a) Scene 1 (b) Scene 2 (c) Scene 3
(d) Scene 4 (e) Scene 5 (f) Scene 6
(g) Scene 7 (h) Scene 8 (i) Scene 9
(j) Scene 10 (k) Scene 11 (l) Scene 12

Fig. 3. Error statistics for scenes 1–12 ((a)–(l)) of Table 1. For each scene only the results for TRWS algorithm with $k_d = 2$, $k_s = 1$ are shown. V_{max} = 1,2,10 and 100 are represented by o, *, □ and ◇ respectively. Only the error rates between 0–40% are shown.

not have any constraints for those pixels and as a result they assign labels in a way that minimizes the energy but does not necessarily find a good estimate of disparity. This clearly suggests that the energy terms are not sufficiently capturing the labeling constraints.

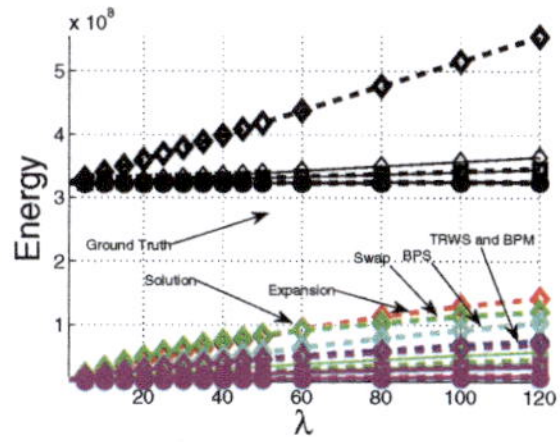

Fig. 4. Energies for computed solutions (lower curves) are typically much lower than energies for ground truth (upper curves), which implies that the energy function does not capture well the structure of cluttered scenes. Data shown are for scene 8, and various values of V_{max} and k_s, and various algorithms (see labels).

6 Conclusion

In this paper, we have looked at how some of the existing MRF based stereo formulations (data and smoothness terms) perform for cluttered scenes. To this end, we first described the model that we use for rendering synthetic scenes, then we used the model to generate synthetic scenes with different underlying statistics. Finally, we empirically observed the set of configurations and optimizers that work best for each type of scene. We found that Expansion, BP-M and TRW-S with $V_{max} = 1,2$ and $k_s = 1,2$ perform well for most of the scenes. In the case of data term, $k_d = 2$ (squared difference) is usually better since it gives better accuracy and has very low error rate for a large range of λ values. We observed that the best error rate does not necessarily correspond to the lowest energy configuration. The energy for ground truth is much higher than what is found by all the algorithms.

The experiments show that even if we have scenes tailored to work well with the form of smoothness term (ie. fronto-parallel surface), there is still room for improvement. Future work will include doing similar analysis with stereo pairs of real scenes with range data and using algorithms that specify additional constraints. Work will also be done on formulating smoothness priors that makes use of scene statistics.

Acknowledgements. We would like to thank the reviewers for their helpful comments and suggestions. We also acknowledge NSERC and FQRNT for their support.

References

1. Birchfield, S., Tomasi, C.: Multiway cut for stereo and motion with slanted surfaces. In: IEEE International Conference on Computer Vision, vol. 1, p. 489 (1999)
2. Scharstein, D., Szeliski, R.: A taxonomy and evaluation of dense two-frame stereo correspondence algorithms. International Journal of Computer Vision 47(1-3), 7–42 (2002)
3. Langer, M.S.: Surface visibility probabilities in 3d cluttered scenes. In: European Conference on Computer Vision, pp. 401–412. Springer, Heidelberg (2008)
4. Chen, J., Cihlar, J.: Plant canopy gap-size analysis theory for improving optical measurements of leaf-area index. Applied Optics 34(27), 6211–6222 (1995)

5. van Gardingen, P.R., Jackson, G.E., Hernandez-Daumas, S., Russell, G., Sharp, L.: Leaf area index estimates obtained for clumped canopies using hemispherical photography. Agricultural and Forest Meteorology 94(3-4), 243–257 (1999)
6. Riao, D., Valladares, F., Condés, S., Chuvieco, E.: Estimation of leaf area index and covered ground from airborne laser scanner (lidar) in two contrasting forests. Agricultural and Forest Meteorology 124(3-4), 269–275 (2004)
7. Li, S.: Markov Random Field Modeling in Image Analysis. Springer Publishing Company, Incorporated, Heidelberg (2009)
8. Szeliski, R., Zabih, R., Scharstein, D., Veksler, O., Kolmogorov, V., Agarwala, A., Tappen, M., Rother, C.: A comparative study of energy minimization methods for markov random fields with smoothness-based priors. IEEE Transactions on Pattern Analysis and Machine Intelligence 30, 1068–1080 (2008)
9. Birchfield, S., Tomasi, C.: A pixel dissimilarity measure that is insensitive to image sampling. IEEE Transactions on Pattern Analysis and Machine Intelligence 20, 401–406 (1998)
10. Boykov, Y., Veksler, O., Zabih, R.: Fast approximate energy minimization via graph cuts. IEEE Transactions on Pattern Analysis and Machine Intelligence 23, 1222–1239 (2001)
11. Pearl, J.: Probabilistic reasoning in intelligent systems: networks of plausible inference. Morgan Kaufmann Publishers Inc., San Francisco (1988)
12. Tappen, M.F., Freeman, W.T.: Comparison of graph cuts with belief propagation for stereo, using identical mrf parameters. In: IEEE International Conference on Computer Vision, Washington, DC, USA, p. 900. IEEE Computer Society, Los Alamitos (2003)
13. Kolmogorov, V.: Convergent tree-reweighted message passing for energy minimization. IEEE Transactions on Pattern Analysis and Machine Intelligence 28(10), 1568–1583 (2006)
14. Wainwright, M., Jaakkola, T., Willsky, A.: Map estimation via agreement on (hyper)trees: Message-passing and linear-programming approaches. IEEE Transactions on Information Theory 51(11), 3697–3717 (2005)
15. Huang, J., Lee, A., Mumford, D.: Statistics of range images. In: IEEE Conference on Computer Vision and Pattern Recognition, vol. 1, pp. 324–331 (2000)
16. Lee, A.B., Mumford, D., Huang, J.: Occlusion models for natural images: A statistical study of a scale-invariant dead leaves model. International Journal of Computer Vision 41(1-2), 35–59 (2001)
17. Belhumeur, P.N., Mumford, D.: A bayesian treatment of the stereo correspondence problem using half-occluded regions. In: IEEE Conference on Computer Vision and Pattern Recognition, pp. 506–512 (1992)

Medial Spheres for Shape Approximation

Svetlana Stolpner, Paul Kry, and Kaleem Siddiqi

Abstract. We study the problem of approximating a solid with a union of overlapping spheres. We introduce a method based on medial spheres which, when compared to a state-of-the-art approach, offers more than an order of magnitude speed-up and achieves a tighter volumetric approximation of the original mesh, while using fewer spheres. The spheres generated by our method are internal to the object, which permits an exact error analysis and comparison with other sphere approximations. We demonstrate that a tight bounding volume hierarchy of our set of spheres may be constructed using rectangle-swept spheres as bounding volumes. Further, once our spheres are dilated, we show that this hierarchy generally offers superior performance in approximate separation distance tests.

1 Introduction

The choice of representation of 3D volumetric data is an important question in motion planning, solid modeling, computer vision, computer graphics, medical imaging and computer aided design. Whereas representations that approximate shape boundaries with triangles are popular, representation of solids as a union of spheres are a valuable alternative. In 1994, Ranjan and Fournier advocated the use of a union of spheres for representing volumetric data and argued that this representation captures the significant features of a shape robustly with respect to noise and resolution [1]. In recent years, a number of applications have made use of sphere-based representations. Particularly, hierarchies of spheres are used for collision detection in [2] because sphere-sphere intersection tests are fast and simple. Sphere approximations are also used for fast soft shadow generation [3] where a low-frequency object representation is sufficient. Additionally, sphere-based representations are used for

Svetlana Stolpner · Paul Kry · Kaleem Siddiqi
School of Computer Science and Centre for Intelligent Machines, 3480 University Street, Room 410, Montréal, Québec, Canada, H3A 2A7
e-mail: {sveta, siddiqi}@cim.mcgill.ca, kry@cs.mcgill.ca

J. Angeles et al. (Eds.): Brain, Body and Machine, AISC 83, pp. 137–148.
springerlink.com

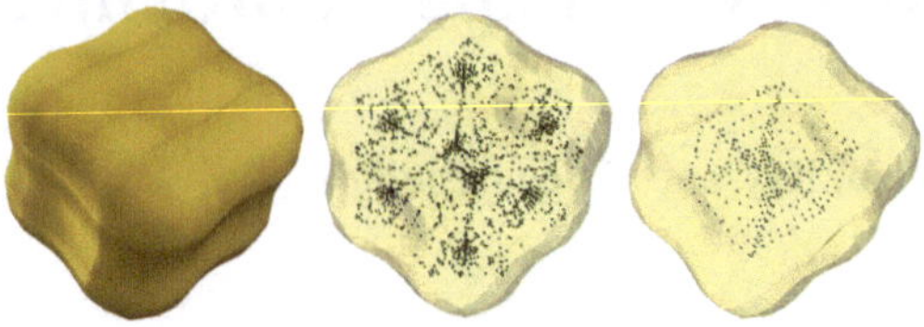

Fig. 1. A mesh (left), sphere centers given by Voronoi vertices (center), sphere centers computed with our method (right). Note that the Voronoi vertices are highly clustered at the centers of the osculating spheres of the 8 corners of the object, while the sphere centers produced with our method are distributed nearly uniformly.

efficient level-of-detail rendering [4], shape matching [5], and shape deformation [6]. Typically, the boundary of the object is first represented as a triangle mesh, and next the interior of the mesh is approximated using a union of spheres.

In the above applications, it is desirable to generate a sphere-based approximation of a solid that 1) has a small number of spheres, 2) provides a tight fit to the original shape, and 3) is fast to compute. To find the minimum number of spheres that cover a set of points on the object boundary is a hard computational problem. In fact, it can be shown to be NP-hard by reduction from Set Cover [7]. Therefore, to find a small set of approximating spheres to an object boundary heuristics are used.

A popular strategy for approximating an object with a small number of well-fitting spheres is to start by computing the Voronoi spheres of a set of boundary points [2, 7, 8]. As we shall see in our subsequent discussion, the centers of these spheres, the Voronoi vertices, are inherently clustered at rounded corners of objects (*cf.* Fig. 1(center)). Subsequently, to generate a small number of well-fitting spheres, such methods require an optimization step to remove and redistribute spheres, making them computationally expensive.

Recently, Stolpner *et al.* [9] proposed a method to compute an alternative representation of a solid as a union of spheres such that the sphere centers are "well-distributed" on the medial surface of the object. The distribution of the sphere centers generated with this method is shown in Fig. 1 (right). In the current article, we observe that by working directly with this distribution, we avoid the expensive post-processing steps of adding, deleting, or redistributing spheres. In comparison with the state-of-the-art, we show that this sphere-generation method works significantly faster, generates fewer spheres, and provides a tighter fit to the original object in terms of volumetric error.

The spheres that we generate have the special property that they are *internal to the bounding surface* of the object. This fact allows us to quickly compute the exact volumetric error of our approximation and to compare the error with that of a sphere approximation that is not necessarily internal. Further, we demonstrate how our spheres may be quickly dilated in order to provide improved boundary coverage for approximate separation distance computation.

2 Background and Previous Work

We begin by providing some necessary definitions. The complete (generically, infinite) set of spheres whose union corresponds to a solid is given by the Medial Surface Transform. Given a solid $\Omega \subset \mathbb{R}^3$ with boundary $\mathscr{B}$, the *medial surface* $\mathscr{MS}$ of Ω is the locus of centers of maximal inscribed spheres in Ω. A maximal inscribed sphere in Ω is called a *medial sphere*; the set of all medial spheres is the *Medial Surface Transform* of Ω.

When approximating an object with spheres, a popular strategy is to start with a subset of the spheres of the medial surface transform of the object. A useful tool for doing this is the following discrete analogue of the medial surface transform.

Given a set of point sites $S = \{\mathbf{s_1}, \mathbf{s_2}, \ldots, \mathbf{s_n}\}$, $\mathbf{s_i} \in \mathbb{R}^3$, the *Voronoi diagram* of S is a partition of $\mathbb{R}^3$ into Voronoi cells $V(\mathbf{s_i})$ with respect to the Euclidean distance d_E, such that

$$V(s_i) = \{\mathbf{x} \in \mathbb{R}^3 \mid d_E^2(\mathbf{s_i}, \mathbf{x}) \leq d_E^2(\mathbf{s_j}, \mathbf{x}),\ \forall j \neq i\}. \tag{1}$$

A Voronoi cell of a site $\mathbf{s_i}$ is, hence, the set of points that are closer to the site $\mathbf{s_i}$ than to any other site in S. A vertex of $V(\mathbf{s_i})$ is called a *Voronoi vertex* and is the center of a *Voronoi sphere* that touches 4 or more sites of S, but does not contain any sites in its interior.

When the points S are sampled on the boundary $\mathscr{B}$ of an object Ω, Amenta et al. [10] show that a subset of the Voronoi vertices of S, the *poles*, converges to the medial surface of Ω as the sampling density of S approaches infinity, and hence, the union of a subset of Voronoi spheres converges to Ω. This property is the basis for several methods that approximate a solid with spheres.

We now overview methods that approximate a solid with spheres with the goal of providing a tight approximation to the object using a small number of spheres. An early method for the approximation of an object with spheres [11], bounds each mesh triangle with a sphere but does not necessarily offer a tight fit. Also, a number of methods have been proposed recently for the simplification of sphere representations [7, 12], but these methods do not explicitly share the same goal of ensuring tightness of fit.

There are two methods which approximate an object Ω with tight-fitting spheres, starting with a set of Voronoi spheres. Hubbard [8] greedily selects adjacent Voronoi spheres for merging when their bounding sphere has the best tightness of fit. Bradshaw and O'Sullivan [2] improve the tightness of fit of Hubbard's algorithm by using an adaptive greedy strategy that adds sample points to the boundary of Ω to generate new Voronoi spheres as needed. Tightness of fit is evaluated as the maximum distance between each sphere and Ω. As this quantity is difficult to compute exactly, an approximation is used. Methods that initialize sphere centers to Voronoi vertices of a set of boundary points face the inherent problem that some of these vertices may be clustered; for example, rounded parts of the object boundary generate many Voronoi vertices near the center of the osculating sphere of such regions, as these are points which are equidistant from 4 boundary points.

In more recent work, Wang *et al.*[3] propose a variational optimization method that achieves tighter fits than the method of [2], but is only feasible for approximations having up to a small number of spheres (around 128 spheres). The error measure used is an approximation to total sphere volume outside the object Ω. Approximate volumetric error is also used in [13] to compare the performance of the methods of [8], [2], and an octree-based method for approximating deforming objects with spheres. In the present article we use also use a volume-based error measure.

3 Computation of Spheres

In this section, we describe how we approximate the object Ω with spheres such that the centers of the spheres are "well-distributed" and lie near a subset of the medial surface.

We will compute a set of *approximate* medial point locations using a method based on the analysis of the gradient of the Euclidean distance transform ∇D of Ω described in [9]. This gradient, $\nabla D : \mathbb{R}^3 \rightarrow \mathbb{R}^3$, is a vector field that assigns each point in Ω the direction to its closest point on $\mathscr{B}$. ∇D is a smooth vector field everywhere inside Ω, except on the medial surface where it is multi-valued. In order to generate a well-distributed set of approximate medial points, we find those voxels of a regular grid that are internal to Ω, analyze the values of the ∇D vector field in each grid cell, and, if the presence of the medial surface is likely, we look for approximate medial point locations in these grid cells. At most one approximate medial point is output per grid cell. This property ensures the "well-distributed" nature of our approximate medial points. To find approximate medial points, we use the property shown in [14], that states that the medial surface intersects a line segment (p,q) interior to Ω, where $q = p + \gamma \cdot \nabla D(p)$, iff $\nabla D(p) \neq \nabla D(p)$, for any scalar value of γ. Refer to Fig. 2, left, for an illustration. This property allows us to find the location of medial points within a user-chosen distance of the medial surface by performing binary search on line segments (p,q) intersected by the medial surface. When the radius of the spheres at the approximate medial points is chosen to be the distance from these points to their nearest boundary points, such spheres are internal and tangent to the boundary $\mathscr{B}$. We compute nearest points on the mesh boundary to query points quickly using a version of the PQP software [15] modified for efficient point-to-mesh distance queries. Additional details of the algorithm can be found in [9].

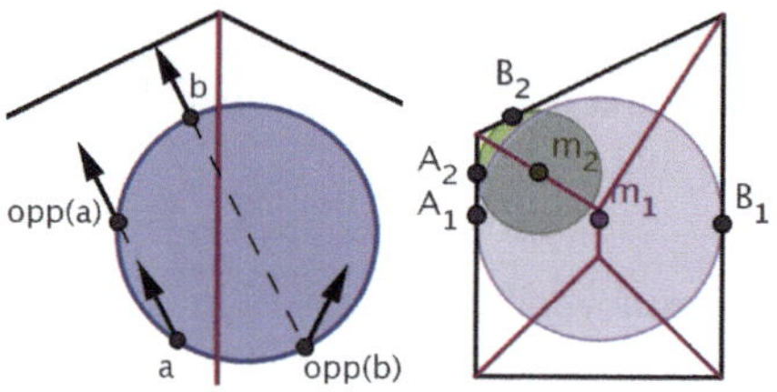

Fig. 2. The boundary of an object is shown in black and its medial surface is in red. Left, arrows show ∇D, the directions to nearest locations on the boundary of the object. In this example, the medial surface intersects the line segment $(b, opp(b))$ because $\nabla D(b) \neq \nabla D(opp(b))$. Right, the object angle of circle m_1 is is greater than that of circle m_2 since $\angle A_1 m_1 B_1 > \angle A_2 m_2 B_2$.

Suppose that a medial point $m \in \mathcal{MS}$ is equidistant from two points $A, B \in \mathcal{B}$. Angle $\angle AmB$ (see Fig. 2, right) is the *object angle*. The complete medial surface of a polyhedron can be a complex structure; for example, it includes bisectors of each pair of mesh faces that meet convexly. Our goal is to locate a few medial spheres such that the volume of their union approximates the volume of Ω well. As proven in [16], removal of spheres having a small object angle has a small impact on the volume of the reconstructed object (refer to Fig. 2, right). As we will be looking for a set of spheres whose volume is close to that of the original solid, we will look for those medial points having a large object angle. For an approximate medial point computed, only one point of tangency is known exactly (the nearest point on $\mathcal{B}$ to the approximate medial point), while the other point is approximated. In our experiments, we use a threshold of 0.6 radians for the object angle.

Figure 1 shows a distribution of approximate medial points computed with our method and that computed with a Voronoi-based method of [10]. Voronoi-based methods approximate the medial surface using a set of boundary samples as input and the density of the medial surface approximation depends on the distribution of the boundary points. In contrast, the distribution of medial points produced by our distance-based method depends on the sampling strategy of the interior of the object and allows one to control the density of the medial points produced.

The larger the grid resolution, the larger the radius of the smallest sphere computed. As small spheres contribute less to the volume of the reconstructed object than large spheres, by choosing a large grid resolution and a large object angle threshold, we generate a small number of spheres that provide good volumetric object coverage, as shown experimentally in Section 4.3. A possible limitation of our approach is that to approximate objects with long narrow parts well we would have to use a fine grid, resulting in a large number of spheres. Such objects may be better approximated with primitives other than spheres, such as oriented bounding boxes or rectangle-swept spheres.

Although fast approximate GPU-based methods for distance field computation exist [17, 18], our use of exact point-to-mesh distance allows us to compute spheres that are guaranteed to be interior to the object boundary. This property is essential to allow efficient comparison of volumetric error in Section 4.3.

4 Volumetric Error for Unions of Spheres

In this section, we evaluate the quality of our sphere-based approximation. In Section 4.1, we compute lower bounds on the volumetric error of the approximation of an object with a union of spheres. We discuss the combinatorial structure of a union of spheres in Section 4.2 and review a formula for the exact computation of the volume of a union of spheres. We use the error bounds and this formula to compare volumetric error of our approximation and that of a leading method in Section 4.3.

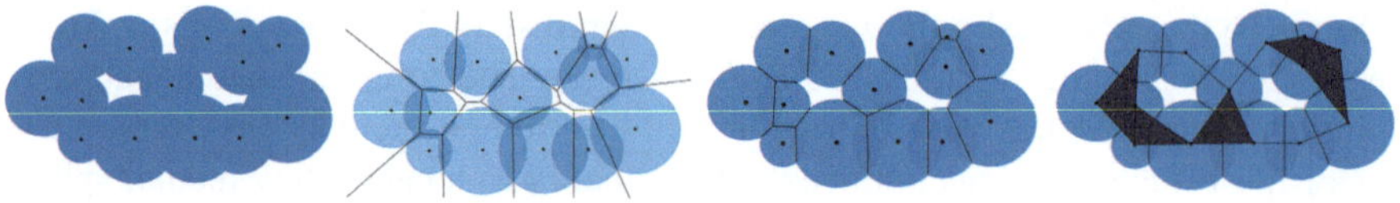

(a) (b) (c) (d)

Fig. 3. (a) A union of a set of disks; (b) its power diagram overlayed; (c) the union of the set of disks decomposed using the power diagram; (d) the dual of the power diagram.

4.1 Volumetric Error: Exact or Lower Bound

We evaluate tightness of fit of an approximation of Ω with a set of spheres S as the volume of S outside Ω plus the volume of Ω outside S. Let $\text{vol}(\cdot)$ denote volume and $\overline{A}$ be the complement of a set A. Then the error of the approximation of Ω with S is given by:

$$\text{Err}_\Omega(S) = \text{vol}(S \cap \overline{\Omega}) + \text{vol}(\Omega \cap \overline{S})$$

It is easy to show that

Lemma 1. $\max\{\text{vol}(S) - \text{vol}(\Omega), \text{vol}(\Omega) - \text{vol}(S)\} \leq \text{Err}_\Omega(S)$ *and when* $\text{vol}(S \cap \overline{\Omega}) = 0, \text{vol}(\Omega) - \text{vol}(S) = \text{Err}_\Omega(S)$.

4.2 Unions of Spheres: Tools

The combinatorial structure of a union of spheres has been the subject of much recent research. We provide a brief overview of the important concepts to be used in this work. Let us define a special space filling diagram for a set of sphere sites, called the *power diagram* [19]:

Given a set of spheres, $S = \{(\mathbf{c_1}, r_1), (\mathbf{c_2}, r_2), \ldots, (\mathbf{c_n}, r_n)\}$, with centers $\mathbf{c_i} \in \mathbb{R}^3$ and radii $r_i \in \mathbb{R}$, the *power diagram* of S, denoted $PD(S)$, is a partition of $\mathbb{R}^3$ into convex *power cells* $P(s_i)$, such that

$$P(s_i) = \{\mathbf{x} \in \mathbb{R}^3 \mid d_E^2(\mathbf{c_i}, \mathbf{x}) - r_i^2 \leq d_E^2(\mathbf{c_j}, \mathbf{x}) - r_j^2, \forall j \neq i\}. \quad (2)$$

Power cells, eq.(2), are identical to Voronoi cells, eq.(1), when all the sphere radii are the same. Figure 3 shows a 2D example of the power diagram of a set of disks.

Consider the dual complex $DC(S)$ of $PD(S)$, which is a simplicial complex that captures the topology of S. As shown in [20], to find the total volume of the union of spheres s_i, $\text{vol}(\cup_i s_i)$, one need only consider the spheres corresponding to vertices, edges, faces and tetrahedra of $DC(S)$: $\text{vol}(\cup_i s_i) = \sum_{i \in DC(S)} \text{vol}(s_i) - \sum_{(i,j) \in DC(S)} \text{vol}(s_i \cap s_j) + \sum_{(i,j,k) \in DC(S)} \text{vol}(s_i \cap s_j \cap s_k) - \sum_{(i,j,k,l) \in DC(S)} \text{vol}(s_i \cap s_j \cap s_k \cap s_l)$ We use this formula to compute the exact volume of a given union of spheres S, $\text{vol}(S)$. Since computing $\text{vol}(\Omega)$ is easy, using the result of Lemma 1, we can compute either the exact volumetric error of an approximation of a solid with spheres or a lower bound on the error.

4.3 Experimental Results

Given a representation of an object using its triangulated boundary, applications such as soft shadow generation can benefit from an alternative representation of this object using a small number of well-fitting spheres. We use our method described in Section 3 to compute a set of internal spheres whose centers are "well-distributed" on the medial surface of the object and then use tools developed in Section 4.3 to evaluate the exact volumetric error of our approximation with respect to the original boundary. Recall that approximate volumetric error has been used in the literature [3, 13] to evaluate the quality of sphere approximations and provides a global measure of fit. Wang *et al.*[3] provides a tighter fit than [2], but is only feasible for generating small sphere sets (approximately 128 spheres). Reference [2] is a state-of-the-art method for approximating objects tightly with a large number of spheres. As our method is able to generate a large number of spheres quickly, we compare it to that of [2].

Let S_D be the approximation to Ω computed with our distance-based method described in Section 3 and let S_V be the approximation to Ω computed with the Voronoi-based AMAA method of [2]. As explained in Section 3, the sphere set S_D is completely contained inside Ω, *i.e.*, $\text{vol}(S_D \cap \overline{\Omega}) = 0$. By Lemma 1, $\text{vol}(\Omega) - \text{vol}(S_D)$ gives the exact error of the approximation of Ω with S_D, while $\text{vol}(S_V) - \text{vol}(\Omega)$ is a lower bound on the error of the approximation of Ω with S_V. We will define a normalized error $\text{nerr}(S) = \text{Err}_\Omega(S)/\text{vol}(\Omega)$. Thus, $\text{nerr}(S_D) = (\text{vol}(\Omega) - \text{vol}(S_D))/\text{vol}(\Omega)$ and $\text{nerr}(S_V) \geq (\text{vol}(S_V) - \text{vol}(\Omega))/\text{vol}(\Omega)$.

We tabulate the exact error of S_D given by $\text{nerr}(S_D)$, and a lower bound on the error of S_V, $\text{nerr}(S_V)$, in Table 1. Note that the set of spheres S_V does not necessarily contain the object Ω being approximated, so the computation of exact error in this case is a challenging computational problem. In finding the appropriate set of spheres S_D, the voxel resolution was iteratively increased until the error of the approximation given by our method was smaller than that of AMAA. Table 1 compares

Table 1. Sphere sets S_D are generated with our distance-based method, while sphere sets S_V are generated with a Voronoi-based method of [4]. Timings are given in seconds. S_D provides a tighter approximation using fewer spheres significantly faster than S_V. Volumetric error evaluates the global quality of fit of the sphere approximation. Shown is the smallest computed set of spheres S_D such that the volumetric error of S_D is smaller than S_V.

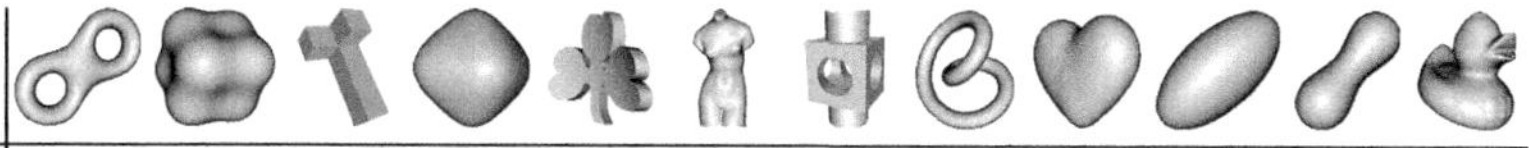

$\lvert S_V \rvert =$	512	482	506	348	496	497	493	498	484	431	474	503
$\lvert S_D \rvert =$	344	367	249	163	393	406	468	436	233	108	179	396
$\text{nerr}(S_V) \geq$	0.035	0.015	0.17	0.021	0.266	0.054	0.284	0.051	0.018	0.016	0.017	0.050
$\text{nerr}(S_D) =$	0.034	0.010	0.133	0.009	0.195	0.052	0.178	0.051	0.016	0.014	0.016	0.046
$\text{Time}(S_V) =$	2342	7428	1890	25698	1930	2672	1515	1581	3078	3421	4049	9479
$\text{Time}(S_D) =$	45	184	15	98	30	29	14	107	75	77	134	51

the performance of the two methods for a number of models for a fixed maximum number of AMAA spheres. Timings are shown on a 3.6 GHz Pentium IV processor with 3 GB of RAM. As can be seen from Table 1, our method generates a set of tighter fitting spheres significantly faster than AMAA.

AMAA construction proceeds top-down by building a hierarchy of a fixed branching factor and depth. For the timings in Table 1, S_V are the leaves of an 8-ary hierarchy of depth 4. Fast construction of a tight binary hierarchy for the spheres S_D is discussed in Section 5.2.

Note that our "well-spaced" distribution of sphere centers is suboptimal in certain cases; for example, for the ellipsoid model in Table 1, a distribution with fewer small spheres and more large spheres would provide better volumetric coverage than our distribution. However, we observe that for this model as well, our method performs better than that of [2].

5 Approximate Separation Distance

For those objects whose surface can be tightly approximated as a surface of a union of a small number of spheres relative to the number of surface triangles, separation distance computations can be accelerated by working with the sphere representation instead of the boundary mesh representation. Further, pairwise distance tests are significantly faster between spheres than triangles. In this section, we show how to quickly grow our set of spheres so as to improve boundary coverage, and how to construct a tight bounding volume hierarchy of the set of spheres. We then evaluate the performance of our sphere approximation against that of the state-of-the-art method AMAA of [2] for approximate separation distance computation.

5.1 Improving Boundary Coverage by Conservative Dilation

In Section 4, we aimed to fill the volume of the object being approximated tightly. However, in this section, we require our set of spheres to provide a tight approximation to the boundary of the object. We show a simple heuristic for conservatively growing and adding spheres that allows us to achieve a more favourable approximation to the boundary.

The surface of the union of spheres S is a non-differentiable object. As explained in Section 4.2, sphere adjacency in the union of spheres S is described by the dual complex $DC(S)$ of the power diagram $PD(S)$. In $DC(S)$, edges correspond to pairs of spheres intersecting along circular arcs on the surface of the union of spheres, while triangles correspond to triplets of spheres intersecting at vertices on the surface of the union of spheres. We can solve for the locations of these vertices, called *v-points*, by considering all triangles (s_i, s_j, s_k) in $DC(S)$, finding the 2 intersection points of spheres s_i, s_j and s_k, and retaining those intersection points that lie on the boundary of the union of spheres. We also note which triplet of spheres contributed to the creation of a v-point. For a v-point v, we consider the *nearest location on the mesh boundary* to v. The radius of each sphere s is increased sufficiently so as to cover the

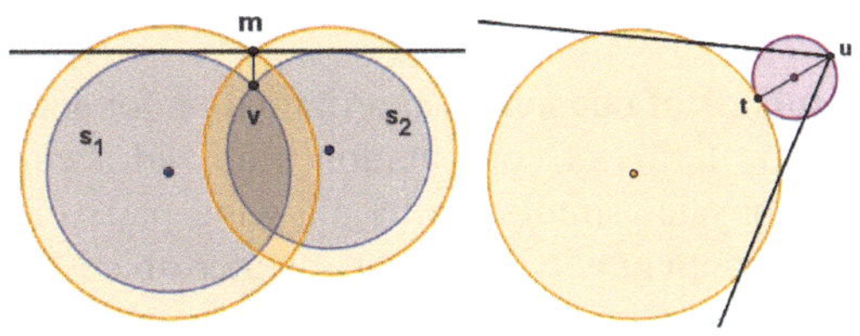

Fig. 4. (a) The blue circles s_1 and s_2 intersect at point v, whose nearest point on the boundary is m. The radii of both blue circles are increased to create orange circles that contain m. (b) To cover the corner of the object at u, we add a new purple circle with center along the line segment (u,t).

nearest mesh point to each of the v-points that is created by s. Figure 4(a) presents a 2D example. For those edges of $DC(S)$ that are not part of any triangles of $DC(S)$, we sample 2 random points on the circle of intersection of the two corresponding spheres and proceed similarly as with v-points.

In order to further improve the fit of the union of spheres S, we introduce new spheres at those vertices of the mesh boundary whose *nearest location on the set of spheres* is more than a tolerance away—see Fig. 4(b). Let $\mathbf{u}$ be a vertex of the mesh boundary $\mathscr{B}$ and $\mathbf{t}$ be the nearest point on S to $\mathbf{u}$. We initially place a new sphere $s = (\mathbf{c}, r)$ with radius $r = \|\mathbf{t}-\mathbf{u}\|_2/2$ centered at $\mathbf{c} = \mathbf{u} + r(\mathbf{t}-\mathbf{u})/\|\mathbf{t}-\mathbf{u}\|_2$.

5.2 Hierarchy Construction Using Rectangle-Swept Spheres

In order to use our set of spheres S to perform fast proximity queries, we fit a bounding volume hierarchy to the set of spheres S. We observe that the medial surface simplified by object angle is often composed of relatively flat sheets, along which the radius of the medial spheres varies smoothly. A Rectangle-Swept Sphere (RSS) (introduced in [15], see Fig. 5) is therefore a suitable bounding volume for medial spheres. The hierarchy is constructed top-down, each parent has 2 children, and each RSS is fitted to the set of leaf spheres it bounds. We find the orientation of the RSS rectangle and the partitioning plane using a linear least square fit. The radius is found by using a procedure similar to that in [15].

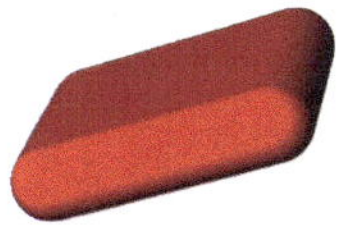

Fig. 5. Rectangle-Swept Sphere.

Alternatively, one may compute a bounding volume hierarchy using spheres as the bounding volume. Given our sphere set approximation, we compute the hierarchy using a similar top-down procedure as when using RSSs as the bounding volume. We compute the minimum bounding sphere of a set of spheres using the algorithm presented in [21].

5.3 Experimental Results

We evaluate the usefulness of our sphere approximation for approximate separation distance computation in terms of computation time and accuracy. In our experiments, we perform a physics simulation by allowing pairs of the same object to collide in a box with random gravity and measure inter-object distance both approximately using sphere approximation and exactly using the meshes at each frame. Such a simulation considers pairs of non-intersecting objects at a variety of separation distances and relative orientations.

We compare the performance of approximating separation distance for the set of spheres computed with our method and that computed with the AMAA method of [2]. We grow and add spheres to our initial sphere set S_D using the strategy outlined in Section 5.1 to create a new sphere set S_D^+. This step takes one to five seconds for the models in Table 2. We compute both an RSS hierarchy and a sphere hierarchy of our set of spheres using the method described in Section 5.2. AMAA constructs a bounding volume hierarchy where bounding elements are spheres. Table 1 lists timings for building a hierarchy with branching factor eight. In proximity query experiments, we consider hierarchies with branching factor two. Computation of a binary AMAA hierarchy for the models shown in Tables 2 takes significantly longer: from one hour (for the knot model) to 11 hours (for the peanut model). In addition to a binary sphere hierarchy, we also construct an RSS hierarchy of the

Table 2. Timing and error results for inter-object distance tests for our updated sphere sets S_D^+ and sphere sets S_V of [2]. Average timings per frame are recorded in milliseconds. SH uses a sphere hierarchy fitted to the sphere set, while RSSH uses a Rectangle-Swept Sphere hierarchy. The maximum dimension of the bounding box for all objects is 10.

			Timings		Error	
		Size	SH	RSSH	Ave.	Max.
	S_D^+	520	0.524	0.085	0.012	0.074
	S_V	498	0.268	0.088	0.024	0.063
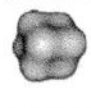	S_D^+	237	0.152	0.061	0.082	0.332
	S_V	397	0.502	0.084	0.104	0.454
	S_D^+	830	0.582	0.139	0.033	0.114
	S_V	831	0.239	0.143	0.053	0.226
	S_D^+	379	0.255	0.067	0.009	0.104
	S_V	379	0.302	0.064	0.067	0.278
	S_D^+	293	1.024	0.171	0.019	0.056
	S_V	420	0.327	0.208	0.036	0.113
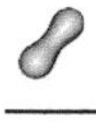	S_D^+	296	0.385	0.066	0.009	0.024
	S_V	439	0.133	0.123	0.020	0.075
	S_D^+	772	0.405	0.109	0.040	0.188
	S_V	695	0.279	0.145	0.106	0.587

AMAA spheres. Constructing an RSS hierarchy for the sphere sets considered takes a fraction of a second on average.

We compare average per-frame time for performing approximate separation distance tests using our sphere approximations and those produced by AMAA, where both sphere hierarchies (SH) and RSS hierarchies (RSSH) are considered. These results are summarized in Table 2. Timings are shown for a 3.6 GHz Pentium 4 CPU with 3 GB of RAM. In our experiments, we find that building an RSS hierarchy of spheres significantly improves distance query time compared to using a sphere hierarchy for both our and AMAA sphere sets, as the RSS offers greater tightness of fit. For most models, we see that the fastest performance is achieved by using our sphere sets and the RSS hierarchy. We believe that because our spheres are nearly medial, RSS bounding volumes provide them with a particularly tight fit.

We evaluate the inter-object distance error as the difference between the exact distance between the meshes and the distance computed using the sphere set approximation. Table 2 presents error statistics for the different sphere sets. Our sphere sets display a smaller average error and generally smaller maximum error than those of AMAA. When exact separation distance is not needed, for those objects that may be approximated well using a small number of spheres relative to the number of boundary triangles, sphere approximations can offer significant speed up at a small cost in quality.

6 Conclusions

We have described a method to compute a tight-fitting union of spheres approximation to a solid without an expensive optimization step that improves the sphere distribution. As such, our method is significantly faster than existing methods and can be used to generate sphere set approximations with a larger number of spheres than previously possible. We have shown experimentally that in comparison with the state-of-the-art AMAA method, our method generates fewer spheres, has a smaller volumetric error, and is significantly faster. Because our spheres are internal to the object, it is possible to evaluate the volumetric error of our approximation exactly and compare the quality of the approximation with a method that generates a set of spheres that are not necessarily internal and do not cover the object boundary. We experimentally demonstrate the benefit of using the rectangle-swept sphere bounding volume for building a hierarchy of approximately medial spheres. Using this hierarchy, we have shown that our sphere sets generally perform faster and more accurately than those of the AMAA method for approximate separation distance computation. There is a wide variety of additional applications where our tight fitting sphere approximations can be valuable, including path planning, point location, shadow generation, shape matching, mesh deformation, shape morphing, shape segmentation, and approximate Minkowski sums.

References

1. Ranjan, V., Fournier, A.: Volume models for volumetric data. Computer 27(7), 28–36 (1994)
2. Bradshaw, G., O'Sullivan, C.: Adaptive medial-axis approximation for sphere-tree construction. ACM Transactions on Graphics 23(1), 1–26 (2004)
3. Wang, R., Zhou, K., Snyder, J., Liu, X., Bao, H., Peng, Q., Guo, B.: Variational sphere set approximation for solid objects. Visual Computer 22(9), 612–621 (2006)
4. Rusinkiewicz, S., Levoy, M.: QSplat: A multiresolution point rendering system for large meshes. In: ACM SIGGRAPH, pp. 343–352 (2000)
5. Shamir, A., Sotzio, A., Cohen-Or, D.: Enhanced hierarchical shape matching for shape transformation. International Journal for Shape Modeling 9(2), 203–222 (2003)
6. Yoshizawa, S., Belyaev, A., Seidel, H.P.: Skeleton-based variational mesh deformations. Computer Graphics Forum 26(3), 255–264 (2007)
7. Aichholzer, O., Aurenhammer, F., Kornberger, B., Plantinga, S., Rote, G., Sturm, A., Vegter, G.: Recovering structure from r-sampled objects. Computer Graphics Forum 28(5), 1349–1360 (2009)
8. Hubbard, P.M.: Approximating polyhedra with spheres for time-critical collision detection. ACM Transactions on Graphics 15(3), 179–210 (1996)
9. Stolpner, S., Whitesides, S., Siddiqi, K.: Sampled medial loci and boundary differential geometry. In: ICCV Workshop on 3-D Digital Imaging and Modeling, pp. 1855–1862 (2009)
10. Amenta, N., Choi, S., Kolluri, R.: The Power Crust, Unions of Balls, and the Medial Axis Transform. Computational Geometry: Theory and Applications 19(2-3), 127–153 (2001)
11. Quinlan, S.: Efficient distance computations between non-convex objects. In: International Conference on Robotics and Automation, pp. 3324–3329 (1994)
12. Giesen, J., Miklos, B., Pauly, M., Wormser, C.: The scale axis transform. In: Symposium on Computational Geometry, pp. 106–115 (2009)
13. Garcia, M., Bayona, S., Toharia, P., Mendoza, C.: Comparing sphere-tree generators and hierarchy updates for deformable objects collision detection. In: International Symposium on Visual Computing, pp. 167–174 (2005)
14. Stolpner, S., Whitesides, S.: Medial axis approximation with bounded error. In: International Symposium on Voronoi Diagrams, pp. 171–180 (2009)
15. Larsen, E., Gottschalk, S., Lin, M.C., Manocha, D.: Fast proximity queries with swept sphere volumes. Tech. rep., University of North Carolina at Chapel Hill (1999)
16. Foskey, M., Lin, M.C., Manocha, D.: Efficient computation of a simplified medial axis. In: Solid Modeling and Applications, pp. 96–107 (2003)
17. Sud, A., Govindaraju, N.K., Gayle, R., Manocha, D.: Interactive 3d distance field computation using linear factorization. In: ACM Symposium on Interactive 3D Graphics and Games, pp. 117–124 (2006)
18. Rong, G., Tan, T.S.: Jump flooding in GPU with applications to Voronoi diagram and distance transform. In: ACM Symposium on Interactive 3D graphics and Games, pp. 109–116 (2006)
19. Aurenhammer, F.: Power diagrams: properties, algorithms and applications. SIAM Journal of Computing 16(1), 78–96 (1987)
20. Edelsbrunner, H.: The union of balls and its dual shape. In: Symposium on Computational Geometry, pp. 218–231 (1993)
21. Gärtner, B.: Fast and robust smallest enclosing balls. In: European Symposium on Algorithms, pp. 325–338 (1999)

A Heuristic Algorithm for Slicing in the Rapid Freeze Prototyping of Sculptured Bodies

Eric Barnett, Jorge Angeles, Damiano Pasini, and Pieter Sijpkes

Abstract. The subject of this paper is a heuristic slicing algorithm for converting STL or PLY CAD data into boundary and fill paths for rapid freeze prototyping (RFP). The algorithm, developed for one commercial robotic system, can also be used to produce toolpaths for other rapid prototyping systems. The algorithm entails five steps: (a) geometry data and other control parameters are imported; (b) the geometry is sliced at several equidistant heights to form bounding paths; (c) contours for the scaffolding material are computed; (d) part and scaffolding paths are buffered in or out to account for deposition path width; and (e) fill paths are computed. A STL file of a 300 mm-high statue of James McGill is used as an example part for demonstrating the capabilities of the algorithm.

1 Introduction

Ice construction has fascinated people for thousands of years. On the practical side, ice roads and shelters have enabled access to and settlement of remote areas. Artistic ice construction has a long tradition of its own; in recent decades, it has become more popular than ever. As new technologies have become available, ice construction is becoming increasingly automated. Computer numerical control (CNC) ice

Eric Barnett · Jorge Angeles
Centre for Intelligent Machines and Department of Mechanical Engineering, McGill University, Montreal, Quebec H3A 2K6, Canada
e-mail: {ebarnett, angeles}@cim.mcgill.ca

Damiano Pasini
Department of Mechanical Engineering, McGill University, Montreal, Quebec H3A 2K6, Canada
e-mail: damiano.pasini@mcgill.ca

Pieter Sijpkes
School of Architecture, McGill University, Montreal, Quebec H3A 2K6, Canada
e-mail: pieter.sijpkes@mcgill.ca

J. Angeles et al. (Eds.): Brain, Body and Machine, AISC 83, pp. 149–162.
springerlink.com

construction is quite well established, with several companies offering the capability to build certain parts on demand[1].

In traditional CNC machining, a part is formed by removing material using techniques such as milling and drilling. An obvious alternative to *material removal* is *material addition*, also known as rapid prototyping (RP) [1]. RP is a relatively new technology, first achieving widespread use in the 1980s. An important distinction exists between RP *deposition* systems such as fused deposition modeling (FDM) and systems such as selective laser sintering (SLS) and stereolithography, which selectively fuse material already in place. In the field of ice construction, Bryant and Leu have developed a deposition rapid freeze prototyping (RFP) system consisting of a valve/nozzle water delivery system positioned by stepper-motor driven axes [2, 3, 4].

There are significant advantages to using RP as opposed to traditional CNC techniques. Extremely complex parts can be built using RP that would either be impossible or require specific, expensive tooling using traditional CNC techniques. Also, the process from design to fabrication is much simpler using RP; it can be as simple as "printing" a 3D part much as one would print a 2D document with a regular printer. Of course, RP also has many drawbacks, the most obvious being fabrication time: a part that measures roughly 100 mm in each dimension could easily take 50 hours to build. Also, RP machines typically cost several hundred thousand dollars, and most build materials cost about $50/kg. Additional advantages of RFP construction include the use of an inexpensive, more environmentally-friendly build material and lower equipment costs, since a valve/nozzle deposition system is used and fume control infrastructure is not needed.

In [5], we reported on the long history of ice construction research at McGill University. In the past few years, we have focused on the development of computer-assisted ice construction techniques. At the small scale, two systems have been retrofitted for rapid prototyping with ice: an Adept Cobra 600 robot and a Fab@home desktop 3D printer [6]. For the Cobra 600 system, only the end-effector and part of the distal link are inside the freezer during construction. Our Cobra 600 RFP system is novel because the Cobra 600 has never before been used for RP, not to mention RFP. The four-axis SCARA architecture is well-suited to RP, although many additional subsystems and considerable software development are needed to retrofit the system for RFP. Fluid is supplied under pressure from pressurized dispensing tanks outside the freezer; inside the freezer, the fluid lines are heated using an on/off temperature controller; the set point for the temperature near the nozzle tips is 20°C.

Elsewhere, we described in [7] the trajectory control algorithm developed for the Cobra 600 RFP system. This scheme is specifically adapted to the strengths and weaknesses of the Cobra architecture and the Adept C40 Compact Controller. The algorithm is an improved version of a previously developed Matlab code [8] and since is not adapted specifically for the Cobra 600, it can be used to generate control trajectories for other rapid prototyping systems. The previous code produced self-intersecting paths for some parts with complex geometries; this problem has

[1] Ice Sculptures Ltd., Grand Rapids, MI ((`www.iceculture.com`)
Ice Culture Inc., Hensall, ON (`www.iceguru.com`)

(a) (b)

Fig. 1. James McGill STL part: (a) One million facet model; (b) Decimated model with 3906 facets

been eliminated with `rpslice` through the implementation of a new path buffering technique. Additionally, the data storage and processing efficiency has improved considerably, allowing parts with a high level of detail to be sliced.

Matlab is preferable as a coding environment to programming languages such as C or C++ because it is a superlanguage, which means that it offers many useful functions, less development time, and ease of debugging. Specifically, many functions available in the mapping toolbox are useful for rapid prototyping. The performance of `rpslice` is measured in terms of the quality of path data generated and the computational time required; specifications for the computer used are: Intel Core 2 Duo processor @ 2.2 GHz, 2 GB RAM, 500 GB drive @ 7200 RPM. Throughout the paper, we use a James McGill[2] CAD model, shown in Fig. 1, to demonstrate the capabilities of `rpslice`.

Many steps are involved in the slicing algorithm; literature on the subject typically focuses on one of the steps [9, 10, 11, 12].

[2] James McGill (1744–1813) bequested an estate and funds in his will for the construction of McGill College (later McGill University), officially founded in 1821. A natural scale original statue of James McGill features prominently in the McGill downtown campus.

2 Data Import and Transformation

The first step in `rpslice` is to import a TXT file containing control parameters; the parameters for the part shown in Fig. 1(a) are included in Table 1.

Table 1. Control parameter input file

Parameter [units][a]	[Data format] Description (Data restrictions)
1	[int] Number of STL and/or PLY files to import (> 0)
jamesmcgill.stl	[char char] STL and/or PLY file names (no spaces within each name)
0	[1/0 (on/off)] *inexact* option: used to indicate if common vertices for adjacent triangular facets are not identical
2	[1/2/3] Slicing option: 1, boundary paths are found; 2, boundary and fill paths are formed; 3, 2D option (indicates the model need only be sliced once and contour data can be copied for all of the part slices)
1	[1/0 (on/off)] scaffolding option
1×10^{-4}	[float] Tolerance value for comparing floating-point data (> 0)
1.5 1.5 [mm]	[float float] [part support] Deposition path width (> 0)
0.4 [mm]	[float] Slice thickness (> 0)
45 [°]	[float] Max. angle between path segments that is considered to be *smooth* (> 0)
0 0 0 0	[1/0 float float float] [(on/off) Sx Sy Sz] Scaling vector (> 0)
1 -90 0 0 [°]	[1/0 float float float] [(on/off) Rx Ry Rz] Rotation vector (1-2-3 Euler angles)
0 0 0 0 [mm]	[(1/0) Tx Ty Tz][b] [(on/off) Tx Ty Tz] Translation vector
0 1	[float float] [min max] Part fraction to slice: [0 1] indicates the entire part (0 <= float <= 1)
0 1	[float float] [min max] Scaffolding subset(0 <= float <= 1)
1 1 1	[int int int] [p1 p2 p3] Part copy vector: p3 part copies will be built, laid out in a grid measuring [p1 × p2] (> 0)
5 [mm]	[float] Scaffolding buffer

[a] The parameters used for the James McGill ice statue STL file are shown here.
[b] There are three options for each axis: Numerical entries [float] indicate the minimum coordinate for the part on an axis; [cent] entries will center the part on an axis; [orig] entries will keep the part centered at its original location on an axis.

2.1 *Facet Data Importation with* `facetread`

PLY or STL facet data are imported with our `facetread` function, which can read binary or ASCII format. Table 2 shows the storage and processing efficiencies for different formats and settings, using the the James McGill statue shown in Fig. 1(a) as an example.

It can be seen that ASCII files are considerably less efficient in both data-storage and processing time. More storage space is required for ASCII files, mainly because approximately twice as many characters are needed to store floating-point numbers in the ASCII format. Additionally, for STL files in the ASCII format, several text labels surround the data for each facet. Both of these factors contribute in making STL ASCII file sizes approximately five times larger than their binary counterparts. Reading in ASCII data also takes much longer, because file sizes are larger, data must be encoded into machine format, and data are stored in variable-width fields. The advantage of using the ASCII format is that it can be read directly by people, which can be useful for debugging. However, since the binary format is significantly more efficient in terms of processing time and data-storage, it should be used in almost all cases.

The PLY format also consists of triangular facets, which are stored much more efficiently. In the STL format, each facet is stored as nine floating point numbers, which represent the three vertices. Since adjacent facets share a common edge, every vertex is repeated several times in the file. In the PLY format, each unique vertex is stored as three floating point numbers. Each facet is stored as three integers, which are the indices of its three vertices. From Table 2, we see that the PLY format is significantly more storage-efficient than the STL format.

In order to exploit Matlab's vectorization capabilities and minimize import time, looping structures should be avoided when importing large amounts of data. For some of the formats described above, this can be difficult, since each facet contains several different data types, and in some cases, extraneous information. However, with Matlab's `fscanf` and `fread` commands for ASCII and binary data respectively, it is possible to skip unwanted information and read in all facet data with a single command.

Throughout `rpslice`, looping structures are avoided whenever possible to reduce processing time; this is especially important when large, detailed models are sliced. Simple computations are vectorized, and more complex operations are accomplished with cell functions. Both of these techniques lead to significant reductions in processing time. However, they can also be highly memory-intensive, especially when cell functions are used. For some especially complex or memory-intensive operations, `for` loops are used.

Table 2. A comparison of different techniques for reading in the facet data for the part shown in Fig. 1(a)

File format	Technique	File size (MB)	Import time (s)
binary PLY	two `fread` statements	18.1	0.7
ASCII PLY	two `fscanf` statements	34.9	7.1
binary STL	one `fread` statement	47.6	0.7
binary STL	`for` loop with one `fread` statement	47.6	32.5
ASCII STL	one `fscanf` statement	257.8	50.0
ASCII STL [a]	`while` loop with multiple `fscanf` statements	16.1	603.8

[a] The previous code [8] was used in this case; A decimated model with only 62500 facets was used because the one million-facet model failed to read in after over ten hours.

2.2 *Transformation of Facet Data*

The `rpslice` algorithm translates, rotates, and scales the triangular facets based on the data imported from the input file. This feature is especially useful when a STL or PLY file is the only CAD model available. If rotation or scaling is desired, the model is initially moved to a location where its minimum value in each dimension is zero. Facets are then rotated, scaled, and translated as described in Table 1. Rotation is performed using the 1-2-3 Euler-angle order, though this order can be easily changed, for example, if the 1-2-3 order leads to an orientation singularity typical of the Euler-angle representation of rotations [13].

3 Part Boundary Paths

The boundary contours for each slice are formed using a function called `facet-slice`, composed of the three subfunctions, `intersectfacet`, `segmerge` and `pathfilter`. First, `intersectfacet` is used to identify the facets that intersect each slice and compute plane-plane intersections, forming an array of disconnected line segments. Next, an iterative procedure is performed to link the line segments in each slice to form part boundary paths. A function called `polymerge` exists in Matlab to perform this task, though it returns an "OUT OF MEMORY" error if the array of line segments to merge is too large. As such, we developed a new iterative linking function, which we call `segmerge`. This function forms a polygon segment-by-segment: at each iteration, the last point of the partially-completed polygon is matched with an endpoint of a segment in the disconnected array. A comparison between `segmerge` and `polymerge` is included in Table 3. For RFP, the model is sliced at increments of 0.4 mm, which corresponds to the thickness of scaffolding slices. However, since ice slices are only 0.2 mm thick, every second water slice has repeated boundary paths.

Boundary paths that have unnecessarily high resolution are decimated using `pathfilter`; a minimum point-spacing of 0.2 mm is imposed to accommodate

Table 3. A comparison of `segmerge` and Matlab's `polymerge` function for forming contours from segments with matching endpoints[a]

Facets	Segments	Computational Time (s)	
		`polymerge`	`segmerge`
3906	126	0.05	0.03
62500	545	0.17	0.06
500000	1604	1.10	0.22
1000000	2154	-[b]	0.35

[a] Different resolutions are tested for a slice 8 mm from the base of the part shown in Fig. 1.
[b] OUT OF MEMORY error.

limitations of the Cobra controller [7]. Additionally, subsequent steps in the algorithm complete much more quickly when there are fewer path points.

4 Scaffolding Boundary Paths

For the Cobra 600 RFP system, scaffolding is needed to support overhanging part features. During deposition, the RFP system alternates between part and scaffolding slices. It is possible to model the scaffolding with CAD software by "subtracting" the part from a block of material. However, this technique is not always straightforward and will usually result substantial waste of scaffolding material, since support is only needed *below* overhanging features of the part. Additionally, if the part is only available in a surface rather than a solid format, this technique frequently fails. These factors motivate the generation of scaffolding contours from the part itself, after the slicing stage.

Chalasani et al. [14] developed a rapid prototyping technique in which they consider only the 2D contours in each slicing plane, while others [9, 15] consider the 3D model, analyzing triangular facets. Our `rpscaf` function is similar to the first technique, with additional options included to accommodate the specific characteristics of the Cobra RFP system.

The `rpscaf` function works in the following manner. A "merged" region is formed for every slice, which consists of the Boolean union between all of the part slices above. The scaffolding region for one slice is defined as the Boolean subtraction of the part region for that slice from the merged region for the slice above. This technique is fast and efficient, though it also produces many thin scaffolding features. Since thin features are built less accurately by the RFP system, the merged regions are buffered outward, by an amount specified in the input TXT file,

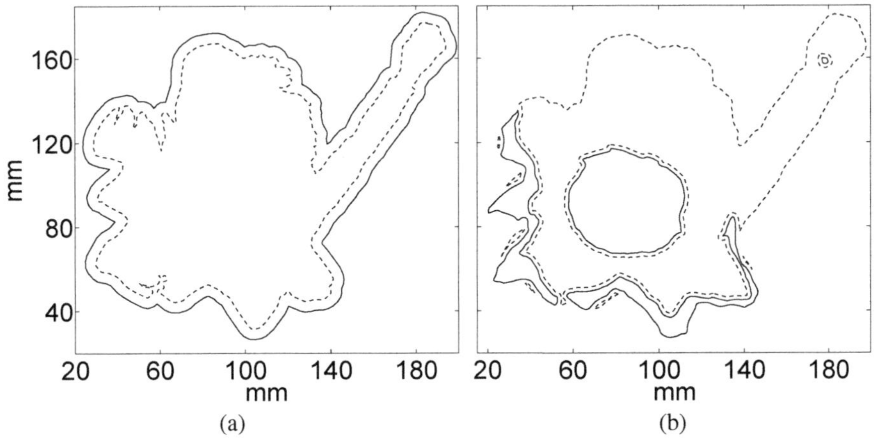

Fig. 2. (a) Merged region for one slice before buffering (dashed) and after buffering (solid); (b) Part (solid) and scaffolding (dashed) regions for the same slice

before performing the Boolean subtraction. This is accomplished using the `bufferf` function, described in Sect. 5. All ice features in every slice are thus completely surrounded by scaffolding; obviously, this increases the amount of scaffolding material and construction time needed. However, it also significantly increases the part accuracy. Figure 2 displays graphical representations of the regions described here for one slice of the part shown in Fig. 1(a). In Fig. 2(b), part and scaffolding regions are shown *after* path buffering, described in Sect. 5.

5 Path Buffering

Path buffering is needed because many of the part and scaffolding paths generated in the previous two steps are identical. Additionally, since the path width pw used with our RFP systems is 1.5 mm, if the path and scaffolding boundaries were used directly as deposition paths, an error of $pw/2 = 0.75$ mm would be introduced. To avoid this error, all boundary paths are buffered inward by $pw/2$ to form the deposition paths. A custom contour-buffering technique was implemented in [8], but it was not computationally efficient, and produced self-intersecting paths for certain complex geometries. For `rpslice`, path buffering is performed with `bufferf`, a modified version of the `bufferm` function available in Matlab's mapping toolbox.

5.1 *The Matlab* `bufferm` *Function*

The syntax of `bufferm` is: `[latb,lonb] = bufferm(lat,lon,...`
`dist,direction,npts,outputformat)`

`[latb,lonb]`	output latitude-longitude region
`[lat,lon]`	input latitude-longitude region
`dist`	buffering distance
`direction`	buffering direction (`in` or `out`)
`npts`	number of points used for vertex buffer circles
`outputformat`	data output format (`cell`, `vector`, or `cutvector`)

A polygon region is composed of several contours, stored in the NaN-delimited vector format or in separate cells of an array. Vertices for external contours are ordered clockwise, while internal contours, or holes, are ordered counterclockwise. Internal contours are buffered in the opposite direction to the commanded buffering direction.

The input contours are buffered one at a time. Rectangles are formed, centered around each contour segment, of width `2dist` and length the segment length. Circles with `npts` points are formed around each point on the contour. The Matlab function `polybool` is then used to form the Boolean union between each circle and the adjacent rectangle on one side. The `polybool` function then merges all of these new polygons, one at a time, to form the buffered contours. A Boolean union or intersection is then performed between the buffered contour group and the original contour, depending on whether the inward or outward buffering is desired and whether or not the contour represents a hole.

In this fashion, a buffered contour group composed of one or more contours is formed for each of the input contours. The groups are then merged, one at a time, to form the output region, using Boolean unions for external contours and subtractions for internal contours.

5.2 `bufferf`, *a Contour Buffering Function for Planar Regions*

For `bufferm`, input data are assumed to be latitude and longitude coordinates, and geometric calculations are thus done for a spherical rather than a planar surface. Therefore, if contour data pertain to a planar surface, this function will introduce some error when computing buffered contours. Data can be scaled down to fit within a region where surface curvature has a minimal effect to minimize this error. However, computations are simpler for planar geometry and it is preferable to work directly with input data. Thus, we have replaced all spherical surface computations with their planar counterparts.

The `bufferm` function is quite robust; as long as input data are in the correct format, it will never produce self-intersecting paths. However, it also has a high computational cost; thus, any modifications that can reduce this cost will significantly impact the overall processing time for `rpslice`. Nearly all of the computational time expended during `bufferm` is during calls to `polybool`, a function in Matlab that performs Boolean operations on polygonal regions. As a result, our modifications consist mostly of techniques for minimizing the number of calls to `polybool`.

Firstly, we replaced the circle-and-rectangle technique used by `bufferm` with a function which computes a single "boundseg" polygon for each line segment, as shown in Fig. 3(a). This reduces the number of polygons to merge by half.

Secondly, we eliminated the iterative procedure whereby the boundseg polygons are merged one-by-one. As with `bufferm`, `polybool` can be used to perform Boolean operations on polygonal regions composed of multiple contours. However, contours within each region must not intersect with either each other or themselves. Therefore, we created a `prebool` function, which forms a handful of contour groups from the boundseg polygons for one contour. The maxima and minima in the Cartesian dimensions are used to form a bounding rectangle around each boundseg polygon, as shown in Fig. 3(a); contour groups are formed composed of contours whose bounding rectangles are non-intersecting. One boundseg contour group, for a slice 8 mm from the base of the STL model shown in Fig. 1(a), is shown in Fig. 3(b). Since the contour is composed of 1018 segments, 2036 Boolean operations would be needed to buffer it using `bufferm`. Using the boundseg polygons and `prebool`, only 10 Boolean operations are needed because 10 groups of segment-buffered polygons are formed. An alternative method would be to detect whether polygons intersect, which would result in even fewer contour groups to merge. However, detecting

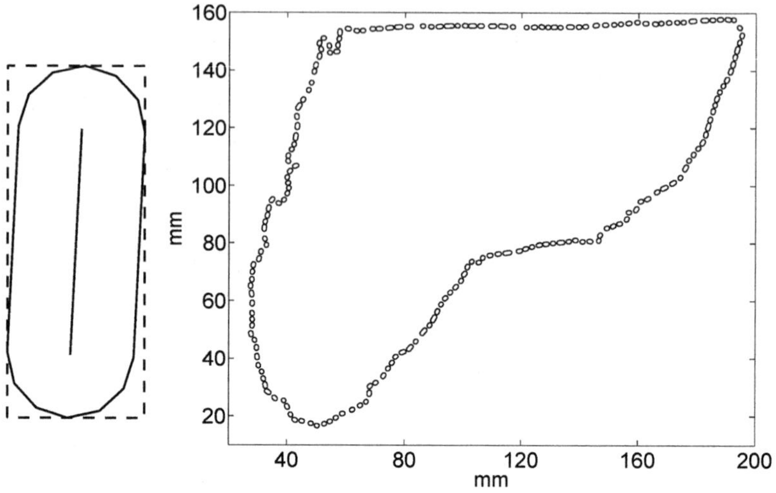

(a) (b)

Fig. 3. (a) One contour segment along with its boundseg polygon and bounding rectangle; (b) One group of boundseg polygons for a slice 8 mm from the base for the part shown in Fig. 1(a)

polygon intersections requires nearly as much computational time as performing Boolean operations, thereby lowering the overall performance.

We also reduced the number of calls to `polybool` needed to merge the buffered versions of the original contours that form the region. Since `polybool` operations are only needed when polygonal regions are intersecting, non-intersecting polygons can simply be grouped with each other in cell arrays or the NaN-separated vector format. In our slicing algorithm, region contours will not intersect as long as the STL input file is properly defined. The nesting among contours, or the number of other contours each contour lies within, is also known. Even levels of nesting indicate external contours and odd levels of nesting indicate internal contours. In `bufferf`, buffered contours are first merged to form groups of contours at each nesting level. All contours that have been buffered inward will also not intersect with each other, so one group of contours at each inwardly-buffered nesting level can be formed with no calls to `polybool`. Contours that have been buffered outward might intersect, however; thus, at each nesting level, groups of non-intersecting contours are found with `prebool` and then merged with `polybool`, if necessary. Once one group of contours has been formed at each nesting level, a series of Boolean operations is performed, alternating between subtraction and union, to form the final buffered region. Usually, there are only 1–3 levels of nesting, which means 0–2 Boolean operations are needed at this step.

Table 4 shows the computational times for `bufferm` and `bufferf` for different slices of the James McGill statue in Fig. 1(a). The `bufferf` function is always

Table 4. Computational times for Matlab's `bufferm` function vs. `bufferf` for the James McGill statue in Fig. 1(a)

Slice height (mm)	Vertices	Computational Time (s) `bufferm`	`bufferf`
8	235	1.55	0.21
8	2155	113.93	4.67
178	235	1.97	0.29
178	2973	49.90	3.76

significantly faster, and the performance gain becomes most apparent when the polygon resolution is high.

6 Fill Paths

Many authors [16, 17, 12] reportedly use zig-zag paths to fill object areas in their rapid prototyping algorithms. With the zig-zag technique, typically a series of parallel lines are intersected with the boundary contours, and adjacent lines are linked to produce paths that are very long and have many abrupt changes in direction. Xu and Shaw [18] consider 2D material gradients within each slice to produce smooth filling paths. We developed an iterative inward path buffering technique to form fill paths, mainly to avoid abrupt changes in direction [8]. However, this technique is computationally very intensive, smooth paths are not produced in all cases, and it fails for certain complex geometries. To address these problems, we have developed a new filling-path technique, called `zzfill`, similar to the zig-zag filling

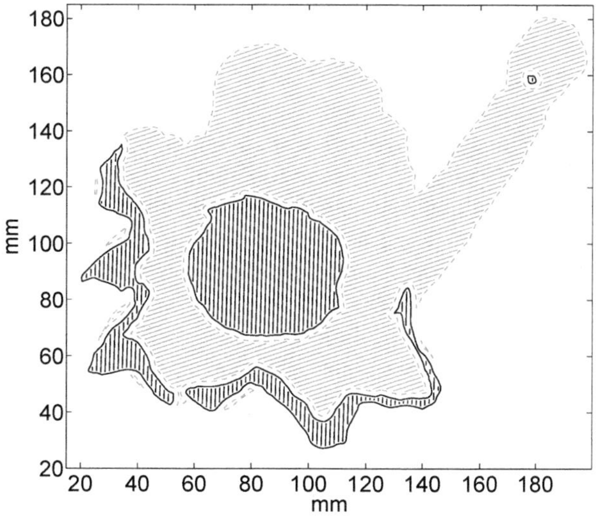

Fig. 4. Fill paths for one slice of the part shown in Fig. 1(a); solid lines indicate the part boundaries and dashed lines indicate the scaffolding boundaries

technique, except that the parallel lines are simply not linked. This eliminates *all* abrupt changes in direction during deposition, since we require that deposition only occurs when the tool is moving at a constant speed [7]. A graphical representation of the part and support contours, with the fill paths, is shown in Fig. 4.

7 Results

Table 5 shows the breakdown of computational times for slicing the coarse and fine James McGill STL models shown in Fig. 1. Additional post-processing steps are needed to produce the trajectory data needed for the Cobra deposition control program [7]. The part resolution should be close to the minimum point spacing imposed during `facetslice` to achieve the optimum balance between computational speed and part accuracy.

Table 5. Breakdown of computational time for the slicing algorithm for the James McGill statue in Fig. 1(a)

Facets	3906	1 million
traj. points[a]	933 858	1 824 094
	Computational Time (s)	
`facetread`	0.03	0.74
`facetslice`	6.59	194.72
`rpscaf`	108.85	135.37
`bufferf`	297.32	1469.80
`zzfill`	12.09	14.85
total	441.25	1844.80

[a] Total trajectory points in Cartesian space.

8 Conclusions

The major steps involved in a slicing algorithm for Rapid Freeze Prototyping are reported in this paper. Some functions used in the algorithm were developed from scratch, while others were produced by either modifying Matlab functions or using them directly. The paths produced with the algorithm were used to generate the control trajectories for building the James McGill ice statue of Fig. 5(b). Figure 5(a) shows the bronze version from which the STL file of Fig. 1 was produced. Future work with the slicing algorithm will involve increasing the efficiency and robustness of some parts of the code. Also, the algorithm will be configured to be accessible to a wide range of potential users. Ultimately, a user will be able to provide a STL file and an input TXT file, execute an EXE file, and produce the trajectory control information for the Cobra 600 RFP system, without any expert assistance.

(a) (b)

Fig. 5. James McGill statue, 306 mm high: (a) Original bronze statue; (b) Ice statue, built on its side, with the Cobra 600 RFP system: 852 slices built in 132 hours

Acknowledgements. The authors gratefully acknowledge the support of The Social Sciences and Humanities Research Council of Canada (SSHRC), the Natural Sciences and Engineering Research Council of Canada (NSERC), Quebec's Fonds québecois de la recherche sur la nature et les technologies (FQRNT), and McGill University's Faculty of Engineering. The generous rebate received from Adept Technology is dutifully acknowledged.

References

1. Crawford, R.H., Beaman, J.J.: IEEE Spectr. 36(2), 34 (1999)
2. Bryant, F.D., Leu, M.C.: Rapid Prototyp. J. 55(1), 317 (2009)
3. Bryant, F.D., Sui, G., Leu, M.C.: Rapid Prototyp. J. 9(1), 19 (2003)
4. Zhang, W., Leu, M.C., Yi, Z., Yan, Y.: IEEE Spectr. 20, 139 (1999)
5. Sijpkes, P., Barnett, E., Angeles, J., Pasini, D.: Archit. Res. Cent. Consort Spring Conf. (ARCC 2009), San Antonio, TX, April 15-18, pages 6 (2009)
6. Barnett, E., Angeles, J., Pasini, D., Sijpkes, P.: IEEE Int. Conf. Robot. Autom., Kobe, JP, May 12-17, pp. 146–151 (2009)
7. Barnett, E., Angeles, J., Pasini, D., Sijpkes, P.: To appear in Proc. ASME 2010 Int. Des. Eng. Tech. Conf., (Montreal, QC, Canada, August 15-18 (2010)
8. Ossino, A., Barnett, E., Angeles, J., Pasini, D., Sijpkes, P.: Trans. Can. Soc. Mech. Eng. 33(4), 689 (2009)
9. Allen, S., Dutta, D., Des, J.: Manuf. 5(3), 153 (1995)

10. Choi, S.H., Kwok, K.T.: Rapid Prototyp. J. 8(3), 161 (2002)
11. Haipeng, P., Tianrui, Z.: Rapid Prototyp. J. 187–188, 623 (2007)
12. Luo, R.C., Pan, Y.L., Wang, C.J., Huang, Z.H.: IEEE Int. Conf. Robot. Autom., Orlando, FL, May 15-19, pp. 883–888 (2006)
13. Angeles, J.: Rotational Kinematics. Springer, New York (1988)
14. Chalasani, K., Jones, L., Roscoe, L.: Solid Freeform Fabr. Symp., Austin, TX, August 7–9, pp. 229-241 (1995)
15. Huang, X., Ye, C., Wu, S., Guo, K., Mo, J.: Int. J. Adv. Manuf. Tech. 42, 1074 (2008)
16. Chen, H., Xi, N., Sheng, W., Chen, Y., Roche, A., Dahl, J.: IEEE Int. Conf. Robot. Autom, Taipei, Taiwan, September 14–19, pp. 3504–3509 (2003)
17. Luo, R.C., Chang, C.L., Tzou, J.H., Huang, Z.H.: IEEE Int. Conf. Robot. Autom., Barcelona, Spain, April 18-22, pp. 584–589 (2005)
18. Xu, A., Shaw, L.L.: Comput. Aided Des. 37, 1308 (2005)

Robust Design of 2nd Order Terminal ILC Using μ-Analysis and a Genetic Algorithm Approach

Guy Gauthier, Mathieu Beauchemin-Turcotte, and Benoit Boulet

Abstract. In the thermoforming industry, the heater temperature setpoints can be automatically tuned with Terminal Iterative Learning Control (TILC). This cycle-to-cycle control is used to adjust the heater temperature setpoints so that the temperature profile at the surface of the plastic sheet converges to the desired temperature. The robustness of a closed-loop system with this TILC algorithm is measured using the μ-analysis approach. A Genetic Algorithm (GA) is used to find the 2nd order TILC controller parameters giving the most robust closed-loop system.

1 Introduction

The reheat phase is an important part of the process of thermoforming, since plastic sheets have to be heated to the right temperature before being molded [1-4]. Up to now, the heater temperature setpoints have been adjusted manually, by trial and error.

To improve heater temperature setpoint tuning, we can use a cycle-to-cycle control approach. Since the plastic sheet temperature measurement is done only once, when the sheet is exited from the thermoforming oven, then the selected cycle-to-cycle control approach is the Terminal Iterative Learning Control (TILC) algorithm [5-8]. This algorithm was introduced first in [9] and then in a PhD thesis [10]. The high-order TILC in [9-12] was proposed to improve robustness. TILC is a variant of Iterative Learning Control (ILC) [13].

To tune a 2nd order TILC algorithm, a Genetic Algorithm (GA) is used to select the best set of controller parameters. The GA selects the best candidates by evaluating the robustness of the closed-loop system with the μ-analysis [14]. When we have completed all generations of the GA, the individual with the genes giving the most robustness is selected to tune the TILC algorithm.

Guy Gauthier · Mathieu Beauchemin-Turcotte
Ecole de Technologie Supérieure, Montreal, Quebec

Benoit Boulet
McGill University, Montreal, Quebec

J. Angeles et al. (Eds.): Brain, Body and Machine, AISC 83, pp. 163–174.
springerlink.com

Some researchers have used μ-synthesis and other robust approaches to do ILC design as shown in the literature review of paper [15]. But μ-synthesis requires a lot of work to simplify the designed controller before its implementation. This becomes cumbersome when the number of input and output of the system is high. The use of μ-analysis with GA removes this simplification step, since the controller structure is defined beforehand. The goal is to find the parameters of the most robust controller.

Section 2 presents the system used to design the TILC. Section 3 introduces a second-order TILC. The μ-analysis concept and parameters are explained in Section 4. Section 5 presents the genetic algorithm used to select the best parameters for the TILC algorithm. Simulation results, using the TILC designed by this approach, are shown in Section 6. Section 7 concludes this paper.

2 Problem Setup

For the design of the TILC algorithm, the following linear system is considered:

$$\begin{aligned}\dot{\mathbf{x}}_k(t) &= \mathbf{A}\mathbf{x}_k(t) + \mathbf{B}\mathbf{u}_k \\ \mathbf{y}_k(t) &= \mathbf{C}\mathbf{x}_k(t)\end{aligned} \tag{1}$$

This linear system comes from the linearized model of the thermoforming machine reheat phase [4-7, 16].

In (1), $t \in \mathbb{R}$ expresses the time within the cycle, from 0 to T. The cycle length T has a fixed duration, because it is an important assumption of the TILC approach. The cycle number is expressed by $k \in \mathbb{N}$. System matrices **A**, **B** and **C** are assumed time-invariant. The state vector $\mathbf{x}_k(t) \in \mathbb{R}^n$ expresses the temperature at n points on the plastic sheet. The input vector $\mathbf{u}_k \in \mathbb{R}^m$ contains the temperature of the m heaters, and those temperatures are maintained constant during the entire reheat cycle. The p measured surface temperatures of the plastic sheet are expressed by the output vector $\mathbf{y}_k(t) \in \mathbb{R}^p$.

The control task is to update the heater temperature setpoint $\mathbf{u}_k$ so that the sheet surface temperatures converge to a desired terminal value vector $\mathbf{y}_d \in \mathbb{R}^p$ at time T. From (1), the terminal output is

$$\mathbf{y}_k(T) = \mathbf{\Gamma}\mathbf{x}_k(0) + \mathbf{\Psi}\mathbf{u}_k \tag{2}$$

In (2), matrix $\mathbf{\Gamma} \in \mathbb{R}^{p \times n}$ is used to obtain the zero-input response

$$\mathbf{\Gamma} := \mathbf{C}e^{\mathbf{A}T} \tag{3}$$

Matrix $\mathbf{\Psi} \in \mathbb{R}^{p \times m}$ is used to obtain the zero-state response

$$\mathbf{\Psi} := \mathbf{C}\int_0^T e^{\mathbf{A}(T-\tau)}\mathbf{B}d\tau \tag{4}$$

To put the emphasis on the cycle domain, the notation can be changed, and so (2) is rewritten as this discrete cycle equation

$$\mathbf{y}_T[k] = \mathbf{\Gamma}\mathbf{x}_0[k] + \mathbf{\Psi}\mathbf{u}[k] \tag{5}$$

where $\mathbf{y}_T[k] := \mathbf{y}_k(T)$, $\mathbf{u}[k] := \mathbf{u}_k$ and $\mathbf{x}_0[k] := \mathbf{x}_k(0)$.

Since the surface temperature of the plastic sheet is measured only at time T, (5) is a discretized system in the cycle domain, the z-transform is a useful tool for analyzing it. Then, the z-transform of (5) from the cycle domain is

$$\hat{\mathbf{y}}_T(z) = \mathbf{\Gamma}\hat{\mathbf{x}}_0(z) + \mathbf{\Psi}\hat{\mathbf{u}}(z) \tag{6}$$

where the z-domain variables have caret above them.

3 Second Order TILC Algorithm

To be able to sustain the slow variation of the ambient temperature that cause a drift on the initial temperature of the plastic sheet, and to increase the robustness of the closed-loop system, a second order TILC is used. This 2nd order TILC is defined in the z-domain as

$$\hat{\mathbf{u}}(z) = \mathbf{\Psi}^+ \left\{ \left((z+1)\mathbf{I} - \mathbf{Q} \right)(z-1) \right\}^{-1} (\mathbf{K}_1 z - \mathbf{K}_2)(\hat{\mathbf{y}}_d(z) - \hat{\mathbf{y}}_T(z)) \tag{7}$$

where $\mathbf{Q} \in \mathbb{R}^{p \times p}$, $\mathbf{K}_1 \in \mathbb{R}^{p \times p}$ and $\mathbf{K}_2 \in \mathbb{R}^{p \times p}$ are a diagonal matrices. The matrix $\mathbf{\Psi}^+ = \mathbf{\Psi}^T \left(\mathbf{\Psi}\mathbf{\Psi}^T \right)^{-1}$ is the pseudo-inverse of the system matrix $\mathbf{\Psi}$. The matrix $\mathbf{Q}$ is diagonal and expressed by

$$\mathbf{Q} = \begin{bmatrix} \alpha_1 & 0 & \cdots & 0 \\ 0 & \alpha_2 & \cdots & 0 \\ \vdots & \vdots & \ddots & \vdots \\ 0 & 0 & \cdots & \alpha_p \end{bmatrix} \tag{8}$$

The two diagonal gain matrices $\mathbf{K}_1$ and $\mathbf{K}_2$ are defined as (with $i \in \{1, 2\}$):

$$\mathbf{K}_i = \begin{bmatrix} k_{i1} & 0 & \cdots & 0 \\ 0 & k_{i2} & \cdots & 0 \\ \vdots & \vdots & \ddots & \vdots \\ 0 & 0 & \cdots & k_{ip} \end{bmatrix} \tag{9}$$

For the following Lemma, the TILC algorithm defined in (7) is rewritten as $\hat{\mathbf{u}}(z) = \hat{\mathbf{C}}(z)(\hat{\mathbf{y}}_d(z) - \hat{\mathbf{y}}(z))$, where:

$$\hat{\mathbf{C}}(z) = \mathbf{\Psi}^+ \left\{ \left((z+1)\mathbf{I} - \mathbf{Q} \right)(z-1) \right\}^{-1} (\mathbf{K}_1 z - \mathbf{K}_2) \tag{10}$$

Lemma 1: The closed loop-system is internally stable if and only if the following matrix is invertible:

$$\begin{bmatrix} \mathbf{I}_m & \hat{\mathbf{C}}(z) \\ -\mathbf{\Psi} & \mathbf{I}_p \end{bmatrix} \tag{11}$$

for all z outside the unit circle.

Proof: The proof can be found in [17]. □

The determinant of (11) with the controller defined in (7) is

$$\begin{aligned} p(z) &= \det\left(\begin{bmatrix} \mathbf{I}_m & \mathbf{C}(z) \\ -\mathbf{\Psi} & \mathbf{I}_p \end{bmatrix}\right) \\ &= (z-1)^{-m} \det\left(\mathbf{I}_p + \mathbf{\Psi}\mathbf{\Psi}^+ \left\{(z+1)\mathbf{I}_p - \mathbf{Q}\right\}^{-1} (\mathbf{K}_1 z - \mathbf{K}_2)\right) \end{aligned} \tag{12}$$

The solution of (12) gives the poles of the closed loop-transfer function. If the system matrix $\mathbf{\Psi}$ is full rank, the product $\mathbf{\Psi}\mathbf{\Psi}^+$ is equal to the identity matrix. Then, the poles are located at the values of z which solve $p(z) = 0$

$$q(z) = \prod_{i=1}^{p} \left(z^2 + (k_{1i} - \alpha_i) z + (\alpha_i - k_{2i} - 1)\right) \tag{13}$$

From (13), the poles are

$$z_i = \tfrac{1}{2}\left(\alpha_i - k_{1i} \pm \sqrt{k_{1i}^2 - 2k_{1i}\alpha_i + \alpha_1^2 - 4\alpha_i + 4k_{2i} + 4}\right) \tag{14}$$

If the system matrix $\mathbf{\Psi}$ is rank deficient, the product $\mathbf{\Psi}\mathbf{\Psi}^+$ in (12) is not equal to the identity matrix. This product $\mathbf{\Psi}\mathbf{\Psi}^+$ gives an idempotent matrix and can be rewritten as $\mathbf{\Psi}\mathbf{\Psi}^+ = \mathbf{V}\mathbf{\Lambda}\mathbf{V}^{-1}$, as shown in the following Lemma.

Lemma 2: Suppose that matrix $\mathbf{A} \in \mathbb{R}^{p\times p}$ is idempotent and its rank is equal to $r < p$. The spectral decomposition of $\mathbf{A}$ can be written as

$$\mathbf{A} = \mathbf{V}\mathbf{\Lambda}\mathbf{V}^{-1} \tag{15}$$

where $\mathbf{V} \in \mathbb{R}^{p\times p}$ is a matrix, each column of which is an eigenvector of $\mathbf{A}$, and $\mathbf{\Lambda} \in \mathbb{R}^{p\times p}$ is a diagonal matrix defined as

$$\mathbf{\Lambda} = \begin{bmatrix} \mathbf{I}_r & \mathbf{0} \\ \mathbf{0} & \mathbf{0} \end{bmatrix} \tag{16}$$

Proof: Since the eigenvalues of an idempotent matrix are either 0 or 1, then the multiplicity of the 1 eigenvalue is equal to the rank of the idempotent matrix [14]. □

Using the Lemma 2, equation (12) can be rewritten

$$\begin{aligned} p(z) &= (z-1)^{-m} \det\left(\mathbf{I}_p + \mathbf{V}\boldsymbol{\Lambda}\mathbf{V}^{-1}\left\{(z+1)\mathbf{I}_p - \mathbf{Q}\right\}^{-1}(\mathbf{K}_1 z - \mathbf{K}_2)\right) \\ &= (z-1)^{-m} \det\left(\mathbf{I}_p + \boldsymbol{\Lambda}\mathbf{V}^{-1}\left\{(z+1)\mathbf{I}_p - \mathbf{Q}\right\}^{-1}(\mathbf{K}_1 z - \mathbf{K}_2)\mathbf{V}\right) \end{aligned} \tag{17}$$

In that case, the poles are the values of z which solve $p(z) = 0$.

4 The mu-Analysis Approach

μ-analysis is a tool used to measure the robustness of the TILC algorithms [17, 18]. Figure 1 shows the detailed block diagram of the system matrix $\boldsymbol{\Psi}$ with its uncertainties. That system is related to the nominal one, $\boldsymbol{\Psi}_0$, with the weighting functions and matrices necessary to define the uncertainty of the system [7].

The system to control is expressed by its nominal matrix $\boldsymbol{\Psi}_0 \in \mathbb{R}^{p\times m}$, and the uncertainty amplitudes on each entry of the system matrix $\boldsymbol{\Psi}$ are expressed by each entry $\Delta_{\Psi,ii} \in \mathbb{R}^+$, $\forall i \in \{1, 2, \cdots, n_u\}$ of the real diagonal matrix $\boldsymbol{\Delta}_\Psi$. The size of $\boldsymbol{\Delta}_\Psi$ depends on n_u, the number of uncertain parameters in the matrix $\boldsymbol{\Psi}$, and the maximum size of $\boldsymbol{\Delta}_\Psi$ is $pm \times pm$.

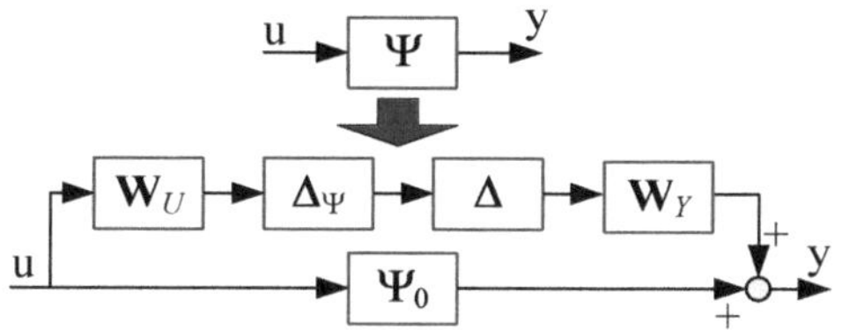

Fig. 1. Representation of system uncertainty.

The real diagonal matrix $\boldsymbol{\Delta}$ is such that each entry on the main diagonal is strictly smaller than 1 ($|\Delta_{ii}| < 1$). Therefore, the size of the matrix $\boldsymbol{\Delta}$ is the same as $\boldsymbol{\Delta}_\Psi$. To associate each entry $\Psi_{0,ij}$ of $\boldsymbol{\Psi}_0$ with the corresponding uncertainty amplitude $\Delta_{\Psi,kk}$, we need two real matrices identified by $\mathbf{W}_U$ and $\mathbf{W}_Y$. Then, the uncertain system can be written as

$$\boldsymbol{\Psi} = \boldsymbol{\Psi}_0 + \mathbf{W}_Y \boldsymbol{\Delta}\boldsymbol{\Delta}_\Psi \mathbf{W}_U \tag{18}$$

When all parameters of the system $\boldsymbol{\Psi}$ are uncertain, the matrix $\mathbf{W}_U \in \mathbb{R}^{pm\times m}$ is

$$\mathbf{W}_U = \begin{bmatrix} \mathbf{I}_m & \mathbf{I}_m & \cdots & \mathbf{I}_m \end{bmatrix} \tag{19}$$

where the identity matrix is repeated p times and matrix $\mathbf{W}_Y \in \mathbb{R}^{p \times pm}$ is

$$\mathbf{W}_Y = \mathbf{I}_p \otimes \begin{bmatrix} 1 & 1 & \cdots & 1 \end{bmatrix} \tag{20}$$

where the 1 is repeated m times.

For the parameters with no uncertainty, the corresponding lines and columns have to be removed from matrices as shown in Figure 2. This is explained in [7].

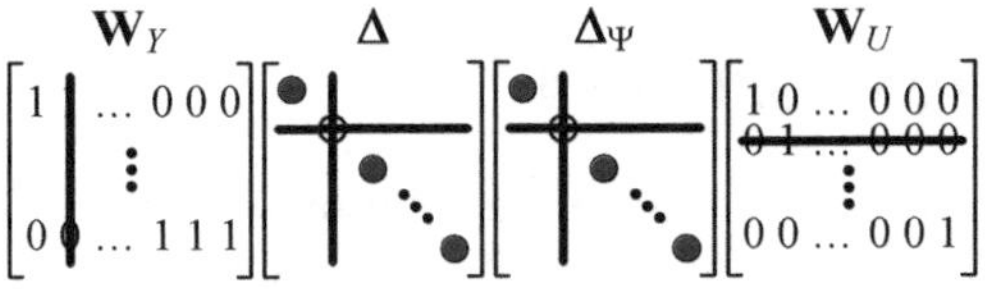

Fig. 2. Removal of 0 uncertainty.

The uncertain system $\mathbf{\Psi}$ is connected to a 2nd order cycle-to-cycle control to close the loop in the cycle domain as shown in Figure 3.

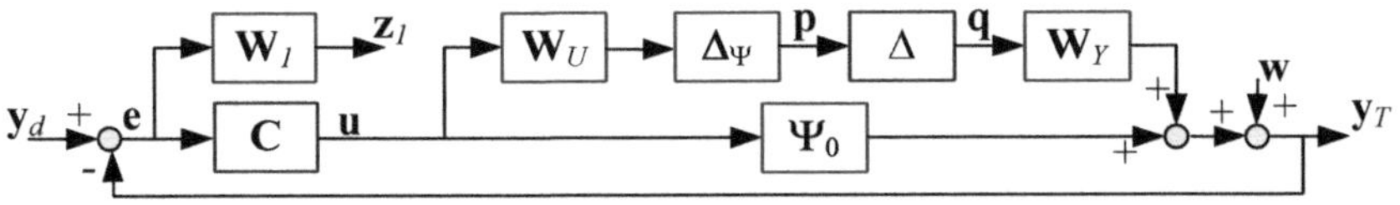

Fig. 3. Block diagram with weighting function and uncertainty for μ-analysis.

Since the robustness of the controller is obtained by the uncertainties expressed earlier, the performance specifications are included in the $\mathbf{W}_1(z)$ matrix containing the weighting function on the main diagonal. Each entries of $\mathbf{W}_1(z)$ have the following transfer function [7]

$$W_{1,ii} = \frac{1}{M_{1,i}} \left\{ \frac{(2 + M_{1,i}\omega_{1,i})z - (2 - M_{1,i}\omega_{1,i})}{(2 + \varepsilon_{1,i}\omega_{1,i})z - (2 - \varepsilon_{1,i}\omega_{1,i})} \right\} \tag{21}$$

where $i \in \{1, 2, \cdots, p\}$.

The parameters of the weighting function are as follows [7]:

- $M_{1,i}$ is the high-frequency gain of the inverse of $W_{1,ii}$, the purpose of which is to limit the high-frequency gain of the sensitivity function of the closed-loop system;
- $\varepsilon_{1,i}$ is the low-frequency gain of the inverse of $W_{1,ii}$;
- $\omega_{1,i}$ is the frequency where the gain of the inverse is equal to 1 - this frequency will determine the speed of convergence of the closed-loop system.

To do the μ-analysis, we start from the closed-loop block diagram shown on Figure 3. This block diagram can be reorganized into the one shown in Figure 4, where the TILC $\mathbf{C}$ and uncertainty $\boldsymbol{\Delta}$ are put in distinct blocks and all other blocks are grouped into a block named $\mathbf{N}$

$$\mathbf{N} := \begin{bmatrix} \mathbf{0} & \mathbf{0} & \boldsymbol{\Delta}_{\Psi}\mathbf{W}_U \\ -\mathbf{W}_1\mathbf{W}_Y & \mathbf{W}_1 & -\mathbf{W}_1\boldsymbol{\Psi}_0 \\ -\mathbf{W}_Y & \mathbf{I}_p & -\boldsymbol{\Psi}_0 \end{bmatrix} \tag{22}$$

The $\mathbf{N}$ matrix is the main component of the following relationship between input and output shown in Figure 4

$$\begin{bmatrix} \mathbf{p} \\ \mathbf{z} \\ \mathbf{e} \end{bmatrix} = \mathbf{N} \begin{bmatrix} \mathbf{q} \\ \mathbf{y}_d \\ \mathbf{u} \end{bmatrix} \tag{23}$$

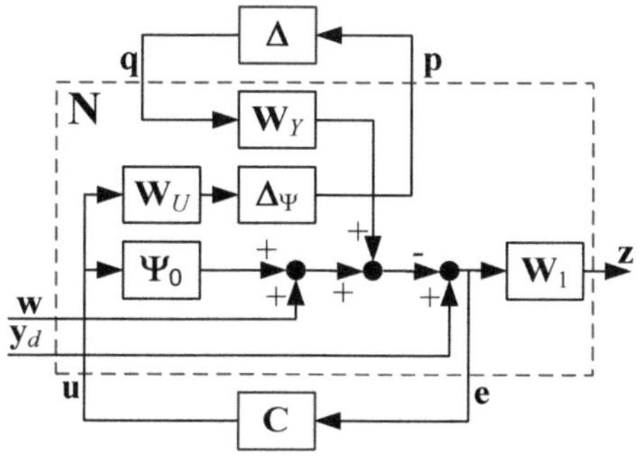

Fig. 4. Reorganized block diagram of the closed-loop system.

From the block diagram in Figure 4 that defines the general formulation of the μ-analysis, we can do the analysis about the robust stability or robust performance of a given controller $\mathbf{C}$. To do so, the $\mathbf{N}$ and $\mathbf{C}$ blocks are combined together into a new block named $\mathbf{P}$.

The matrix corresponding to the block $\mathbf{P}$ is expressed by the following lower linear fractional transformation (LFT) [7, 14]:

$$\mathbf{P} := \mathcal{F}_l(\mathbf{N},\mathbf{C}) = \begin{bmatrix} -\boldsymbol{\Delta}_{\Psi}\mathbf{W}_U\mathbf{U}_0\mathbf{W}_Y & \boldsymbol{\Delta}_{\Psi}\mathbf{W}_U\mathbf{U}_0 \\ \mathbf{W}_1\mathbf{S}_0\mathbf{W}_Y & \mathbf{W}_1\mathbf{S}_0 \end{bmatrix} \tag{24}$$

where $\mathbf{U}_0$ and $\mathbf{S}_0$ are respectively the nominal input and output sensitivities of the closed-loop system.

The performance specification is expressed by a fictitious uncertainty matrix $\boldsymbol{\Delta}_P \in \mathbb{C}^{p\times p}$ connecting the output $\mathbf{z}$ of the weighting function $\mathbf{W}_1$ to the input $\mathbf{y}_d$. This fictitious uncertainty matrix $\boldsymbol{\Delta}_P$ is such that

$$\|\mathbf{\Delta}_P\| = \bar{\sigma}(\mathbf{\Delta}_P) < 1 \tag{25}$$

at all frequencies. The matrix $\mathbf{\Delta}_P$ is a full matrix of complex values. Both uncertainty matrices are combined into this block diagonal matrix:

$$\mathbf{\Delta}_c = \begin{bmatrix} \mathbf{\Delta} & 0 \\ 0 & \mathbf{\Delta}_P \end{bmatrix}. \tag{26}$$

Since the norm of $\mathbf{\Delta}_c$ is smaller than 1 [14], combining the system expressed by matrix **N** and the controller **C**, using the small gain theorem, must give a matrix **P** having a gain (or norm) smaller than 1 at all frequencies.

The structured singular value $\mu_\Delta(\mathbf{P})$ is used to measure the robustness [17, 18]

$$\mu_\Delta(\mathbf{P}) := \frac{1}{\min\{\bar{\sigma}(\mathbf{\Delta}_c) : \det(\mathbf{I} - \mathbf{P}\mathbf{\Delta}_c) = 0\}} \tag{27}$$

The smaller the value obtained in (27), the more robust is the closed-loop system. The value $\mu_\Delta(\mathbf{P})$ of must be strictly below 1.

5 Genetic Algorithm

The GA is used to find the values of the 2nd order TILC controller parameters that minimize the value of $\mu_\Delta(\mathbf{P})$ and make the closed-loop system robust.

An initial population, of size N_X, is generated with a number of randomly generated controller parameters to cover the space of all possible stable controllers. Since there is three controller parameters for each of the p loops (since there is p inputs), each chromosome has a length equal to $3p$ genes and contains the real values corresponding to the main diagonal of matrices **Q**, $\mathbf{K}_1$ and $\mathbf{K}_2$. Here is a chromosome and its genes

$$Chromosome = \begin{bmatrix} \alpha_1 & \cdots & \alpha_p & k_{11} & \cdots & k_{1p} & k_{21} & \cdots & k_{2p} \end{bmatrix} \tag{28}$$

At each generation, the robustness measure $\mu_\Delta(\mathbf{P})$ is evaluated for all chromosomes, and the fitness function is equal to the inverse of $\mu_\Delta(\mathbf{P})$. Once the fitness value has been determined, the choice of which of the N_X individuals of the population will be used for mating is performed by roulette-wheel selection [19].

The roulette-wheel is weighted according to the fitness values of all individuals. An individual having a bigger fitness value will have more area assigned to it on the wheel. Then, this individual has a higher probability to be selected when the biased roulette wheel is spun.

Each pair of selected individuals generates an offspring by real-value recombination. The algorithm for recombination is the intermediate recombination [20] defined by

$$\alpha_i^{Offspring} = a_i \alpha_i^{Parent\_1} + (1 - a_i)\alpha_i^{Parent\_2} \tag{29}$$

with $a_i \in [-0.25, 1.25]$ an uniform random variable, for $i \in \{1, 2, \cdots, p\}$. Same applies to k_{i1} and k_{i2}.

Following real-value recombination, every genes of the chromosome of all offspring can have a mutation with a probability of 10%. The mutation is necessary to insure that the GA will not converge to local maxima of the fitness value. The real-value mutation is defined by [20]:

$$\alpha_i^{Mutated} = \alpha_i^{Offspring} + s_i \cdot 2^{-4u} \tag{30}$$

with $u \in [0,1]$ a uniform random variable and $s_i \in (-1, +1)$ another uniform random variable. Note that all random variables are independent. Again, same applies to k_{i1} and k_{i2}.

The N_o offsprings are reinserted into the population and the complete population is evaluated with the fitness function. By elitism selection, the N_X individuals having the highest fitness value (or the N_X best individuals) are kept, and we restart the process with this population for a new generation.

The GA stops when a certain number of generations have been achieved. The number of generations of the GA was selected arbitrarily after some trials. The genetic diversity seems to decrease rapidly when the GA converges to a solution. The number of generations has been selected from this observation about the loss of genetic diversity.

The best individual of the last generation is selected and the contents of its chromosome become the main diagonal of the **Q**, $\mathbf{K}_1$ and $\mathbf{K}_2$ matrices used for the TILC algorithm.

This approach can be compared with the μ-synthesis that is usually used to obtain the most robust controller using D-K iterations. With the μ-synthesis, there is a need to simplify the controller before implementing it. For a big MIMO system, this simplification can be cumbersome. With the proposed approach, using μ-analysis and GA, the controller structure is defined beforehand and can be implemented directly, even with a big MIMO system.

6 Simulation Results

To show the effectiveness of the design approach, the thermoforming nonlinear model defined in [5-7] is used. We use a four heaters and four sensors configuration of the thermoforming oven (Figure 5). This is done by combining heater in four groups: T_{T1}-T_{T2}-T_{T3}, T_{T4}-T_{T5}-T_{T6}, T_{B1}-T_{B2}-T_{B3} and T_{B4}-T_{B5}-T_{B6}. The temperature measurement is done with the four following sensors: IR_{T2}, IR_{T5}, IR_{B2} and IR_{B5}.

The GA algorithm will start with an initial population of 40 individuals. The number of generations is 20. At each generation, 35 offsprings are generated. Each chromosomes has 3 genes: α, k_1 and k_2; since we choose that all gain values are

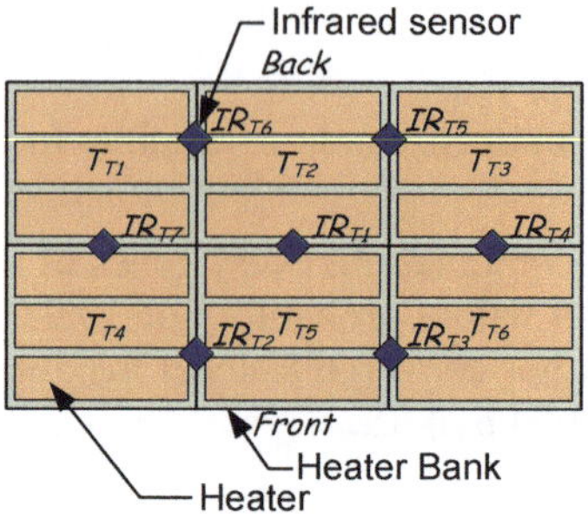

Fig. 5. Heaters and sensors location of the top of oven (bottom identical with subscript B).

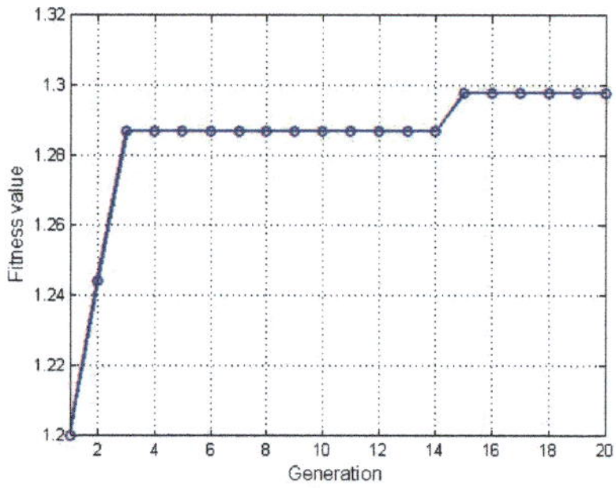

Fig. 6. Plot of the fitness value

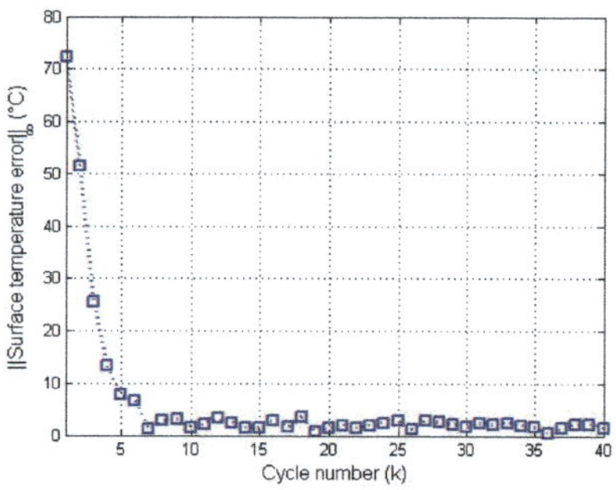

Fig. 7. Maximum error for a four heaters and four sensors oven.

equal. The parameters of the weighting function $\mathbf{W}_1$ are: $M_{1,i} = 2$, $\varepsilon_{1,i} = 0.01$ and $\omega_{1,i} = 0.5$. Figure 6 shows the fitness value evolution during 20 generations.

As shown in Figure 7, the maximum temperature error drops below 5°C in seven cycles. The error remains below 5°C for the remaining of the simulation despite the noise (white noise with standard deviation of 1°C), the initial condition change (a sinusoidal of 20°C peak to peak with a period of 60 cycles) and a drift of the air temperature inside the oven (1°C per cycle). Furthermore, the plastic sheet parameters are shifted by about 10% from the parameters used for the design of the 2nd order TILC algorithm.

Repeating the GA a second time, with an initial population including the best individual obtained in the previous run of the GA, leads to a better result (see fitness value in Figure 8).

Under the same conditions as the previous simulation, the convergence is a little bit faster than the previous design, as shown in Figure 9.

In each design, the duration to do the 20 generations is about 17 minutes. The design using the μ-analysis with the GA seems to work well up to second order TILC and need to be generalized for a high order TILC.

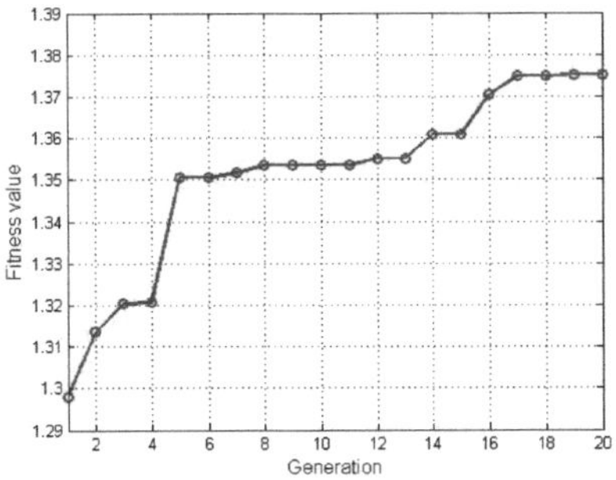

Fig. 8. Plot of the fitness value (2nd run of the GA).

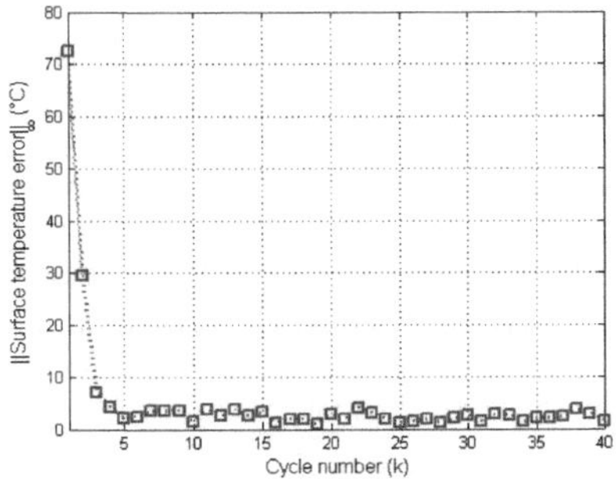

Fig. 9. Maximum error for a four heaters and four sensors oven.

7 Conclusion

The design of a TILC, by tuning the 2nd order controller parameters, leads to a robust closed-loop system as shown by the simulation results. The GA is used to find the system giving the lowest $\mu_\Delta(\mathbf{P})$. The initial population of the GA has to cover the space of the possible $\mathbf{Q}$, $\mathbf{K}_1$ and $\mathbf{K}_2$ matrices to be able to find the minimal $\mu_\Delta(\mathbf{P})$. With the proposed approach, the closed-loop robustness analysis and TILC design are automated.

In future work, we will have to determine whether or not this approach can be generalized for higher-order TILCs.

Acknowledgments. The authors wish to thanks the reviewers for their constructive comments that had help to improve this paper.

References

[1] Throne, J.L.: Technology of thermoforming. Hanser Publishers (1996)

[2] Throne, J.L.: Thermoforming. Hanser Publishers (1987)

[3] Moore, B.: In-cycle control of the thermoforming reheat process, Master's thesis, McGill University, Montreal (2002)

[4] Ajersch, M.: Modelling and real-time control of sheet reheat phase in thermoforming, Master's thesis, McGill University, Montreal (2004)

[5] Gauthier, G., Ajersch, M., Boulet, B., Haurani, A., Girard, P., DiRaddo, R.: A new absorption based model for a sheet reheat in thermoforming. In: Annual Technical Conference on SPE 2005, Boston, MA, pp. 1192–1196 (2005)

[6] Gauthier, G., Boulet, B.: Terminal iterative learning control applied to thermoforming reheat phase. In: 2006 International Symposium on Industrial Electronics, Montreal, Canada, pp. 353–357 (2006)

[7] Gauthier, G.: Terminal iterative learning for cycle-to-cycle control of industrial processes, Ph.D. thesis, McGill University, Montreal (2008)

[8] Girard, P., DiRaddo, R., Thompson, V., Boulet, B.: Advanced in-cycle and cycle-to-cycle on-line adaptive control for thermoforming of large thermoplastic sheets. In: SAE Advances in Plastic Components, Processes and Technologies 2005 (2005), paper id: 2005-01-1520

[9] Chen, Y., Xu, J.-X., Wen, C.: A High-Order Terminal Iterative Learning Control Scheme. In: Proceedings of the 36th IEEE Conference on Decision and Control, San Diego, California, pp. 3771–3772 (1997)

[10] Chen, Y.: High-Order Terminal Iterative Learning Control: Convergence, Robustness and Applications, Ph.D., Nanyang Technological University, Singapore (1997)

[11] Chen, Y., Xu, J.-X.: High-Order Terminal Iterative Learning Control with an Application to a Rapid Thermal Process for Chemical Vapor Deposition. In: Iterative Learning Control – Convergence, Robustness and Applications. Lecture Notes in Control and Information Science, vol. 248, pp. 95–104. Springer, Heidelberg (1999)

[12] Chen, Y., Xu, J.-X., Lee, T.H., Yamamoto, S.: An Iterative Learning Control in Rapid Thermal Processing. In: Proceedings of the IASTED International Conference on Modeling, Simulation and Optimization, Singapore, pp. 189–192 (August 1997)

[13] Moore, K.L.: Iterative Learning Control: An Expository Overview. In: Datta, B.N. (ed.) Applied and Computational Control, Signals, and Circuits, pp. 151–214. Birkhauser, Basel (1999)

[14] G. Gauthier and B. Boulet, Robust Design of Terminal ILC with an Internal Model Control Using μ-analysis and a Genetic Algorithm Approach, Proceedings of the 2010 American Control Conference, Baltimore, Maryland, 2010, 2069-2075.

[15] van de Wijdeven, J., Donkers, T., Bosgra, O.: Iterative Learning Control for uncertain systems: Robust monotonic convergence analysis. Automatica 45, 2382–2391

[16] Gauthier, G., Boulet, B.: Robust design of terminal ILC with $H\infty$ mixed sensitivity approach for thermoforming oven. In: Fifth International Conference on Industrial Automation, Montreal, Canada (2007)

[17] Gu, D.W., Petkov, P.H., Konstantivov, M.M.: Robust control design with MATLAB. Springer, New York (2005)

[18] Zhou, K., Doyle, J.C., Glover, K.: Robust and Optimal Control. Prentice Hall, Upper Saddle River (1996)

[19] Baker, J.E.: Reducing Bias and Inefficiency in the Selection Algorithm. In: Proceedings of the Second International Conference on Genetic Algorithms and their Application, Hillsdale, New Jersey, pp. 14–21 (1987)

[20] Mühlenbein, H., Schlierkamp-Voosen, D.: Predictive Models for the Breeder Genetic Algorithm: I. Continuous Parameter Optimization. Evolutionary Computation 1(1), 25–49 (1993)

Development of an Anthropomorphic Saxophone-Playing Robot

Jorge Solis, Atsuo Takanishi, and Kunimatsu Hashimoto

Abstract. Our research aims to develop an anthropomorphic saxophone-playing robot; as an approach to understand the human motor control from an engineering point of view. In this paper, we present the Waseda Saxophonist Robot No. 2 (WAS-2) which is composed of 22 degrees of freedom (DOF). In particular, he functioning of the lips, fingers, tongue, oral cavity and lungs have been mechanically simulated to enable WAS-2 to play an alto saxophone. Furthermore, in order to ensure the accuracy of the air pressure control, a feed-forward control system with dead time compensation has been implemented. A set of experiments were carried out to verify the effectiveness of the proposed system. From the experimental results, the range of sound pressure was increased and the air pressure control was improved.

1 Introduction

The development of anthropomorphic robots is inspired by the ancient dream of humans replicating themselves. However, human behaviors are difficult to explain and to model. Owing to the evolution of computers, electronics, and signal-processing, this ancient dream is becoming a reality. In fact, current humanoid robots are able to perform activities such as dancing and playing musical instruments. However, these mechanical devices are still far from understanding and processing emotional states as humans do. Research on musical robots seems like a particularly promising path toward helping to overcome this limitation because music is a universal communication medium, at least within a given cultural context. Furthermore, research into robotic musical performance can shed light on aspects of expression that traditionally have been hidden behind the rubric of "musical intuition". In 1984, at

Jorge Solis · Atsuo Takanishi
Waseda University, 2-2 Wakamatsu-cho, Shinjuku-ku, Tokyo, Japan
e-mail: solis@ieee.org, contact@takanshi.mech.waseda.ac.jp

Kunimatsu Hashimoto
Partner Robot Division, Toyota Motor Corporation, Aichi, Japan

J. Angeles et al. (Eds.): Brain, Body and Machine, AISC 83, pp. 175–186.
springerlink.com

Waseda University, WABOT-2 was the first attempt at developing an anthropomorphic musical robot able of playing a concert organ. Then, in 1985, the WASUBOT built also at Waseda, could read a musical score and play a repertoire of 16 tunes on a keyboard instrument [1]. The late Prof. Ichiro Kato argued that an artistic activity such as playing a keyboard instrument would require human-like intelligence and dexterity [2].

Nowadays, different kinds of musical robots and robotic musicians have been developed. Musical robots are designed to closely reproduce the required motor skills displayed by humans in order to play musical instruments ([3], [4], [5], [6], etc.). From this approach, musical robots are used as benchmarks to study the human motor control from an engineering point of view as well as to better understand the human-robot interaction. For this purpose, anthropomorphic robots are designed to closely reproduce the required motor skills displayed by humans to play musical instruments. In contrast, robotic musicians are conceived as automated mechanisms designed to create new ways of musical expression from the musical engineering point of view.

Some examples of musical robots are described below. Shibuya is developing an anthropomorphic arm which reproduces the movement required to play a violin [3]. In particular, this violin robot is designed to produce expressive sounds by considering kansei (sensitivity). The arm has a total of 7-DOF actuated by DC motors. From experimental results, the violin robot is able of playing notes with a high level of repetitiveness. Takashima has been developing different music robots that are able of playing wind instruments such as [7]: saxophone, trumpet, etc. In particular, the saxophone-playing robot has been developed under the condition that the musical instrument played by robots should not be changed or remodeled at all. This robot is composed of an artificial mouth, finger mechanisms and air-supplying system. Due to the complexity of replicating the motion of human fingers, the finger mechanism is composed of 23 fingers so that each finger can press each key of the saxophone. Shimojo has worked on a violin-playing robot, which is composed of a commercial 7-DOF manipulator and a 2-DOF fingering mechanism [4]. The end-effector of the manipulator has been designed to hold a bow. A wrench sensor was attached to the end-effector to control the bowing pressure. As a result, the violin-playing robot is able of performing simple musical scores.

Other researchers work on automatic musical instruments from the music technology point of view. Weinberg and Driscoll are developing an improvisational robotic marimba player which is composed by four robotic arms designed to hold the mallets [8]. Typically, human players hold four mallets (two in each hand). The robotic arms are arranged in pairs with overlapping workspace to allow various combinations of chords to be played. Each of the arms has 4-DOF, actuated by a linear guide and servomotors. Goto and Yamasaki are developing different kinds of automated percussionist mechanical instruments which can be commanded through a gesture controller [9]. Performers wear the gesture controller (data-suit) whose movements are identified and transformed into musical commands. Hayashi developed an automatic piano capable of producing soft tones [10]. Such an automatic piano employs feedback control to operate an actuator to strike a key. Singer introduced the LEMUR musical robots. LEMUR is a Brooklyn-based group of artists and technologists who create robotic musical instruments [11].

In summary, the research on musical robots and robotic musicians has been particularly intensified in recent decades. In fact, we may distinguish four different research approaches [12]: Enabling the Human-Robot interaction, understanding the human motor control, introducing new ways of art/entertainment and new methodologies of musical teaching. Even though the above anthropomorphic musical robots have achieved promising results; up to now, only few of them are able to perform as human musicians (in terms of perception and motor dexterity). Moreover, none of the above robots is able of playing different kinds of musical instruments (i.e. flute, saxophone, etc.) which could be useful to improve our understanding of the nature of human musicians. Therefore, we are aiming as a long-term goal at two basic issues: enabling the interaction between two human-like robots (by developing two different robots able of performing different wind instruments) and enabling a single human-like robot to play different kinds of wind instruments (our ability to enable a single human-like robot to play different kinds of wind instruments can be studied in detail).

This paper is organized as follows: In Section 2, we present details of the mechanical design and controls system implementation of the Waseda Saxophonist Robot No. 2 (WAS-2) designed to reproduce the human organs involved during saxophone-playing. In particular, the functioning of the lips, fingers, tongue, oral cavity and lungs are presented. Moreover, the implementation of a feed-forward control system with dead-time compensation is detailed. Finally, in Section 3, a set of experiments with the WAS-2 are proposed to verify the effectiveness of the designed mechanisms to increase the sound pressure range (related to the lip mechanism) and the control accuracy of the air pressure.

2 Anthropomorphic Saxophone-Playing Robot

As a result of our research, we have developed the Waseda Saxophonist Robot No. 2 (WAS-2) which has been designed to increase the range of sound by improving the design of the artificial lips and increase the range of finger motion by designing a human-like hand. In particular, WAS-2 is composed by 22-DOF that reproduce the physiology and anatomy of the organs involved during saxophone-playing as follows (Figure 1): 3-DOF (from which 1-DOF is passively controlled) to control the shape of the artificial lips, 16-DOF for the human-like hand, 1-DOF for the tongue driving mechanism and 2-DOF for the lung system (1-DOF for the air pump and 1-DOF for the valve mechanism).

2.1 Mechanism Design

The artificial lip of the WAS-1 mouth mechanism was designed with 1-DOF in order to control the vertical motion of the lower lip [13]. Based on the up-and-down motion of the lower lip, it became possible to control the pitch of the saxophone sound. However, it is difficult to control the sound pressure by means on 1-DOF. In addition, in the near future, the authors are planning to design a saxophone-playing robot with the capability of standing-up as a human player does. Therefore; in such a condition, it is difficult to hold the instrument with the artificial mouth. For this purpose, the mouth mechanism of WAS-2 has designed to expand the range of sound as well as to reduce the movement of the instrument when the robot holds the instrument with its mouth (Figure 2a). As we may observe, the lip mechanism consists of 2-DOF designed to control the up/down motion of both lower and upper lips. In addition, a passive 1-DOF has been implemented to modify the shape of the side-way lips. Basically, the lower lip is used to correct the vibrations produced on the reed to produce the desired pitch. The upper and side-way lips are used to avoid any possible leak of air (if only the lower lip motion is controlled; when its position is slightly changed, the air pressure may drastically changes). The material of the artificial lips is a thermoplastic elastometer, termed septon, which reproduces the elasticity and stiffness of human lips.

In addition to the mechanism design improvement, we have embedded an array of sensors to determine the displacement of the instrument while the saxophone is hold with the robot's mouth. For this purpose, four photo-interrupters were attached as shown in Fig. 2b. The photo interrupter combines a GaAs IRED with a high-sensitivity photo-transistor in a super-mini package with dimensions of 2.7 mm × 3.2 mm. The light emitted by the photo-interrupter is then reflected by the surface of the mouthpiece of the instrument (Figure 2c).

In order to produce the saxophone sound, it is required to control the motion of each of the fingers to push the correspondent keys. The finger mechanism of WAS-1 was composed of a link connected directly to the RC motor axis. In particular, 11

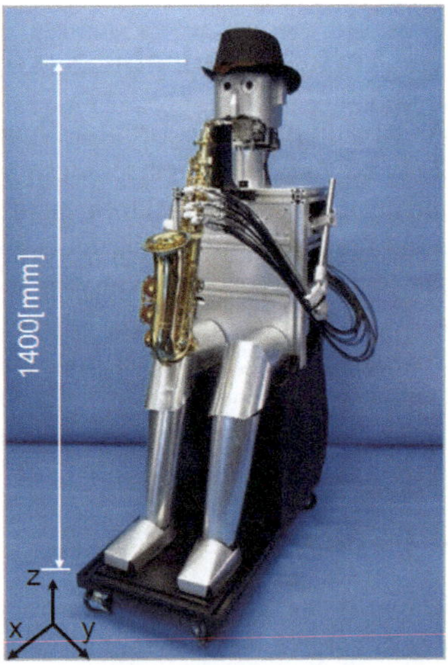

Fig. 1. The Waseda Saxophonist Robot No.2 (WAS-2.eps) is composed of artificial lips, tongue, fingers, lungs and oral cavity

motors were used in order to push each of the keys required to play from the C3 to C♯5 [13]. However, with the alto saxophone it is possible to play from A♯2 to F♯5. For this purpose, a human-like hand was designed for the WAS-2, which it is composed of 16-DOF (Figure 3). In order to reduce the load on the hand part, the actuation mechanism is composed of a wire and pulley connected to the RC motor axis. In order to control the motion of each single finger, the RS-485 communication protocol was used.

The oral cavity of WAS-2 was also designed using septon. Its capacity and size are nearly similar to the size of a human adult male (Fig. 4a). In addition, the strength of the oral cavity has been modeled to support pressures upper 8kPa (which are the maximum pressure measured on human saxophone players [14]). On the other hand, the tonguing mechanism is shown in Fig. 4b. The motion of the tongue tip is controlled by a DC motor which is connected to a link attached to the motor axis.

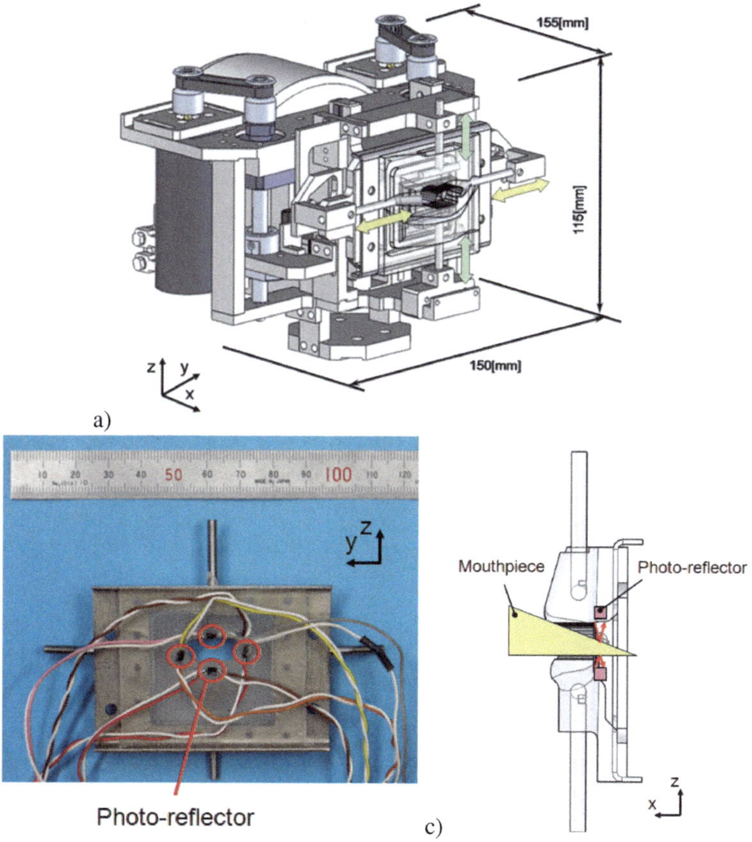

Fig. 2. a) Detail of the mouth mechanism designed for WAS-2; b) Four photo-switch sensors were embedded in the artificial lip; c) the photo-reflector senses the displacement of the instrument by measuring the light reflected by the surface of the saxophone mouthpiece

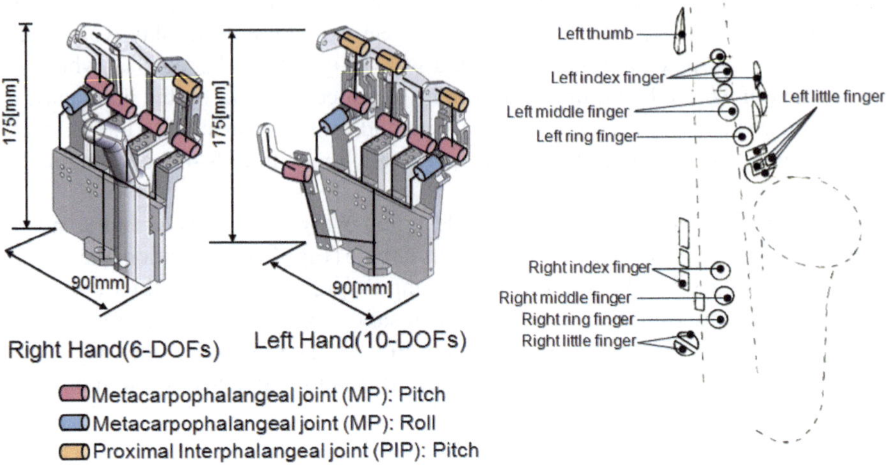

Fig. 3. Detail of the human-like hand designed for WAS-2

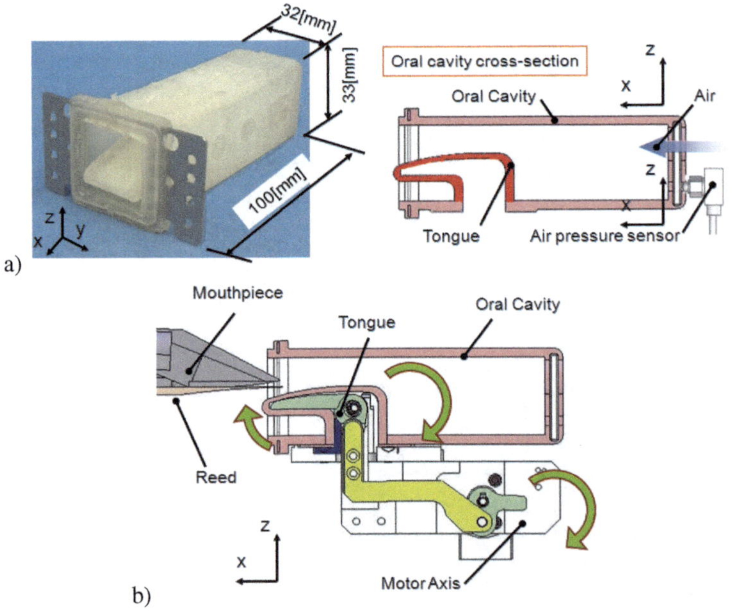

Fig. 4. Detail of the oral cavity and tonguing mechanism of the WAS-2: a) the oral cavity is made of septon and its capacity is nearly similar to a human adult male; b) the tonguing mechanism has 1-DOF to apply pressure on the saxophone reed

In this way, the air flow can be blocked by controlling the motion of the tongue tip. Thanks to the tongue mechanism of the WAS-2, the attack and release of the note can be reproduced.

Regarding the air source of WAS-2, a DC servomotor was used to control the motion of the diaphragm of the air pump; which is connected to an eccentric crank mechanism (Fig. 5a). This mechanism was designed to provide a minimum air flow of 20 L/min and a minimum pressure of 30kPa. Basically, some experimental information can be found in the literature of music terminology [15]. For our research purpose; we experimentally determine such required quantities by developing a mock-up. In particular, we have programmed the mock-up to strike different kinds of notes so that by fixing the desired pitch, the air pressure was changed from 0 kPa to 50 kPa (which corresponds to a variation from 0 L/min to 40 L/min).

Thus, we recorded air flow and pressures that ensured the production of sound. By changing the rotational speed of the motor axis, the air flow quantity can be accurately controlled by measuring it with a flow meter. The noise and vibration produced by the air pump was reduced by installing a case and vibration insulation gel. Thanks to this design, the size of the mechanism was considerably reduced (150 × 133 × 150mm) as well as its weight (1.8kg). Moreover, a DC servo motor has been designed to control the motion of an air valve so that the air delivered by the air pump is effectively rectified (Fig. 5b).

2.2 Control System Implementation

In our previous research, a cascade feedback control system was implemented to ensure the accuracy of the air pressure during a musical performance [13]. Based on the measurements of the pressure sensed at the output of the air pump and the position of the lower lip, the air pressure was controlled. However, during the attack time the target air pressure is reached around 100ms later during a musical performance. This effect is related mainly to the way the musical performance control is

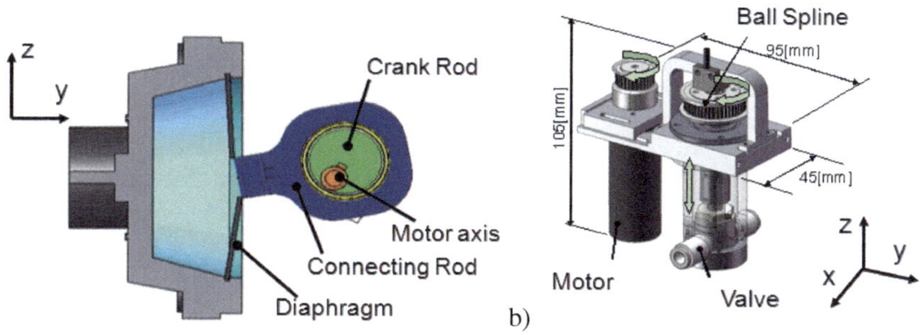

a) b)

Fig. 5. Detail of the lung mechanism designed for WAS-2: a) the air pump is composed of 1-DOF to control the air source; b) the air valve was designed to compensate the ripple effect of the air supplied by the air pump

implemented. Basically, the signal of the note to be played is sent to the control system through a MIDI message. As soon as the message of a note change is received, the air pressure as well as the position of the lower lip are adjusted. Thus, a delay on the control of the air pressure is observed.

In this paper, a modified version of feedback-error learning has been used. Feedback error learning is a computational theory of supervised motor learning proposed by Kawato [16] which is inspired by the central nervous system theory. In addition, Kawato extended that the cerebellum, by learning, acquires an internal model of inverse dynamics of the controlled object. From this extension, feedback-error learning can be also used as training signal to acquire the inverse dynamics model of the controlled system based on Artificial Neural Networks (ANN). On the other hand, the dead-time compensation is used to control devices that take a long time to respond to any input to change. A dead-time compensation control uses an element to predict how changes made now by the controller will affect the controlled variable in the future [17].

In particular, we have proposed the implementation of a feed-forward error learning control system with dead-time compensation as shown in Fig. 6. The inputs of the ANN are defined as follows (the input is based on the difference with the previous note played): pressure reference, note, and lower/upper lip positions. In this case, a total of six hidden units were used (experimentally determined while varying the number of hidden units). As an output, the position of the air valve is controlled to ensure the accurate control of the required air pressure to blow a sound. In addition, a dead-time factor (referred as e^{sL}) is introduced to compensate the delay during the attack time.

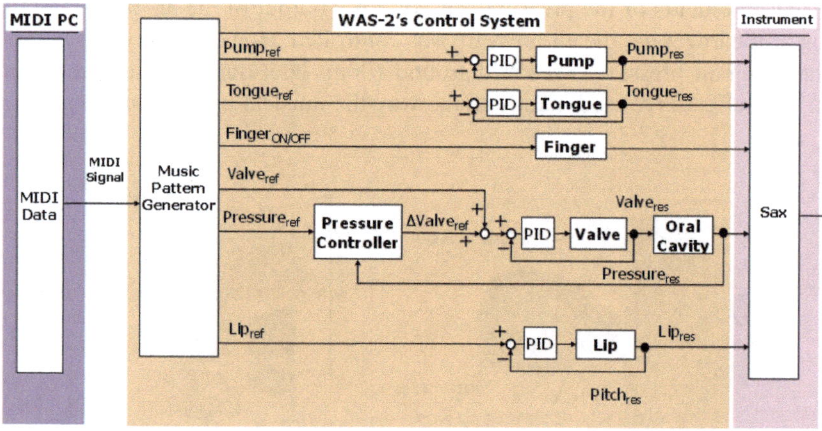

Fig. 6. Block diagram of the proposed feed-forward control system with dead-time compensation implemented to ensure the accuracy of the air pressure control during a performance

3 Experiments and Results

3.1 Mouth Mechanism

In order to verify whether the designed new mouth mechanism enables the extension of the range of sound pressure; we compared the previous mechanism with the new one while playing the notes from C3 to C5. The experimental results are shown in Fig. 7a. As we may see, the new mechanism has effectively increased the range of sound pressure (an average increment of 33%). Even though the range of sound pressure was expanded, there are still differences with the one measure by an intermediate level saxophonist.

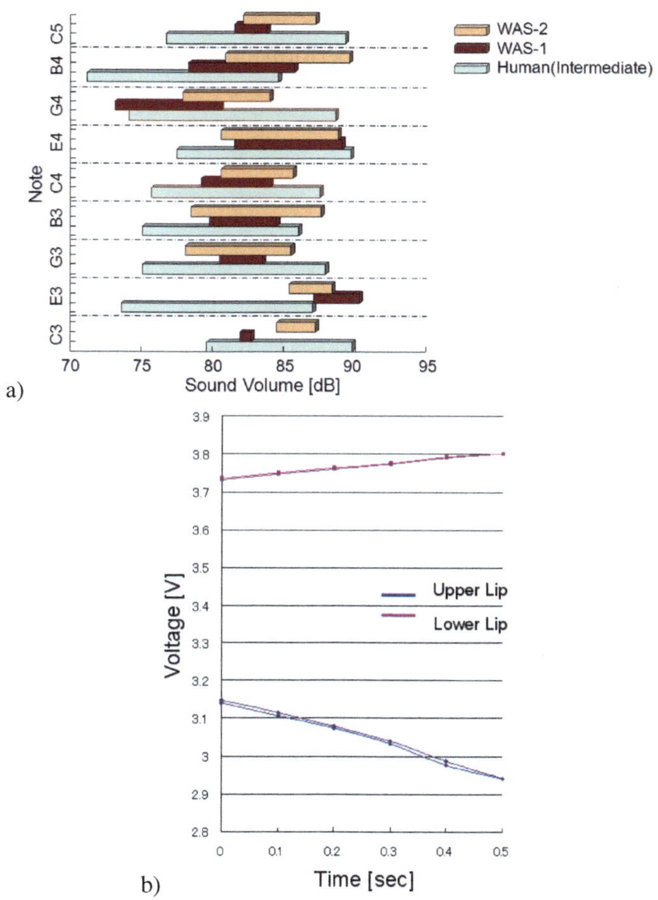

Fig. 7. a) Experimental results to verify the effectiveness of the new mouth mechanism designed to expand the sound pressure range; b) Experimental results while sensing the movement of the upper and lower lips with the proposed embedded sensors

On the other hand, an experiment has been carried out to verify if the proposed array of sensors can detect the movement of the lips while playing the saxophone. For this purpose, WAS-2 was programmed to move up-and-down (0.5 mm) the upper and lower lips. The experimental results are shown in Fig. 7b. As we may see, a linear response has been obtained. Based on the sensed information of the lips movement; in the future, we consider to adjust the position of the saxophone by adding an actuated arm mechanism. Moreover, there is also the possibility to measure the vibration of the reed by analyzing the frequency content of the signal from the sensor placed on the lower lip.

3.2 Control System

In order to determine the effectiveness of the proposed control system implemented on WAS-2, we programmed the robot to perform the Glenn Miller's "Moonlight Serenade". In order to train the ANN, a total of 523 learning steps were conducted. The

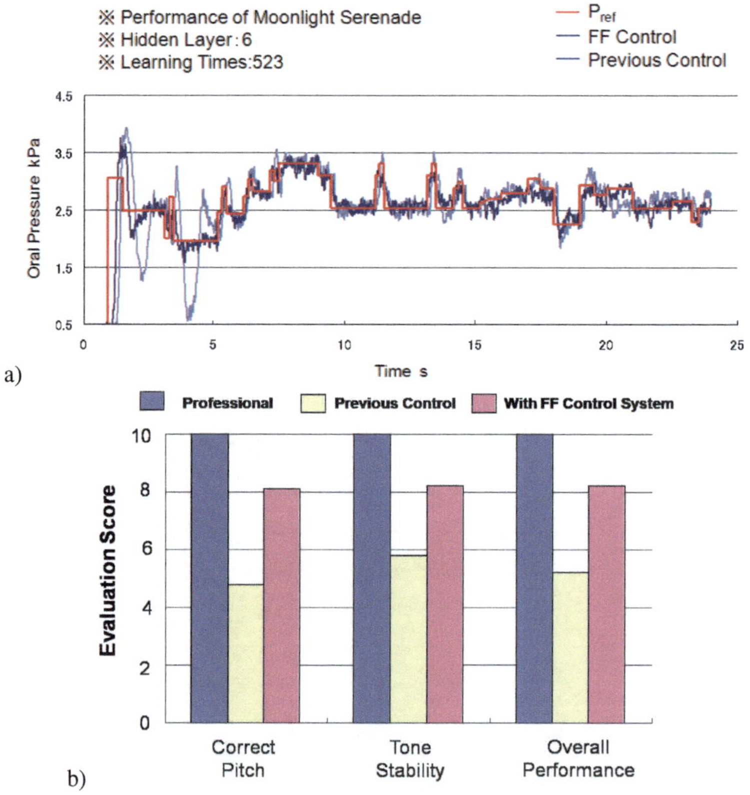

Fig. 8. a) Experimental results with the feed-forward control system with dead-time compensation; b)Experimental results of the subjective evaluation of the improvements of WAS-2

experimental results are shown in Fig. 8a. We can clearly observe that the proposed feed-forward control system with dead-time compensation produced a more stable dynamic response to the air pressure reference. In order to compare both dynamic responses, we computed the correlation coefficient with respect to the target signal (P_{ref}). The correlation coefficient is a quantity that gives the quality of a least-square fitting to the original data. As a result, we found that the air pressure with the feed-forward control system with dead-time compensation became closer to the target signal (correlation coefficient of 0.636) than the previous control system (correlation coefficient of 0.459). With these results, we could confirm the improvements achieved with respect to the previous control system (cascade feedback control).

Finally, we performed a subjective analysis of the improvements done for WAS-2. For this purpose, we have recorded the performance of WAS-2 while playing the "Moonlight Serenade" with the previous (cascade feedback) and the improved (feed-forward) control systems. For this purpose, 12 subjects were asked to compare the above recordings with the recording performance of a professional saxophone player. Subjects were asked to compare each performance in terms of the pitch quality, tone stability and overall performance; where the maximum score (10) was considered the professional one. The experimental results are shown in Fig. 8b. As we may see, a higher evaluation was given to the performance in all the evaluation parameters when the feed-forward control system is used.

4 Conclusions and Future Work

In this paper, the details of the mechanical improvements on WAS-2 were presented. Moreover, an air pressure feed-forward control system with dead-time compensation has been implemented to improve the dynamic response of the air pressure control. From the experimental results, we could confirm the effectiveness of the mechanical and control improvements.

As future work, the proposed feed-forward control system will include pitch information as well (acoustic feedback) as we still detect pitch deviations during the performance.

Acknowledgements. A part of this research was done at the Humanoid Robotics Institute (HRI), Waseda University and at the Center for Advanced Biomedical Sciences (TWINs). This research was supported (in part) by a Gifu-in-Aid to the WABOT-HOUSE Project by Gifu Prefecture. This work was supported in part by Global COE Program "Global Robot Academia" from the Ministry of Education, Culture, Sports, Science and Technology of Japan. WAS-2 was designed by using software SolidWorks. Special thanks to SolidWorks Japan K.K. for the software contribution.

References

1. Sugano, S., Kato, I.: WABOT-2: autonomous robot with dexterous finger-arm coordination control in keyboard performance. In: Proceedings of the International Conference on Robotics and Automation, pp. 90–97 (1987)

2. Kato, I., Ohteru, S., Kobayashi, H., Shirai, K., Uchiyama, A.: Information-power machine with senses and limbs. In: Proceedings of the CISM-IFToMM Symposium on Theory and Practice of Robots and Manipulators, pp. 12–24 (1973)
3. Shibuya, K.: Toward developing a violin playing robot: bowing by anthropomorphic robot arm and sound analysis. In: Proceedings of the 16th IEEE International Conference on Robot and Human Interactive Communication, pp. 763–768 (2007)
4. Kuwabara, H., Seki, H., Sasada, Y., Aiguo, M., Shimojo, M.: The development of a violin musician's robot. In: Proceedings of the IROS Workshop on Musical Performance Robots and Its Applications, pp. 18–23 (2006)
5. Solis, J., Taniguchi, K., Ninomiya, T., Takanishi, A.: Understanding the mechanisms of the human motor control by imitating flute playing with the waseda flutist robot WF-4RIV. Mechanism and Machine Theory 44(3), 527–540 (2008)
6. Toyota Motor Corporation, Overview of the partner robots, `http://www.toyota.co.jp/en/special/robot/`
7. Takashima, S., Miyawaki, T.: Control of an automatic performance robot of saxophone: Performance control using standard MIDI files. In: Proceedings of the IROS Workshop on Musical Performance Robots and Its Applications, pp. 30–35 (2006)
8. Weinberg, G., Driscoll, S.: The design of a perceptual and improvisational robotic marimba player. In: Proceedings of the 16th IEEE International Conference on Robot and Human Interactive Communication, pp. 769–774 (2007)
9. Goto, S., Yamasaki, F.: Integration of percussion robots robot- music with the data-suit bodysuit: Technological aspects and concepts. In: Proceedings of the 16th IEEE International Conference on Robot and Human Interactive Communication (2007)
10. Hayashi, E.: Development of an automatic piano that produce appropriate touch for the accurate expression of a soft tone. In: Proceedings of the IROS Workshop on Musical Performance Robots and Its Applications, pp. 7–12 (2006)
11. Singer, E.: LEMUR's musical robots. In: Proceedings of the Conference on New Interfaces for Musical Expression, pp. 181–184 (2004)
12. Solis, J., Taniguchi, K., Suefuji, K., Takanishi, A.: The development of the anthropomorphic flutist robot at Waseda university. International Journal of Humanoid Robots 3(2), 127–151 (2006)
13. Solis, J., Takeshi, N., Petersen, K., Takeuchi, M., Takanishi, A.: Development of the anthropomorphic saxophonist robot WAS-1: Mechanical design of the simulated organs and implementation of air pressure. Advanced Robotics Journal 24, 629–650 (2010)
14. Fucks, L., Sundberg, J.: Blowing pressures in reed woodwind instruments. TMH-QPSR 37(3), 41–56 (1996)
15. Guillemain, P., Vergez, C., Ferrand, D., Farcy, A.: An instrumented saxophone mouthpiece and its use to understand how an experienced musician plays. Acta Acoustica 96(4), 622–634 (2010)
16. Kawato, M., Gomi, H.: A computational model of four regions of the cerebellum based on feedback-error-learning. Biological Cybernetics 68, 95–103 (1992)
17. Kim, H., Kim, K., Young, M.: On-line dead-time compensation method based on time delay control. IEEE Transactions on Control Systems Technology 11(2), 279–286 (2003)

Human Safety Algorithms for a Parallel Cable-Driven Haptic Interface

Martin J.-D. Otis, Sylvain Comtois, Denis Laurendeau, and Clément Gosselin

Abstract. A parallel cable-driven haptic interface is designed to allow interaction with any type of virtual object. This paper presents and analyzes computational methods for addressing the issues regarding human safety and control reliability using such an interface, thereby ensuring safe operations inside the virtual world. Four strategies are explored: sensor reliability, mechanical interference management, workspace management and human-robot interaction. This paper focuses mainly on the sensors' reliability and workspace management algorithms for a parallel cable-driven haptic interface that imposes special requirements on the control architecture design. One challenging task is to develop efficient computational algorithms for hard real-time processes included in haptic display applications which improve safety without compromising performance.

1 Introduction

A Cable-Driven Locomotion Interface (CDLI) has been proposed as a haptic interface to be used in applications such as training in virtual environments [8]. This CDLI is composed of two cable-driven haptic platforms (or, in the following, the cable-driven haptic interface), one for each foot, controlled with eight cables acting as mechanical transmission. A user can walk inside a virtual environment and interact with any objects in it, including the environment itself. Due to the geometry of cable-driven haptic platforms described in [13] and the trajectories of the feet, cable interference can occur, and the platform can leave its designated workspace. Furthermore, when in use, cables can snap, wind up incorrectly on their reels, or damage their tension sensors, all potentially leading to failure.

Martin J.-D. Otis · Sylvain Comtois · Denis Laurendeau
Computer Vision and Systems Laboratory, University Laval, Québec

Clément Gosselin
Robotics Laboratory, University Laval, Québec
e-mail: gosselin@gmc.ulaval.ca

J. Angeles et al. (Eds.): Brain, Body and Machine, AISC 83, pp. 187–200.
springerlink.com

In the context of a human and a mechanism interacting within the same workspace, the safety of the user is an issue of utmost importance to avoid accidents and injuries. Because it isn't always safe to power-down the mechanism in hazardous situations, the main control algorithm is coupled with a safety manager using an error handler that was designed based on a risk analysis study. For example, the HapticWalker locomotion interface with permanent foot fixture has a safety manager [17]. However, its safety manager could be improved for a cable-driven haptic interface, since cables can interfere with one another or even sever. An analysis of cable interferences is presented in [19] and fault tolerance for a kinematically redundant manipulator is presented in [15]. Yet avoiding cable interference with path-planning such as in [7] is not always feasible since the user movement could be completely random. These two studies are used for managing cable fatigue and interference for enhancing safety.

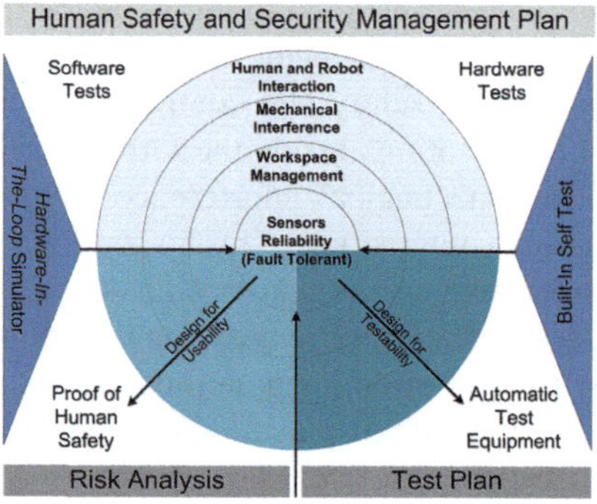

Fig. 1. Human safety and Security Management Plan

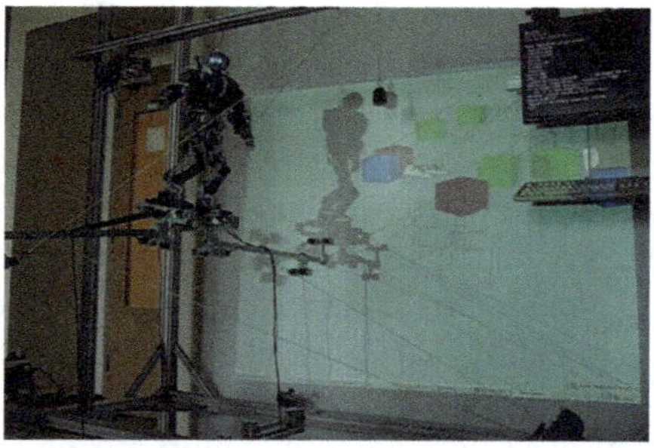

Fig. 2. Full view of the scaled version of the CDLI showing the platforms, the cables and the Virtual Reality screen displaying the scene

Each component of the software must have self-testing capabilities labelled *Built-In Self Test* (BIST) describing a general system test planning for the purpose of quality assurance and safety management. A Hardware-in-the-loop (HIL) simulator could be implemented as a mean of running some parts of the BIST and partially control the platform. Fig. 1 explains the test plan implemented inside the controller for the Cable-Driven Haptic Interface presented in this paper.

This paper addresses modern human safety concerns applied to a parallel cable-driven haptic interface. The approach is itemized as four simple algorithms to enhance human safety. The safety manager defines new modes of operations for a cable-driven haptic mechanism which are investigated thoroughly: 1) critical run-time error (which should never occur); 2) soft run-time error; 3) sensor fault; 4) released cable from an active actuation state; and 5) normal mode. Furthermore, three articular control levels are implemented with an hybrid position/tension control in open and closed-loop.

The main issue presented in this paper is the management of the platform when it leaves its workspace (generated by sensor faults, cable interference and when the platform is forced to go beyond the boundary of its usual workspace) since a defective reel that must be released from an active actuation will modify the workspace size. Indeed, the haptic application is not suitable for this critical situation where safety is more important than rendering virtual objects.

The first section of this paper presents the overall control algorithm of the parallel cable-driven haptic interface with the definition of an Optimal Tension Distriution algorithm designed for redundant actuation. The second section presents the strategies for avoiding user injury. They provide some simple methods of reducing the hazards of human-robot interaction. Finally, since the system is to provide a locomotion interface as a haptic device, the algorithms must be transparent to the user. Thus, the last section presents some results on the scaled-down $1:3$ prototype of the Cable-Driven Locomotion Interface which is shown in Fig. 2. For demonstration, a Kondo KHR-1HV humanoid robot is installed on the platforms.

2 Control Algorithm

In a gait training application, the mechanism can be used to render various ground properties which can potentially reach fatal levels. Some safety algorithms and hardware systems are designed inside the mechanism gait trainer named HapticWalker [17]. The controller presented in this section comes from the idea of Lafourcarde [6] and is an extension of the HapticWalker for a parallel cable-driven interface. Other safety algorithms are presented in [18] which allow human operators and robots to work in close proximity for assisted collaboration. They use vision systems to measure body posture and position and to avoid collisions contrary to the project in [3] where collisions are allowed such that the user can push the robot intuitively (even though the robot is still under position control). In mobile robotics, the safety manager detect hazardous characteristics, evaluate them and act to avoid them [11]. Such a manager could be applied for a gait trainer where cable interference could be a hazardous characteristic.

The software architecture is divided in seven stages as defined in Fig. 3: 1) the testing simulator (BIST) with the safety manager; 2) the haptic rendering [8]; 3) the 4-DOF washout algorithm such as those suggested in [5, 12, 20]; 4) the analysis of stability [4]; 5) the cable interference management algorithm with the Wrench-Closure Workspace [10]; 6) the Optimal Tension Distribution algorithm; and 7) the cable tension controllers including the motorized reels and the power amplifiers. The Wrench-Closure Workspace (WCW), defined in [2], gives the set of poses of the moving platform for which any external wrench can be balanced by positive tensions in the cables. Vector $\mathbf{f}$ represents a flag indicating when the evaluated Cartesian pose $\mathbf{p}_{pf}$ is inside this workspace which will be used in the algorithm 4.

Fig. 1 presents the architecture of the safety design implemented in the safety manager presented in Fig. 3. The design priority goes from the center to the outside of Fig. 1. A fault in the smaller circles could generate a fault in the bigger ones

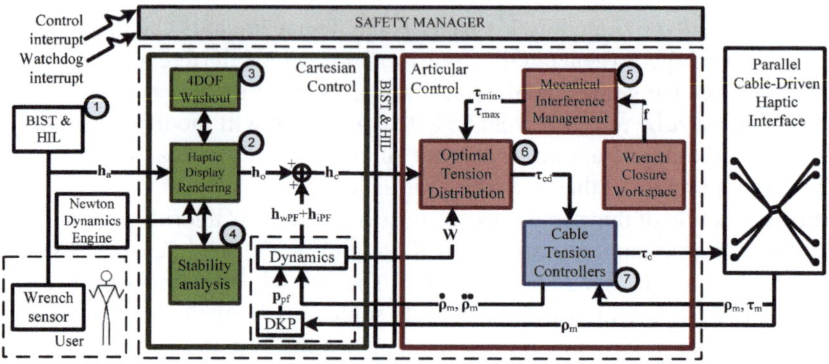

Fig. 3. Simplified control system of the parallel cable-driven haptic interface

(e.g., a damaged reel changes the platform workspace dimensions). However, a fault in the outer circles does not affect the inner ones. Also, the algorithms are not competitive: whether or not cable interference occurs and the platform leaves its workspace, the software continues its safe operation.

An implementation level is shown in Fig. 4 for each safety algorithms with their associated execution priority. Algorithm 1 is the main loop inside the controller where $\mathbf{T}_{cd}$ is a circular buffer of length l containing the previous cable tension values τ_{cd}, m is the index of the actual values and $\mathbf{E}$ represents any errors generated inside the algorithms and inside any functions. This main algorithm presents the calling order for each stage inside the control system which corresponds to a particular case of the Fig. 4 where each safety algorithms with execution priority levels are described.

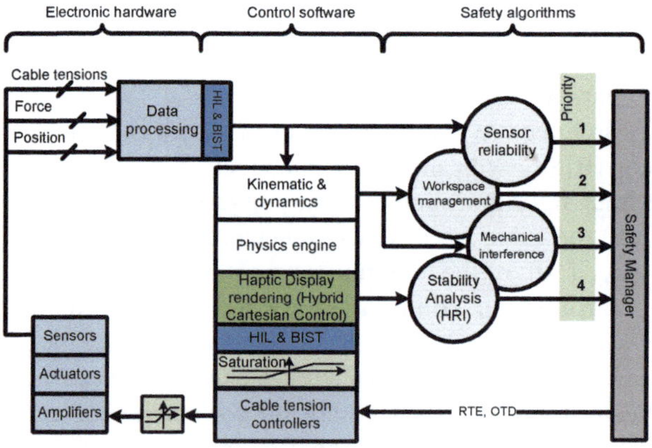

Fig. 4. Implementation of the safety algorithms

Algorithm 1. SafetyManager

repeat
    wait control interrupt
    SensorFaultDetection$\{\tau_m, \dot{\rho}_m\}$
    $[\tau_{min}\, \tau_{max}] \leftarrow$MechanicalInterferenceManagement
    $\mathbf{h}_{wPF} + \mathbf{h}_{iPF} \leftarrow$ DynamicWrenchCompensation$\{\rho_m\}$
    $\mathbf{h}_r \leftarrow$ HapticDisplayRendering$\{\mathbf{h}_a\}$ (see [8])
    $\mathbf{h}_o \leftarrow$ StabilityAnalysis$\{\mathbf{h}_r, \dot{\mathbf{p}}_{PF}\}$ (see [16])
    $\tau_{cd} \leftarrow$ OptimalTensionDistribution$\{\mathbf{W}, \mathbf{h}_c\}$
    $\mathbf{T}^m_{cd} \leftarrow \tau_{cd} \wedge m \leftarrow (m+1)\%l$
    if Watchdog is waiting **then**
        WorkspaceManagement$\{\tau_{cd}\}$
    else
        overrun, soft run-time error, $\mathbf{E} \leftarrow$error
        $\mathbf{T}^{m-1}_{cd} \leftarrow \tau_{min}$, open-loop with τ_{min} for all cables
    end if
until Critical run-time error is detected
Send ending proxy signal

2.1 *Optimal Tension Distribution (OTD)*

Optimal Tension Distribution (OTD) generates a set of equilibrium tension values called the *setpoints* that the cable tension controllers then attempt to follow [21]. Cable sagging is reduced by maintaining the tension in the cables above a minimum threshold value τ_{min}. Indeed, values of τ_{min} are not constants and can be computed with the cable length and the attachement points on the platform. The actuation system (reels, motors and cables) is also limited by a maximum tension τ_{max}. Hence, the following force distribution method in (1) is proposed to avoid cable sagging as well as excessive mechanical deformation of the reel:

$$\begin{cases} \text{minimize} & \frac{1}{2}\tau_{cd}^T \mathbf{G} \tau_{cd} \\ \text{under} & \left| \begin{array}{l} \mathbf{W}\tau_{cd} = \mathbf{h}_c \text{ , with} \\ \tau_{max} \geq \tau_{cd}[i] \geq \tau_{min} \end{array} \right. \end{cases} \tag{1}$$

$$\begin{aligned} \mathbf{G} &= \operatorname{diag}(g_i) \\ \mathbf{h}_c &= \mathbf{h}_{wPF} + \mathbf{h}_{iPF} + \mathbf{h}_o \\ \tau_{cd} &= \begin{bmatrix} \tau_1 & \tau_2 & \dots & \tau_n \end{bmatrix}^T \\ i &= 0,\ 2,\ \dots,\ n-1 \text{ cables,} \end{aligned} \tag{2}$$

where $\mathbf{h}_c$ represents the forces and torques that are applied on a single platform (i.e. the total wrench applied by the cables at the centre of mass of that platform), $\tau_{cd}[i]$ is the tension vector of the ith (of n) cable, $\mathbf{W}$ is the pose-dependent wrench matrix computed by the platform Jacobian matrix that links Cartesian to articular velocities, $\mathbf{G}$ is a weighting matrix with its diagonal elements such that $g_i = 1$ for all i. The mathematical derivation of (1) is presented in [1].

The effect of a broken cable, a defective sensor or a released cable is equivalent to removing a column in the matrix **W**, a value in the vector τ_{cd} and a value in the vector ρ such that $\mathbf{W} \in \mathbb{R}^{k\times(n-j)}$ and $\tau_{cd} \in \mathbb{R}^{(n-j)}$ where k and j are respectively the number of DOF to control and the number of defective sensors. Instead of adjusting the size of the matrices and vectors, the algorithms presented in this paper adjust the values of τ_{min} and τ_{max} in such a way that the respective actuators do not receive enough power from the amplifier. This behaviour implements redundancy on parallel cable-driven haptic interfaces.

Adjusting the size of matrices and vectors has consequences on the Cartesian pose evaluation $\mathbf{p}_{pf}$ of the platform with the use of cable length ρ. Indeed, the measurement of cable length is obviously no longer effective. But, as with the previous Cartesian pose, it is possible to compute the theoretical length of the cable and replace the measurement by this theoretical value. Such an evaluation does not provide an accurate cartesian pose but it sufficiently protects the user and avoids resizing all the matrices and vectors in real-time. The results will show that the computed pose drifts when a cable is severed while the true pose stays constant.

3 Safety Management Algorithm

The following sections detail the safety strategies shown in the Fig. 1. These strategies are ordered from the low-level control (the centre of the circle) to high level haptic interaction control. This figure presents the four main strategies for human safety implemented in the control system shown in Fig. 3 and in the algorithm 1.

Section 3.1 presents the algorithm designed for detecting a broken cable or a defective sensor. Section 3.2 presents the cable tension controllers used inside the control system for allowing the platform to safely go beyond its designated workspace. Finally, section 3.3 develops the control strategy for recovering from crossed cables (or interference), or when a cable intersects with a vertex of the mechanical part of the haptic interface.

3.1 Sensor Reliability

Sensor reliability (for a fault tolerant system) is relevant to the case when a cable or a sensor can be damaged while the user is operating the haptic interface or when a sensor returns null values as shown in Fig. 5, for example. Algorithm 2 is specialized for detecting a broken cable. Two conditions are checked: the tension ($\tau_m[i] \leq 0.5\tau_{min}[i]$) and velocity ($\dot{\rho}_m[i] \geq 0.7\dot{\rho}_{max}$) where τ_m is the cable tension and ρ_m is the measured cable length. A sensor could also become saturated, indicated by the fact that the acquisition board returns its maximum or minimum value for a prolonged period of time. In such conditions, the affected cable must be released from an active actuation state. Vector $\mathbf{S}_c[i]$ represents the state of all cables, ε is the accuracy of the OTD (ε should be near zero) and vector $\mathbf{e}$ indicates a faulty sensor.

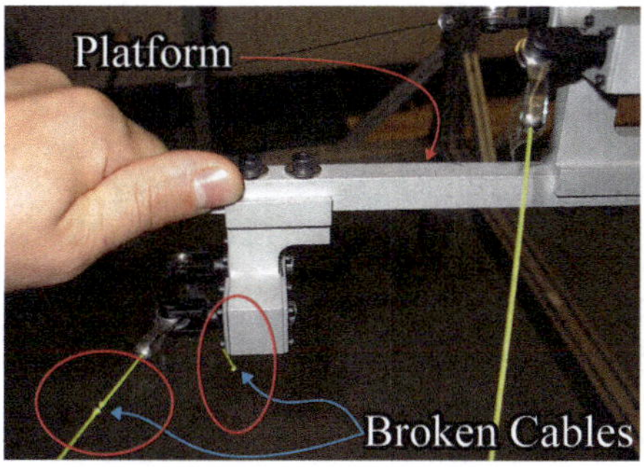

Fig. 5. Management of broken cables

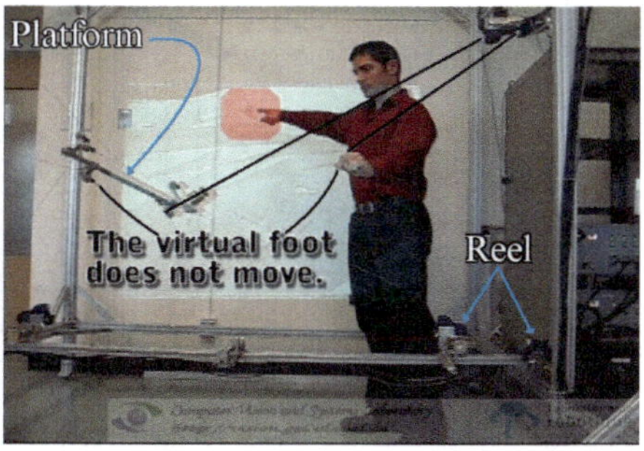

Fig. 6. Cable interference management

Algorithm 2. SensorFaultDetection

Require: $0 \leq \rho[i] \leq \rho_{max}$
  for all i such that $0 \leq i \leq n-1$ **do**
    if not $\mathbf{e}[i]$ **then**
      if $\tau_m[i] \leq 0.5\tau_{min}[i]$ **and** $\dot{\rho}_m[i] \geq 0.7\dot{\rho}_{max}$ **then**
        $\mathbf{e}[i] \leftarrow 1$
      end if
      if Sensor i is saturated over 0.1 second **then**
        $\mathbf{e}[i] \leftarrow 1$
      end if
      if $\mathbf{e}[i] == 1$ **then**
        $\tau_{min}[i] \leftarrow 0,\ \tau_{max}[i] \leftarrow \varepsilon,\ j \leftarrow j+1$
        $\mathbf{S}_c[i] \leftarrow$ cable must be released
      end if
    end if
  end for

3.2 Workspace Management

A runtime error (RTE) switch is implemented as an open-loop controller for ensuring the security of the walker as shown in Fig 7. This RTE, controlled by vector $\mathbf{E}[i]$ for each cable tension controller i, allows the mechanism to go beyond its designated workspace and return to a safe position afterwards. The OTD is deactivated and all cable tensions are maintained in an open loop system with the values $\mathbf{T}_{cd}$ at the position $(m+1)\%l$ previously computed inside the WCW. When the platform ventures outside the WCW, a resulting wrench $\mathbf{h}_w$ is applied on the platform that can be computed with $\mathbf{h}_w = \mathbf{W}\mathbf{T}_{cd}^{(m+1)\%l}$.

The hybrid control presented in Fig. 7 could maintain the platform at the boundary of ρ_{max} by applying cable length control. Otherwise, cable tension control is always used whenever cables are in interference, broken, or sensors are defective. The strategy used for choosing the cable length control considers three conditions:

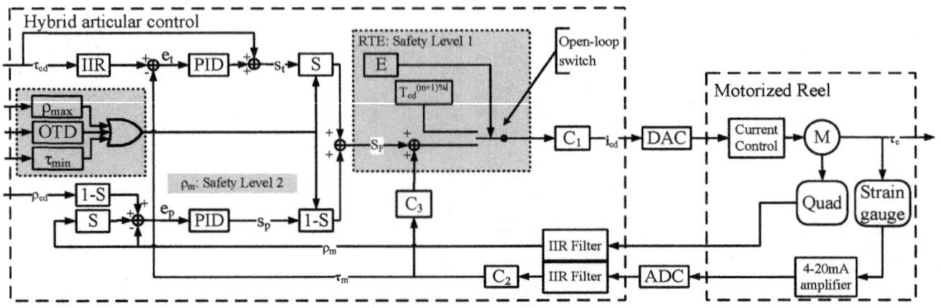

Fig. 7. Cable tension controller

Algorithm 3. WorkspaceManagementAlgorithm

Ensure: $\mathbf{S}_c[i] \wedge \tau_{cd}$
  if $\mathbf{W}\tau_{cd} = \mathbf{h}_c \wedge \mathbf{p}_{pf} \in$ WCW **then**
    $\mathbf{E} \leftarrow$ no error
    for all i **if** $\rho[i] \geq \rho_{max}$ **then** $\mathbf{S}[i] = 0$ **end if**
  else if $\mathbf{W}\tau_{cd} \neq \mathbf{h}_c \vee \mathbf{p}_{pf} \notin$ WCW **then**
    $\mathbf{E} \leftarrow$ error, apply open-loop with $\mathbf{T}_{cd}^{(m+1)\%l}$
  else
    $\mathbf{T}_{cd}^{(m+1)\%l} \leftarrow \tau_{min}$
    $\mathbf{E} \leftarrow$ error, apply open-loop with τ_{min} for all cables
  end if

---

1) maximum cable length is reached; 2) $\mathbf{W}\tau_{cd} \neq \mathbf{h}_c$; and 3) $\tau_m \leq \tau_{min}$. Note that this strategy is a second safety backup in the situation where the runtime error is not activated by the algorithms for allowing articular open-loop control.

3.3 *Mechanical Interference*

Interference between two cables can occur at any time in the course of a random trajectory such as those generated in a haptic application. This interference must be accounted for by choosing which cable needs to be released, or which type of control should be applied in the situation where the interfering cables cannot be released from an active actuation state. The integrity of the workspace must be verified by checking whether or not releasing a given cable keeps the platform inside the WCW. This can result in four distinct cases:

- any released cable generates a pose inside the workspace (case 1);
- cable 1 released: platform is inside the workspace (case 2);
- cable 2 released: platform is inside the workspace (case 3) and
- any released cable generates a pose outside the workspace (case 4).

Indeed, when a particular cable interferes with a mechanical component of the locomotion interface, only this cable can be released. Moreover, case 4 is still not managed. However, it is now known that the direction of the plane described by the two interfering cables can be controlled with a hybrid impedance/admittance approach by combining the developments in [8] and [14]. Currently, when case 4 is detected, the cable tensions are maintained at the $\mathbf{T}_{cd}^{(m+1)\%l}$ values in the open loop controller. Algorithm 4 thus resumes the strategy for choosing which cable must be released between two interfering cables m or n. In this algorithm, $\mathbf{f}[i]$ flags if the platform pose is inside the WCW when cable i is released.

Algorithm 4. MechanicalInterferenceManagement

```
Require: Interference detected between cables m and n [10]
  if m is released then f[m] ← WCW endif
  if n is released then f[n] ← WCW endif
  if CASE 1: f[m] and f[n] then
    if τ_m[m] ≥ τ_m[n] then
      S_c[n] ← Release n
    else
      S_c[m] ← Release m
    end if
  end if
  if CASE 2: not f[n] and f[m] then
    S_c[m] ← Release m
  end if
  if CASE 3: not f[m] and f[n] then
    S_c[n] ← Release n
  end if
  if CASE 4: not f[m] and not f[n] then
    Unstable Interference Domain detected
    Apply control in admittance at this pose
  end if
  if Cable Interference is detected then
    if i is the cable to release then
      τ_max[i] ← τ_min[i] + ε, j ← j + 1
    end if
  end if
```

Fig. 6 shows that when a cable is released, it no longer controls the pose of the platform and thus its measured length is no longer used for computing the pose. Thus, the cable can be reeled in because it is controlled at a minimum tension since the condition $\tau_{max}[i] \leftarrow \tau_{min}[i] + \varepsilon$ is reached and the OTD computes tension set-points τ_c with this condition. This minimum tension is computed to minimize the effect of two ambiguous criteria: reducing both cable sagging and friction at the interference point [10].

4 Results and Analysis

The aforementioned safety algorithms were applied to the control of the Cable-Driven Locomotion Interface prototype. This section presents some results obtained from the final control architecture for improving its reliability. The trajectories of the platforms are imposed by the user without the sensor ($\mathbf{h}_a = 0$) and only the weight of the platform is compensated for ($\mathbf{h}_{wPF} \neq 0$ and $\mathbf{h}_{iPF} = 0$). Indeed, there is only Cartesian impedance control and the platform is maintained in the free space using the OTD algorithm.

The first section described the management of sensor faults and the system reaction to a severed cable. This particular situation generates cable tension discontinuities that must be managed. The second section presents the workspace management algorithm when the user over-strains the platform outside of the workspace. Finally, additional experiments are carried out with the cable interference management algorithm in order to gain further insight on this issue presented first in [10].

4.1 Sensor Reliability

Cable tension sensors could feasibly be damaged during system use. The defective sensor could then return the maximum or minimum value allowed by the acquisition system. It can be ignored and compensated by the other sensors. The solution to when a cable is severed can be seen in Fig 8. This figure shows that before the detection of the event, both conditions in velocity and tension are met. Some oscillations occur around zero after the cable is cut. This vibration is the characteristic Heaviside response of the strain gauge sensor used for measuring the tension inside the reel. The peak tension that is measured also results from the cantilever strain gauge response. It is possible to use this characteristic to detect the tension conditions by evaluating the exponential decreasing sinusoid coefficients with a curve fitting technique. But, this condition is only valid for the actual reel design presented in [9].

The detection delay of the event is approximately 0.1 second and the variation of the estimated pose of the platform $\mathbf{p}_{pf}$ is kept under 5 centimeters as shown in Fig 10. This variation comes from the use of the cable length for the evaluation of the platform pose. The key point is that the control system continues without failure and that the real pose of the platform is constant before and after the event. For reducing the adverse effect of a tension discontinuity shown in Fig. 9, the cable tension controllers use a setpoint IIR filter that smoothes the setpoint tensions, at the cost of reducing the bandwidth of the cable tension controllers.

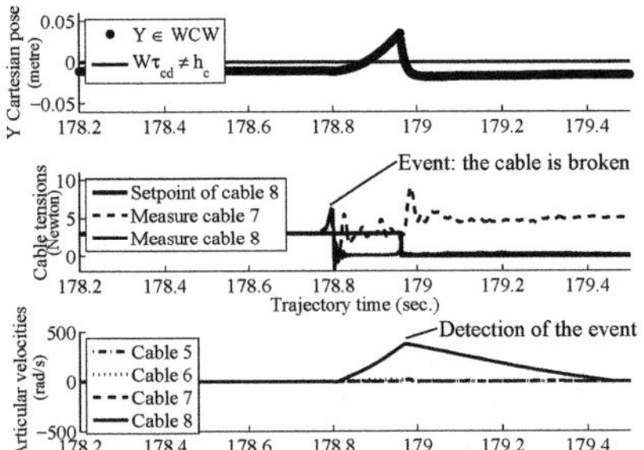

Fig. 8. Cable tensions with the Y movement of the platform when cable 8 is broken

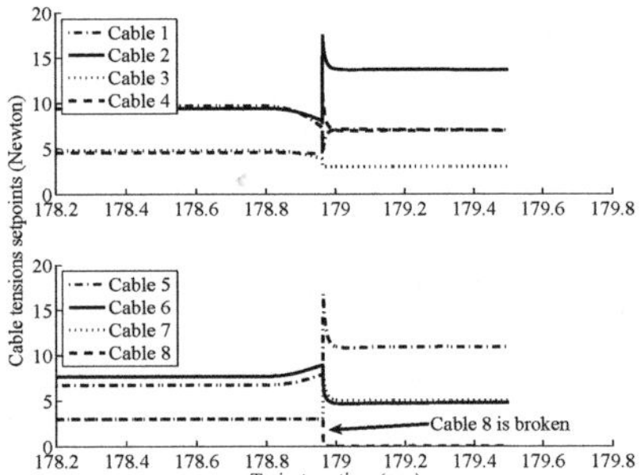

Fig. 9. Computed tension setpoints from the Optimal Tension Distribution algorithm when cable 8 is broken

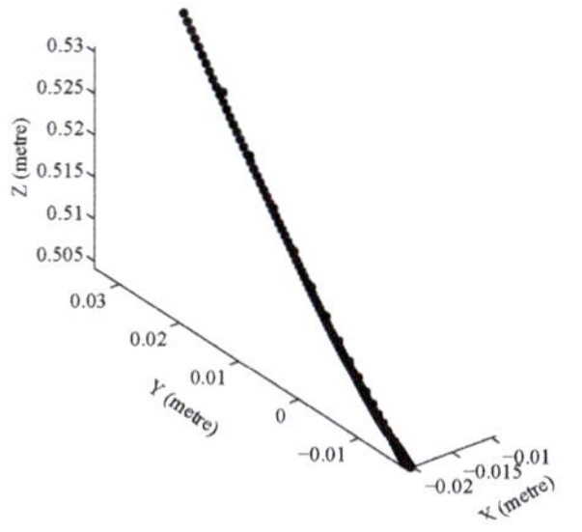

Fig. 10. Variation of the platform pose when cable 8 is broken

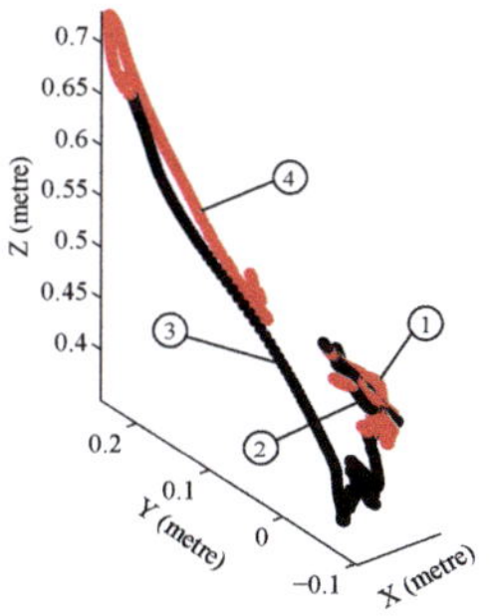

Fig. 11. Cartesian pose with interference between cables 2 and 6

4.2 Workspace Management

Fig. 12 shows the movement of the platform for the evaluation of the workspace management algorithm. When the platform goes outside the WCW, the cable tensions are kept at the constant values $\mathbf{T}_{cd}^{(m+1)\%l}$ generating a wrench $\mathbf{h}_w$, as seen in Fig 13, in the opposite direction of movement, ensuring that the platform will automatically return inside its workspace. Note that the cable tension discontinuities on τ_{cd} at 85.5 seconds in Fig. 12 are the result of the difference between the poses at the moment when the platform goes outside and when the platform comes back inside the WCW. Using a circular buffer $\mathbf{T}_{cd}$ to preserve past cable tensions is useful to minimize the risk that the platform should remain on the boundary of the workspace. Otherwise, when the event occurs, the cable tension setpoints are kept at their value 0.01 second before the event that is defined by the sampling frequency and the length l of the buffer $\mathbf{T}_{cd}$. This timing is chosen experimentally to smoothe the transition of the solution given by the OTD for the user.

4.3 Mechanical Interference between Cables

Fig. 11 shows the trajectory used by the cable interference management algorithm. This trajectory allows one cable interference between cables number 2 and 6. During cable interference, the platform can sometimes venture outside the WCW as shown in Fig. 14 and the cable tension setpoints are kept constant. In part 1 of the figure, between 143 and 145 sec, there is no interference and the cable tensions balance the weight of the platform. In part 2, between 145 and 149 sec, an interference is detected and cable 6 is released from its active actuation state. The setpoint of cable 6 does not drop to its minimal value τ_{min}[5] since the OTD cannot find a valid solution for the actual pose $\mathbf{p}_{pf}$. Indeed, the workspace management algorithm maintains the cable tensions at $\mathbf{T}_{cd}^{(m+1)\%l}$. When the pose $\mathbf{p}_{pf}$ returns inside the workspace, the OTD computes new setpoints and the tension in cable 6 drops to its minimum value,

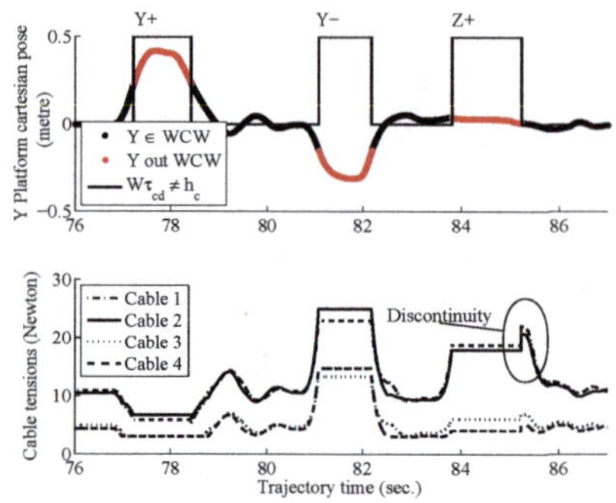

Fig. 12. Cable tensions with the +Y, -Y and +Z movements when the platform goes out of the workspace

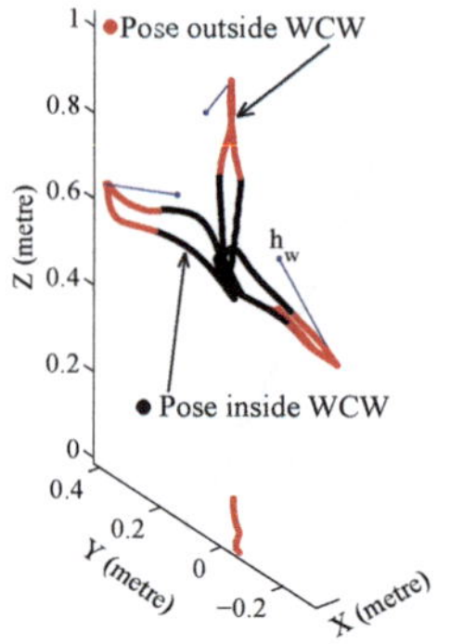

Fig. 13. Cartesian pose when the platform goes outside the WCW

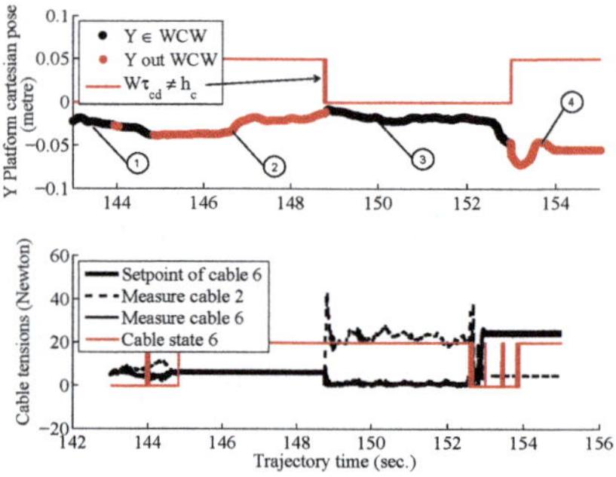

Fig. 14. Cable interference management results

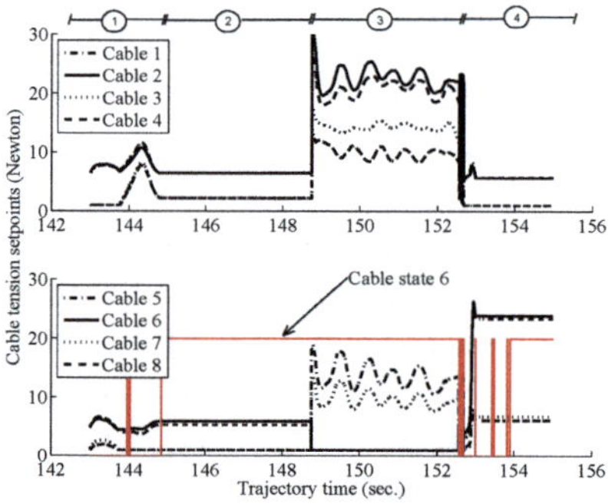

Fig. 15. Cable tension setpoints from OTD with interference

as seen in the third part of the figure. Fig. 15 gives the cable tension setpoints over this trajectory.

5 Conclusion

When a human interacts with an active, semi-autonomous robotic mechanism, safety is a prime concern in any application to avoid injury. Sensor reliability, workspace management and mechanical interference are all critical factors within the system. The results and demonstrations presented here concern the proof that the mechanism will stay stable and safe whenever a fault occurs during operation. Thus, this paper aims to develop a strategy that allows a safer operation at all times, since the use of an emergency shutdown is not always appropriate with a cable-driven mechanism. Of course, such a mechanism does possess major problems of implementation as a haptic device. The safety of this device is a major issue since mechanical transmission via a cable behaves differently compared to a rigid link.

The proposed software architecture includes two levels of safety implemented in Cartesian and articular hybrid controllers with different strategies for testing the overall system. However, the mechanical design plays the most important role, as the software should not be used to increase the reliability of the hardware design. In fact, these algorithms could decrease the performance of the mechanism and thus reduce the transparency of the platform's moving parts. When they are used in haptic applications, special care should be taken in investigating the trade-off between performance and safety. This paper gives further insight into the definition of a safe control algorithm applied to haptic interfaces. However, it could be extended to other systems where human-robot interaction exists.

Finally, in section 3.3, further development could be conducted for managing case 4: when both interfering cables cannot be released from their active actuation state because the platform pose is outside the WCW. Currently, in this situation, the cable tension is maintained at $\mathbf{T}_{cd}$ value.

References

1. Barrette, G., Gosselin, C.: Determination of the dynamic workspace of cable-driven planar parallel mechanisms. Journal of Mechanical Design 127(2), 242–248 (2005)
2. Gouttefarde, M., Gosselin, C.M.: Wrench-closure workspace of six-dof parallel mechanisms driven by 7 cables. Transactions of the Canadian Society for Mechanical Engineering 29(4), 541–552 (2005)
3. Haddadin, S., Albu-Schaffer, A., De Luca, A., Hirzinger, G.: Collision detection and reaction: A contribution to safe physical human-robot interaction. In: International Conference on Intelligent Robots and Systems, pp. 3356–3363. IEEE/RSJ, Nice, France (2008)
4. Hannaford, B., Ryu, J.H.: Time-domain passivity control of haptic interfaces. Transactions on Robotics and Automation 18(1), 1–10 (2002)
5. Iwata, H., Yano, H., Fukushima, H., Noma, H.: Circulafloor. Computer Graphics and Applications 25(1), 64–67 (2005)
6. Lafourcade, P.: Étude des manipulateurs parallèles à câbles, conception d'une suspension active pour soufflerie. École Nationale Supérieur de l' Aéronautique et de l'Espace (2004)
7. Lahouar, S., Ottaviano, E., Zeghoul, S., Romdhane, L., Ceccarelli, M.: Collision free path-planning for cable-driven parallel robots. Robotics and Autonomous Systems 57(11), 1083–1093 (2009)
8. Otis, M.J.D., Mokhtari, M., Du Tremblay, C., Laurendeau, D., De Rainville, F.M., Gosselin, C.M.: Hybrid control with multi-contact interactions for 6dof haptic foot platform on a cable-driven locomotion interface. In: Symposium on Haptics Interfaces for Virtual Environment and Teleoperator Systems, pp. 161–168. IEEE, Piscataway (2008)
9. Otis, M.J.D., Nguyen-Dang, T.L., Laliberté, T., Ouellet, D., Laurendeau, D., Gosselin, C.: Cable tension control and analysis of reel transparency for 6-dof haptic foot platform on a cable-driven locomotion interface. International Journal of Electrical, Computer, and Systems Engineering 3(1), 16–29 (2009)

10. Otis, M.J.D., Perreault, S., Nguyen-Dang, T.L., Lambert, P., Gouttefarde, M., Laurendeau, D., Gosselin, C.: Determination and management of cable interferences between two 6-dof foot platforms in a cable-driven locomotion interface. Transactions on Systems, Man and Cybernetics, Part A: Systems and Humans 39(3), 528–544 (2009)
11. Pace, C., Seward, D.W.: A model for autonomous safety management in a mobile robot. In: International Conference on Computational Intelligence for Modelling, Control and Automation, vol. 1, pp. 1128–1133. IEEE, Vienna (2005)
12. Park, J., Yoon, J., Lim, Y.A., Ryu, J.: A smooth planar walking algorithm for virtual walking machine (k-walker). In: SICE-ICASE International Conference, Piscataway, NJ, USA, pp. 4786–4791 (2006)
13. Perreault, S., Gosselin, C.: Cable-driven parallel mechanisms: application to a locomotion interface. Journal of Mechanical Design, Transactions 130(10), 102, 301–308 (2008)
14. Perreault, S., et al.: Analysis of the interference-free constant-orientation workspace of parallel cable-driven manipulators. Journal of Mechanical Design, Transactions 130(10), 102, 301–308 (2010)
15. Roberts, R.G., Graham, T., Lippitt, T.: On the inverse kinematics, statics, and fault tolerance of cable-suspended robots. Journal of Robotic Systems 15(10), 581–597 (1998)
16. Ryu, J.H., Kim, Y.S., Hannaford, B.: Sampled- and continuous-time passivity and stability of virtual environments. Transactions on Robotics 20(4), 772–776 (2004)
17. Schmidt, H., Hesse, S., Bernhardt, R.: Safety concept for robotic gait trainers. In: International Conference on Engineering in Medicine and Biology, San Francisco, CA, United States, vol. 261, pp. 2703–2706 (2004)
18. Tan, J.T.C., Duan, F., Zhang, Y., Kato, R., Arai, T.: Safety design and development of human-robot collaboration in cellular manufacturing. In: Automation Science and Engineering, pp. 537–542. IEEE, Bangalore (2009)
19. Wischnitzer, Y., Shvalb, N., Shoham, M.: Wire-driven parallel robot: Permitting collisions between wires. International Journal of Robotics Research 27(9), 1007–1026 (2008)
20. Yoon, J., Park, J., Ryu, J.: A symmetric walking cancellation algorithm of a foot-platform locomotion interface. In: International Conference on Humanoid Robots, pp. 668–673. IEEE-RAS, Piscataway (2008)
21. Zheng, Y., Liu, X.: Optimal tension distribution of wire-driven parallel manipulators. Jixie Gongcheng Xuebao/Chinese Journal of Mechanical Engineering 41(9), 140–145 (2005)

Hybrid Stabilizing Control for the Spatial Double Inverted Pendulum

Xinjilefu, Vincent Hayward, and Hannah Michalska

Abstract. The spatial double inverted pendulum actuated at the hip, but not at the foot, may be considered to be a model of standing creatures and robots. Moving in-space, as opposed to in-plane, poses new control problems which, for the most part, are still open. In this paper, a hybrid approach where an energy-shaping, passivity-based swing-up controller hands off the control to a linear-quadratic-regulator in the vicinity of the unstable upright equilibrium is proposed. A direct approach and a pre-compensated approach are described, discussed, and illustrated by means of examples in simulation.

1 Introduction

We propose that any adequate model for studying the active stabilization of articulated structures standing on a small footprint, that is not relying on torques exerted on the ground, should be spatial. The underlying motivation being that to study the stabilization of a multi-body system in the neighborhood of an unstable equilibrium, in addition to the forces due to acceleration, including gravity, the analysis should also include centrifugal and Coriolis terms. The corresponding terms entering in the system dynamics are, of course, nonlinear in essence. The simplest multi-body system that can account for these terms is the double spatial inverted pendulum which is seen in Fig. 1.

Our objective is to describe a control strategy for such systems that are characteristically underactuated and nonholonomic of order two. By the later property we mean that accelerations constraints are not integrable. The spatial double inverted pendulum is a nonlinear, underactuated mechanical system; see related

Xinjelifu · Hannah Michalska
McGill University, Department of Electrical and Computer Engineering, Montréal, Canada
e-mail: xxinji@cim.mcgill.ca, hannah.michalska@mcgill.ca

Vincent Hayward
UPMC Univ. Paris 06, UMR 7222, Institut des Systèmes Intelligents et de Robotique, Paris, France
e-mail: vincent.hayward@isir.upmc.fr

J. Angeles et al. (Eds.): Brain, Body and Machine, AISC 83, pp. 201–215.
springerlink.com

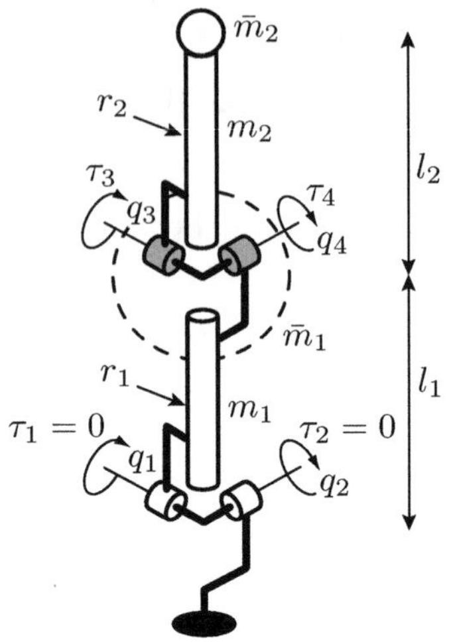

Fig. 1. Two bodies are articulated at the hip and at the ankle. Masses at the head and hip provide a plausible mass distribution for a real machine. Only the hip is actuated.

works in [1, 2]. A popular approach to achieve stabilization of such systems uses swing-up control to a neighborhood of the desired equilibrium and local linear control to maintain balance within that neighborhood. The swing-up control was demonstrated for mechanical systems in the plane by using passivity results while the local linear controller can be designed by making use of one the many available techniques.

The difficulty in employing passivity-based control resides in the fact that any such design must usually be accompanied by energy shaping methods since a given energetic level of the system usually corresponds to multiple equilibria. Energy shaping not only requires a suitable storage function, but also the solution of shaping partial differential equations whose explicit solution can be very difficult. It is perhaps the reason for which passivity-based control has not yet been applied to stabilize spatial mechanical systems such as the double inverted pendulum in space. In this context, the results presented in this paper are a pioneering attempt in this direction which effectively by-passes the need for the solution of the energy shaping equations.

We first present an optimal linear controller which is effective in the neighborhood the upright unstable equilibrium. A second controller based on a hybrid energy shaping approach, is able to steer the system from a much larger set of initial conditions to a state suitably close to the unstable equilibrium where the linear controller completes the stabilization. Simulation results are presented.

2 Model of the Spatial Double Inverted Pendulum

To capture the essential kinematic and dynamic features of a standing machine, the model, Fig. 1, has two rigid cylindrical links of lengths, l_j, and radii, r_j, with masses,

m_j, $j = 1, 2$, that could represent the legs and upper body of a standing machine. Additional point masses $\bar{m}_1$ and $\bar{m}_2$ are attributed to the hip and to the head. The motion of the links are restrained by universal joints at the hip and at the unactuated ankle. The derivation of the model is similar to that in [3], but is reproduced here for completeness.

2.1 Model Derivation

Taking the joints angles as generalized coordinates, $q \triangleq [q_1, \cdots, q_4]^\top$, in the absence of dissipation, the model is a simple Lagrangian system,

$$L(q,\dot{q}) = K(q,\dot{q}) - V(q) = \tfrac{1}{2}\dot{q}^\top M(q)\dot{q} - V(q), \tag{1}$$

where $L(q,\dot{q})$ is the Lagrangian function, $q \in S^2 \times S^2$ is the configuration vector, and $K(q,\dot{q})$ and $V(q)$ are the kinetic and potential energies of the system, respectively. The expression for $M(q)$ can be found in [3]. If $F(q) : S^2 \times S^2 \mapsto \mathbb{R}^{4\times 2}$, represent the selection matrix of the external forces applied to the system then the Euler-Lagrange equations for the system are,

$$\frac{\mathrm{d}}{\mathrm{d}t}\frac{\partial L}{\partial \dot{q}} - \frac{\partial L}{\partial q} = F(q)\boldsymbol{\tau}_a,$$

where $\boldsymbol{\tau}_a \triangleq [\tau_3, \tau_4]^\top \in \mathbb{R}^2$ (subscript a means "actuated"), and $F(q) = [e_3, e_4]$ where e_k is the kth standard basis vector in $\mathbb{R}^4$. Hence, for $k = 1, \cdots, 4$, the system is governed by,

$$\sum_j m_{kj}(q)\ddot{q}_j + \sum_{i,j} \Gamma_{ij}^k(q)\dot{q}_i\dot{q}_j + g_k(q) = e_k^\top F(q)\boldsymbol{\tau}_a,$$

where the gravity terms and the Christoffel symbols are given by [4],

$$g_k(q) = \frac{\partial}{\partial q_k}V(q), \quad \Gamma_{ij}^k(q) = \frac{1}{2}\left(\frac{\partial M_{ij}(q)}{\partial q_k} + \frac{\partial M_{ki}(q)}{\partial q_j} - \frac{\partial M_{jk}(q)}{\partial q_i}\right). \tag{2}$$

In vector form,

$$M(q)\ddot{q} + \dot{q}^\top Q(q)\dot{q} + G(q) = F(q)\boldsymbol{\tau}_a, \tag{3}$$

where Q is a matrix such that $C(q,\dot{q})\dot{q} \triangleq \dot{q}^\top Q(q)\dot{q} \in \mathbb{R}^4$. The terms involving $\dot{q}_i\dot{q}_i$ represent the centrifugal forces and the terms involving $\dot{q}_i\dot{q}_j, i \neq j$, stand for Coriolis forces. Also, $G(q) = [g_1(q), \ldots, g_4(q)]^\top$ contains the gravity terms. Using (2), it is then possible to show that the matrix $\frac{\mathrm{d}}{\mathrm{d}t}M(q) - 2C(q,\dot{q})$ is skew-symmetric. Recalling that $M(q)$ is positive definite, hence invertible, and introducing the Legendre transformation with respect to $\dot{q}$ [5],

$$\boldsymbol{p} = \frac{\partial L}{\partial \dot{q}} = M(q)\dot{q},$$

then allows one to rewrite the system (3) in the Legendre normal form,

$$\dot{q} = M^{-1}(q)\boldsymbol{p},$$
$$\dot{\boldsymbol{p}} = -G(q) + \boldsymbol{p}^{\top}\tilde{Q}(q)\boldsymbol{p} + \tilde{F}(q)\boldsymbol{\tau}_a.$$

Stacking up q and $\boldsymbol{p}$ into $x \triangleq [q;\boldsymbol{p}]$ allows one to see that the Legendre normal form of the model takes the form of a smooth nonlinear system which is affine in the control,

$$\begin{pmatrix}\dot{q}\\ \dot{\boldsymbol{p}}\end{pmatrix} = \begin{pmatrix}M^{-1}(q)\boldsymbol{p}\\ -G(q)+\boldsymbol{p}^{\top}\tilde{Q}(q)\boldsymbol{p}\end{pmatrix} + \begin{pmatrix}0\\ \tilde{F}(q)\end{pmatrix}\boldsymbol{\tau}_a$$
$$\triangleq f(x) + F(x)\boldsymbol{\tau}_a = f(x) + f_1\tau_3 + f_2\tau_4, \tag{4}$$

where $\tilde{Q}(q) \triangleq \left(M^{-\top}(\partial M/\partial q - Q)M^{-1}\right)(q)$, $f : x \rightarrow \mathbb{R}^8$ is the drift vector field related to the gravity field, and $f_1, f_2 \in \mathbb{R}^8$ are constant vector fields with $f_1 = e_7$ and $f_2 = e_8$ where e_k is the k-th standard basis vector in $\mathbb{R}^8$.

2.2 Model Properties

The expression for the system drift that includes the calculation of the Coriolis and centrifugal forces appearing in $C(q,\dot{q})$ expressed through the Christoffel symbols (2) fill very many lines. It is apparent that an exact analysis of the structure of the controllability Lie algebra for the system is practically impossible as it requires the evaluation of repeated Lie brackets of the vector fields $f(x)$, f_1, and f_2.

System (3) is underactuated with control deficiency which is determined by the difference between the rank of $F(q)$ and the dimension of the configuration manifold. Also, the first two equations in (3) constitute a nonlinear motion constraint on the accelerations $\ddot{q}_1$, $\ddot{q}_2$ which cannot be integrated even partially, i.e., the constraints cannot be transformed into an equivalent form that contains only velocities and positions. The relation between integrability and the the gravity term is discussed in [6] and [7] where sufficient conditions for integrability of second order constraints on the system accelerations are given. Non-integrability of the acceleration constraints puts the system in the category of nonholonomic systems of order two and precludes the dimension of the configuration manifold to be reduced by direct integration of the constraints. A further implication is the lack of existence of diffeomorphic state-feedback transformations that can linearize the system globally.

The local linearization of the system in the neighborhood of the unstable upright standing position, is still controllable due to the presence of the gravity term, see [6] and [7]. This allows one to consider constructing linear controllers. Nevertheless, as verified by simulations, the region of attraction for this type of stabilizing feedback is small [8, 9].

The study of small time local controllability (STLC) of the system at every configuration point away from the equilibrium would require a detailed analysis of the structure of controllability Lie algebra of the system [10].

3 Control of the Spatial Double Inverted Pendulum

Since the approach adopted here is of a hybrid type, a linear controller is first constructed to stabilize the linearized system. The linear controller is able to stabilize the nonlinear system in a small neighborhood of the unstable equilibrium. Next, a nonlinear controller is designed to swing-up the system to a region of the state space from which the linear controller can take over.

3.1 LQR Controller for the Linearized System

Denoting $x \triangleq [q; \boldsymbol{p}]$, the linearization of (4) into $\dot{x} = Ax + Bu$ is first derived,

$$A = \left.\frac{\partial f(x)}{\partial x}\right|_{x=0} = \left.\begin{bmatrix} 0 & M^{-1}(q) \\ -\nabla G(q) & 0 \end{bmatrix}\right|_{q=0,\boldsymbol{p}=0},$$

where the matrix $\nabla G(q)$ is the gradient of the vector field $G(q)$, which is also the Hessian of the potential energy. The matrix B is given by $B = F(x) = [e_7, e_8]$ and the control vector is $u = [\tau_3; \tau_4] = \boldsymbol{\tau}_a$. Using the following values of parameters: $m_1 = m_2 = \bar{m}_1 = \bar{m}_2 = 1.0$ kg, $l_1 = l_2 = 1.0$ m, $r_1 = r_2 = 0.1$ m, $g = 9.81$ m/s^2, the cost function is

$$J = \int_0^\infty \left(q^\top Q_1 q + \boldsymbol{p}^\top Q_2 \boldsymbol{p} + \boldsymbol{\tau}_a^\top R \boldsymbol{\tau}_a\right) \mathrm{d}t,\ Q_1 = 50\mathbf{I}_4,\ Q_2 = \tfrac{1}{2}M(0),\ R = 100\mathbf{I}_2.$$

Solving the algebraic Riccati equation using MATLAB delivers the following feedback law for the stabilization of the linearized system,

$$\boldsymbol{\tau}_a = \begin{pmatrix} -606 & 0 & -182 & 0 \\ 0 & -606 & 0 & -182 \end{pmatrix} q - \begin{pmatrix} -37 & 0 & 17 & 0 \\ 0 & -37 & 0 & 17 \end{pmatrix} \boldsymbol{p}.$$

Several initial conditions were used to probe the region of convergence (ROC), see Table 1, with several initial positions of the center of mass (COM).

As seen in Fig. 2a, for Example 1, the system stabilizes in about 3 seconds and in about 5 seconds for Example 2. These examples might lead to believe that the region of convergence of the LQR controller is relatively large; however, Example 3 shows that the same controller cannot stabilize the system, a fact from which we can infer

Table 1. Simulation results with LQR control for different initial conditions.

Example	$q_0^\top$	$\dot{q}_0^\top$	COM eccentricity	x_0 in ROC
1	$[-4°, 2°, 8°, -5°]$	0	3.69	Yes
2	$[10°, -10°, -30°, 30°]$	0	8.26	Yes
3	$[-4°, 0, 0, 0]$	0	8.72	No
4	$[\pm 2.7°, 0, 0, 0]$	0	5.84	Yes
5	$[0, \pm 2.6°, 0, 0]$	0	5.63	Yes
6	$[0, 0, \pm 7.7°, 0]$	0	5.02	Yes
7	$[0, 0, 0, \pm 7.5°]$	0	4.89	Yes

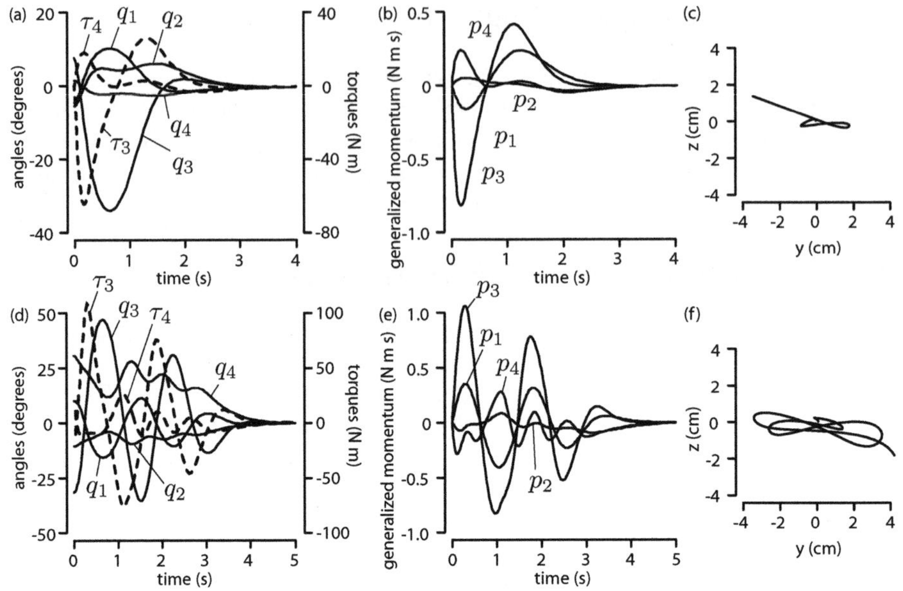

Fig. 2. LQR control Examples 1 (a,b,c) and 2 (d,e,f). a,d: Joint angles and torques through time. b,e: Generalized momenta. c,f: Stabilograms.

that the region of convergence does not have the shape of a ball in the configuration space.

3.2 Energy Shaping and Passivity Based Control

The fact that the generalized inertia matrix is positive definite allows one to employ the collocated partial linearization of the system as proposed in [11]. It leads to system representation in the form of a parallel connection of sub-systems that in turn facilitates the design of passivating system outputs.

Referring to (3), the Euler-Lagrange equations of the underactuated spatial double inverted pendulum can be written,

$$M_{11}(q)\ddot{q}_u + M_{12}(q)\ddot{q}_a + C_u(q,\dot{q}) + G_u(q) = 0, \tag{5}$$

$$M_{21}(q)\ddot{q}_u + M_{22}(q)\ddot{q}_a + C_a(q,\dot{q}) + G_a(q) = \boldsymbol{\tau}_a, \tag{6}$$

where the matrices and vectors are partitioned into the actuated and un-actuated parts,

$$M(q) = \begin{pmatrix} M_{11} & M_{12} \\ M_{21} & M_{22} \end{pmatrix}, \qquad C(q,\dot{q})\dot{q} = \begin{pmatrix} C_u(q,\dot{q}) \\ C_a(q,\dot{q}) \end{pmatrix}$$

$$G(q) = \begin{pmatrix} G_u(q) \\ G_a(q) \end{pmatrix}, \qquad q_u = \begin{pmatrix} q_1 \\ q_2 \end{pmatrix}, \quad q_a = \begin{pmatrix} q_3 \\ q_4 \end{pmatrix}.$$

In (5), the 2×2 block matrix $M_{11}(q)$ is invertible, as $M(q)$ is positive definite, thus $\ddot{q}_u$ can be solved for,

$$\ddot{q}_u = -M_{11}^{-1} M_{12} \ddot{q}_a - M_{11}^{-1} C_u(q,\dot{q}) - M_{11}^{-1} G_u(q) \tag{7}$$

Substituting (7) into (6) yields

$$(M_{22} - M_{21} M_{11}^{-1} M_{12}) \ddot{q}_a + (C_a - M_{21} M_{11}^{-1} C_u) + (G_a - M_{21} M_{11}^{-1} G_u) = \boldsymbol{\tau}_a \tag{8}$$

where the Schur complement of M_{22} in $M(q)$ is $\overline{M}_{22} = M_{22} - M_{21} M_{11}^{-1} M_{12}$ and where $\overline{C}_a = C_a - M_{21} M_{11}^{-1} C_u$ and $\overline{G}_a = G_a - M_{21} M_{11}^{-1} G_u$. The matrix $\overline{M}_{22}$ is positive definite since $M(q)$ is positive definite. Eq. (8) is thus written

$$\overline{M}_{22}\, \ddot{q}_a + \overline{C}_a + \overline{G}_a = \boldsymbol{\tau}_a. \tag{9}$$

If a new control input v is chosen such that,

$$\boldsymbol{\tau}_a = \overline{M}_{22}\, v + \overline{C}_a + \overline{G}_a, \tag{10}$$

then Eq. (9) is partially feedback linearized to read $\ddot{q}_a = v$. Together with Eq. (5), the complete system is then described by,

$$M_{11}(q)\ddot{q}_u + C_u(q,\dot{q}) + G_u(q) = -M_{12}(q) v, \quad \ddot{q}_a = v, \tag{11}$$

with the new control input v defined in Eq. (10). At this point, it is convenient to introduce new state variables,

$$\boldsymbol{\eta} = \begin{pmatrix} q_u \\ q_a \\ \dot{q}_u \end{pmatrix}, \quad z = \begin{pmatrix} q_a \\ \dot{q}_a \end{pmatrix} = \begin{pmatrix} z_1 \\ z_2 \end{pmatrix}.$$

Then, the second order system in (11) is expressed as a first order system,

$$\dot{\boldsymbol{\eta}} = \boldsymbol{\omega}(\boldsymbol{\eta}) + h(\boldsymbol{\eta}, z) + \boldsymbol{\rho}(\boldsymbol{\eta}) v, \tag{12}$$

$$\dot{z} = \boldsymbol{\Phi} z + \boldsymbol{\Psi} v, \tag{13}$$

where the linearized sub-system that represents two double integrators has state space matrices, $\boldsymbol{\Phi} = \begin{pmatrix} 0 & \mathbf{I} \\ 0 & 0 \end{pmatrix}$, $\boldsymbol{\Psi} = \begin{pmatrix} 0 & \mathbf{I} \end{pmatrix}$, and where the functions $\boldsymbol{\omega}(\boldsymbol{\eta})$, $h(\boldsymbol{\eta}, z)$, $\boldsymbol{\rho}(\boldsymbol{\eta})$ are defined by,

$$\boldsymbol{\omega}(\boldsymbol{\eta}) = \begin{pmatrix} \dot{q}_u \\ 0 \\ -M_{11}^{-1}(G_u(q) + \widetilde{C}_u(q,\dot{q}_u)\dot{q}_u) \end{pmatrix},$$

$$h(\boldsymbol{\eta}, z) = \begin{pmatrix} 0 \\ \dot{q}_a \\ -M_{11}^{-1} \widehat{C}_u(q,\dot{q}) \end{pmatrix}, \quad \boldsymbol{\rho}(\boldsymbol{\eta}) = \begin{pmatrix} 0 \\ 0 \\ -M_{11}^{-1} M_{12} \end{pmatrix}.$$

Since the vector $C_u(q,\dot{q})$ is a quadratic function of $\dot{q}$, it can be decomposed into a sum of two terms, $C_u(q,\dot{q}) = \widetilde{C}_u(q,\dot{q}_u)\dot{q}_u + \widehat{C}_u(q,\dot{q})$, where the first term, $\widetilde{C}_u(q,\dot{q}_u)\dot{q}_u$, contains the quadratic terms in $\dot{q}_u$, and the second term, $\widehat{C}_u(q,\dot{q})$, contains the cross terms involving both $\dot{q}_u$ and $\dot{q}_a$ and also the quadratic terms in $\dot{q}_a$. We conclude that

$$h(\boldsymbol{\eta},0) = 0.$$

Introducing the energy-like function associated with the η-sub-system as

$$E_\eta = \tfrac{1}{2}\dot{q}_u^\top M_{11}(q)\dot{q}_u + V(q),$$

it is easy to verify that its time derivative is given by

$$\begin{aligned}\dot{E}_\eta &= \frac{\partial E_\eta}{\partial \boldsymbol{\eta}}\dot{\boldsymbol{\eta}} = \frac{\partial E_\eta}{\partial \boldsymbol{\eta}}(\boldsymbol{\omega}(\boldsymbol{\eta}) + h(\boldsymbol{\eta},z) + \boldsymbol{\rho}(\boldsymbol{\eta})v) \\ &= L_\omega E_\eta + L_h E_\eta + L_{\rho_1} E_\eta v_1 + L_{\rho_2} E_\eta v_2. \end{aligned} \tag{14}$$

The following calculation shows that $L_\omega E_\eta = 0$,

$$\begin{aligned} L_\omega E_\eta = \frac{\partial E_\eta}{\partial \boldsymbol{\eta}}\boldsymbol{\omega}(\boldsymbol{\eta}) =& \frac{1}{2}\dot{q}_u^\top \left(\frac{\partial M_{11}(q)}{\partial q_u}\dot{q}_u\right)\dot{q}_u + G_u(q)^\top \dot{q}_u \\ & - \dot{q}_u^\top M_{11}M_{11}^{-1}(G_u(q) + \widetilde{C}_u(q,\dot{q}_u)\dot{q}_u = 0 \end{aligned}$$

since $\frac{1}{2}((\partial M_{11}(q)/\partial q_u)\dot{q}_u) - \widetilde{C}_u(q,\dot{q}_u)$ is skew-symmetric and appears in a quadratic expression. Hence, $E_\eta = L_h E_\eta + L_{\boldsymbol{\rho}} E_\eta v$, where $L_{\boldsymbol{\rho}} E_\eta = (L_{\rho_1} E_\eta, L_{\rho_2} E_\eta)$. The term $L_h E_\eta$ in (14) vanishes when $z = \dot{q}_a = 0$. It is easy to verify that

$$L_{\boldsymbol{\rho}} E_\eta = -\dot{q}_u^T M_{11} M_{11}^{-1} M_{12} = -M_{12}\dot{q}_u^\top.$$

It is now possible to attempt the local passivation of the parallel interconnection (12,13) in two ways, leading to the same zero dynamics analyses but different design conditions.

3.2.1 Direct Passivity Based Approach

A double integrator sub-system (13) is passive for any output of the form $y_z = Cz = (0; K_2)z$, $K_2 > 0$, with a storage function $S_1 = \frac{1}{2}z^\top Q_1 z$, as it is readily verified that $\boldsymbol{\Phi}^\top Q_1 + Q_1\boldsymbol{\Phi} = 0$ for $Q_1 = (0;0,0;K_2) \geq 0$ while $B^\top Q_1 = C$. Such an input-output system is however not zero state detectable since $y_z \equiv 0$ does not imply that $z_1 \equiv 0$. A possible "energy shaping" output function for the nonlinear sub-system is

$$y_\eta \triangleq (E_\eta - V_{\max}) L_{\boldsymbol{\rho}} E_\eta$$

in which $V_{\max}$ is the largest potential energy of the system at the unstable equilibrium. If not for the presence of the term $h(\boldsymbol{\eta},z)$ in (12), the corresponding storage function for the nonlinear sub-system would be given by,

$$S_\eta(\boldsymbol{\eta}) \triangleq \frac{1}{2}(E_\eta - V_{\max})^2,$$

as then

$$\begin{aligned}\dot{S}_\eta &= (E_\eta - V_{\max})\dot{E}_\eta = (E_\eta - V_{\max})L_h E_\eta + (E_\eta - V_{\max})L_\rho E_\eta v\\ &= (E_\eta - V_{\max})L_h E_\eta + y_\eta v.\end{aligned}$$

The presence of the term associated with $L_h E_\eta$ is an obstacle to the design of a passivity-based stabilizing controller since setting $v = -K_3 y_\eta^\top; K_3 > 0$ does not immediately ensures that $\dot{S}_\eta \leq 0$ at all times. It is useful to investigate the rate of change of the storage function for the parallel interconnection of the linear and nonlinear sub-systems in the hope that the lack of passivity in the nonlinear sub-system can be compensated by passivity in the linear sub-system. The storage function for the parallel interconnection is the sum of the storage functions for the individual sub-systems:

$$S(z,\boldsymbol{\eta}) = S_1(z) + S_\eta(\boldsymbol{\eta}).$$

Thus,

$$\begin{aligned}\dot{S}(z,\boldsymbol{\eta}) &= \tfrac{1}{2}[\dot{z}^\top Q_1 z + z^\top Q_1 \dot{z}] + [E_\eta - V_{\max}]\dot{E}_\eta\\ &= \tfrac{1}{2}v^\top \boldsymbol{\Psi}^\top Q_1 z + \frac{1}{2}z^\top Q_1 \boldsymbol{\Psi} v + [E_\eta - V_{\max}]L_h E_\eta + y_\eta v\\ &= y_z v + y_\eta v + [E_\eta - V_{\max}]L_h E_\eta.\end{aligned}$$

Setting

$$v \triangleq -K_3[y_z + y_\eta]^\top = -K_3 K_2 \dot{q}_a - [E_\eta - V_{max}]K_3 L_{\boldsymbol{\rho}} E_\eta, \tag{15}$$

yields

$$\dot{S}(z,\boldsymbol{\eta}) = -[y_z + y_\eta]K_3[y_z + y_\eta]^\top + [E_\eta - V_{\max}]L_h E_\eta. \tag{16}$$

For the parallel interconnection to be passive one should have $\dot{S} \leq 0$ in some neighborhood of the desired equilibrium. The two terms in (16) contain quadratic as well as linear terms in $\dot{q}_a$. These terms have to be balanced so that $\dot{S} \leq 0$ by adequate selection of the gains K_2 and K_3 while the gain K_3 must be positive definite and bounded, and could be state dependent, if necessary. The following assumption is made to ensure the passivity of the parallel connection.

Assumption 1. *There exist gain matrices $K_2 > 0$, $K_3 > 0$ such that in some neighborhood $\Omega \subset \mathbb{R}^8$ of the unstable equilibrium, for some constant $c > 0$,*

$$\dot{S}(z,\eta) \leq -c[y_z + y_\eta][y_z + y_\eta]^\top. \tag{17}$$

The zero dynamics of the system with control (15) is obtained by observing that if $\dot{S}(z,\boldsymbol{\eta}) \equiv 0$ then $v \equiv 0$ so that the equations of motion reduce to

$$\dot{\boldsymbol{\eta}} = \boldsymbol{\omega}(\boldsymbol{\eta}), \quad \begin{pmatrix} \dot{z}_1 \\ \dot{z}_2 \end{pmatrix} = \begin{pmatrix} z_2 \\ 0 \end{pmatrix}.$$

since $h(\boldsymbol{\eta},z) \equiv 0$ due to $z_2 \equiv 0$. The latter follows from the fact that $z_2 \equiv \text{const} \neq 0$ is impossible in free fall: when $v \equiv 0$ the machine moves only under the action of the gravity field. It follows that also $q_a \equiv$ const, and that $y_z \equiv 0$ so $y_\eta \equiv 0$ for all times. Based on rigid body momentum considerations it can then be shown that:

Proposition 1. *The zero dynamics of the system with control (15) is the dynamics of the system moving only under the gravitational forces (since $v \equiv 0$). The trajectories can only exhibit three types of ω-limit sets:*

1. $\{0;0\}$ *— the unstable equilibrium point,*
2. $\{q_{down};0\}$ *— the stable equilibrium point,*
3. *periodic trajectories where the system rotates in the vertical plane such that $q_a \equiv 0$ at all times.*

The downward equilibrium is the only undesirable ω-limit point which must be avoided and which clearly restricts the region of convergence of the swing-up controller. In view of the above discussion it then follows that

Theorem 1. *Let $\varepsilon > 0$ be such that $B(0;\varepsilon)$ is contained in the region of attraction of a* LQR *controller and let the gain matrices $K_2, K_3 > 0$ be such that (17) holds in some set Ω which contains $B(0;\varepsilon)$. Then every trajectory of the system with the swing-up controller that starts and remains in Ω eventually enters $B(0;\varepsilon)$ where the* LQR *controller stabilizes the system asymptotically to the unstable equilibrium. The hybrid control employing these gains is hence stabilizing the system with region of convergence Ω.*

3.2.2 Passivity Based Approach with Pre-compensation

Prior to passivation of the parallel interconnection, the linear sub-system (13) can first be made asymptotically stable (therefore passive) by introducing a feedback transformation

$$v = -K_1 q_a - K_2 \dot{q}_a + u, \qquad \text{with } K_1 > 0, K_2 > 0,$$

leading to a parallel interconnection with the new control variable u:

$$\begin{aligned} \dot{\boldsymbol{\eta}} &= \boldsymbol{\omega}(\boldsymbol{\eta}) + h(\boldsymbol{\eta}, z_2) - \boldsymbol{\rho}(\boldsymbol{\eta})(K_1 q_a + K_2 \dot{q}_a) + \boldsymbol{\rho}(\boldsymbol{\eta}) u, \\ \dot{z} &= (\tilde{\boldsymbol{\Phi}} - \tilde{\boldsymbol{\Psi}}(K_1;K_2)) z + \tilde{\boldsymbol{\Psi}} u, \end{aligned}$$

where the linearized sub-system is equivalent to two periodic systems (double oscillator) with state space matrices $\tilde{\boldsymbol{\Phi}} = \begin{pmatrix} 0 & \mathbf{I} \\ -K_1 & -K_2 \end{pmatrix}, \qquad \tilde{\boldsymbol{\Psi}} = (0 \; \mathbf{I})$.

An output of the form $y_2 = [Q_{12}^T; Q_{22}]z$, renders the linear sub-system passive, this time, for a strictly positive definite matrix $Q_2 > 0$, satisfying

$$\tilde{\boldsymbol{\Phi}}^\top Q_2 + Q_2 \tilde{\boldsymbol{\Phi}} < 0; \;\; \tilde{\boldsymbol{\Psi}}^\top Q_2 = [Q_{12}^\top; Q_{22}].$$

Clearly, one such matrix Q_2 is of the form

$$Q_2 = \begin{pmatrix} Q_{11} & Q_{12} \\ Q_{12}^\top & Q_{22} \end{pmatrix}.$$

The quadratic form $S_2(z) = \frac{1}{2} z^\top Q_2 z$ then obviously satisfies,

$$\dot{S}_2 \leq y_2 u.$$

and hence is a storage function for the linear sub-system. The bound for the rate of change in the storage function for the parallel connection, $S \triangleq S_2 + S_\eta$, while employing the same output y_η and storage function S_η for the nonlinear sub-system as before, is now:

$$\begin{aligned}
\dot{S}(z, \boldsymbol{\eta}) &= \tfrac{1}{2}[\dot{z}^\top Q_2 z + z^\top Q_2 \dot{z}] + (E_\eta - V_{\max})\dot{E}_\eta \\
&= \tfrac{1}{2} u^\top \boldsymbol{\Psi}^\top Q_2 z + \tfrac{1}{2} z^\top Q_2 \boldsymbol{\Psi} u \\
&\qquad + (E_\eta - V_{\max})(L_h E_\eta - L_\rho E_\eta K_1 q_a - L_\rho E_\eta K_2 \dot{q}_a) + y_\eta u \\
&= y_z u + y_\eta u + (E_\eta - V_{\max})(L_h E_\eta - L_\rho E_\eta K_1 q_a - L_\rho E_\eta K_2 \dot{q}_a).
\end{aligned}$$

Selecting the swing-up control u as before while remembering about the pre-compensation yields

$$\begin{aligned}
v &\triangleq -K_1 q_a - K_2 \dot{q}_a - K_3 (y_z + y_\eta)^\top \\
&= -K_1 q_a - K_2 \dot{q}_a - K_3 Q_{12}^\top q_a - K_3 Q_{22} \dot{q}_a - (E_\eta - V_{\max}) K_3 L_{\boldsymbol{\rho}} E_\eta.
\end{aligned} \tag{18}$$

Hence, the new bound becomes

$$\dot{S}(z, \boldsymbol{\eta}) = -(y_z + y_\eta) K_3 (y_z + y_\eta)^\top + (E_\eta - V_{\max})(L_h E_\eta - L_\rho E_\eta K_1 q_a - L_\rho E_\eta K_2 \dot{q}_a).$$

The counterpart of Assumption 1 is needed.

Assumption 2. *There exist gain matrices $K_1 > 0$, $K_2 > 0$ and $K_3 > 0$ such that in some neighborhood $\Omega \subset \mathbb{R}^8$ of the unstable equilibrium, for some constant $c > 0$*

$$\dot{S}(z, \boldsymbol{\eta}) \leq -c (y_z + y_\eta)(y_z + y_\eta)^\top. \tag{19}$$

It should be noted that the gains must satisfy: $-K_1 - K_3 Q_{12}^\top < 0$ and $-K_2 - K_3 Q_{22} < 0$. Only the first inequality requires attention since $Q_{12} < 0$. The zero dynamics of the system with control (18) is obtained by observing that if $\dot{S}(z, \boldsymbol{\eta}) \equiv 0$ then $v \equiv 0$ so that the equations of motion reduce to

$$\dot{\boldsymbol{\eta}} = \boldsymbol{\omega}(\boldsymbol{\eta}) + h(\boldsymbol{\eta}, z_2),$$
$$\begin{pmatrix} \dot{z}_1 \\ \dot{z}_2 \end{pmatrix} = \begin{pmatrix} z_2 \\ -K_1 z_1 - K_2 z_2 \end{pmatrix}.$$

Clearly, $z_1 \to 0$ and $z_2 \to 0$ since K_1 and K_2 are stabilizing for the linearized sub-system. It follows that $h(\boldsymbol{\eta}, z_2) \to 0$ which implies that the zero dynamics is reduced to

$$\dot{\boldsymbol{\eta}} = \boldsymbol{\omega}(\boldsymbol{\eta}), \quad z_1 \equiv 0, \quad z_2 \equiv 0.$$

Since the zero dynamics is also equivalent to the system moving under the sole action of gravity, Proposition 1 still holds. Like before, the downward equilibrium is the only undesirable ω-limit point which must be avoided. It follows that

Theorem 2. *Let $\varepsilon > 0$ be such that $B(0;\varepsilon)$ is contained in the region of attraction of a* LQR *controller and let the gain matrices $K_1, K_2, K_3 > 0$ be such that (19) holds in some set Ω which contains $B(0;\varepsilon)$, then every trajectory of the system with the swing-up controller that starts and remains in Ω eventually enters $B(0;\varepsilon)$ where the* LQR *controller stabilizes the system asymptotically to the unstable equilibrium. The hybrid control employing these gains is hence stabilizing the system with region of convergence Ω.*

The passivity approach with pre-compensation enjoys advantages compared to the un-compensated approach. The zero dynamics is easier to isolate and $S_2 \to 0$ so that $S \to \frac{1}{2}E_\eta^2 \equiv$ const. The latter delivers a simple criterion by which to tune the controller gains. It is also possible to consider a non-smooth output function $y_\eta \triangleq |E_\eta - V_{\max}| L_\rho E_\eta$ leading to the controller

$$v \triangleq -K_1 q_a - K_3 K_2 \dot{q}_a - |E_\eta - V_{\max}| K_3 L_{\boldsymbol{\rho}} E_\eta \tag{20}$$

3.3 Simulation Results

The switching algorithm between the two controllers is designed as follows.

1. Activate swing-up controller (20);
2. If $V_{\max} + L(q,\dot{q}) < 3\% V_{\max}$, switch to 4;
3. If $V_{\max} + L(q,\dot{q}) > 10\% V_{max}$, switch to 1;
4. Employ LQR to upright position.

where $L(q,\dot{q})$ is the Lagrangian (1) of the system. For the balance phase, the parameters of the LQR controller are as in in Section 3.1. In Example 8 the initial conditions were $q_0 = [10°, -10°, -30°, 30°]^\top$, $\dot{q}_0 = 0$. The following parameters were employed,

$$K_2 K_3 = \begin{pmatrix} 20 & 0 \\ 0 & 20 \end{pmatrix}, \quad K_1 = \begin{pmatrix} 100 & 0 \\ 0 & 100 \end{pmatrix}, \quad K_3 = \begin{pmatrix} 1 & 0 \\ 0 & 1 \end{pmatrix}.$$

As shown in Fig. 3a below, the controller switches at $t = 0.182$ s, and the system stabilizes to its unstable equilibrium configuration in about 6 seconds. In Example 9

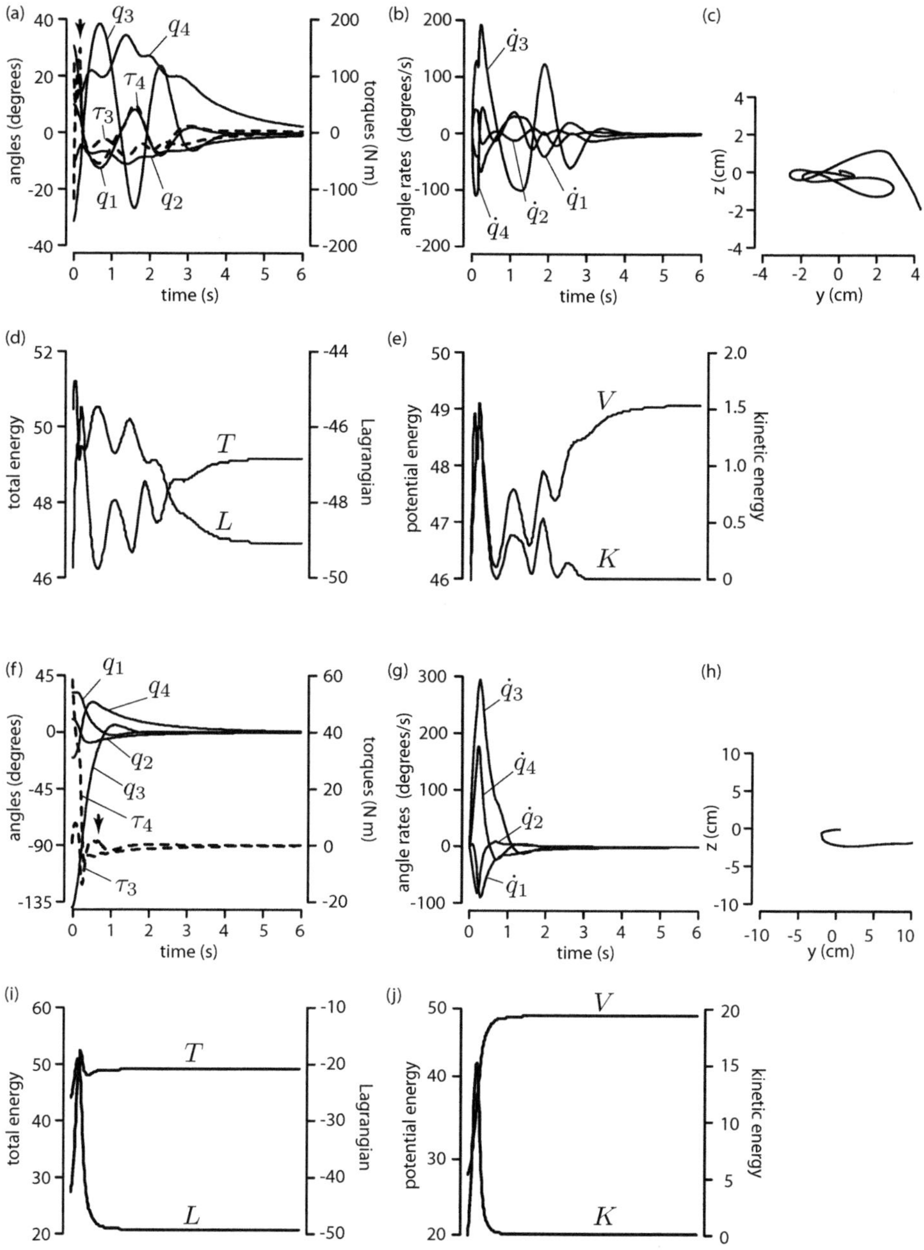

Fig. 3. Examples 8 and 9. (a,f): Joint angles and torques. Arrows shows where the controller switches. (b,g): Joint velocities. (c,h): Stabilograms of COM projected on the ground. (d,i): Total energy and Lagrangian. (e,j): Potential and kinetic Energy.

the initial condition is $q_0 = [30°, 10°, -135°, -20°]^\top$, $\dot{q}_0 = 0$ which is very far from the unstable equilibrium. The following parameters for the swing-up controller were successful,

$$K_2 K_3 = \begin{pmatrix} 8 & 0 \\ 0 & 8 \end{pmatrix}, \quad K_1 = \begin{pmatrix} 16 & 0 \\ 0 & 16 \end{pmatrix}, \quad K_3 = \begin{pmatrix} 2.1 & 0 \\ 0 & 2.1 \end{pmatrix}.$$

The controller switches at $t = 0.688$ s, and the system stabilizes in about 2 seconds. Its trajectories are shown in Fig. 3f above. The hybrid algorithm provides for a much larger ROC but one of the difficulties in applying the passivity approach is to have to tune the parameters.

4 Conclusion

We have employed a passivity-based approach to stabilize the spatial double inverted pendulum actuated at the hip that can be viewed as a model of standing creatures. An energy-shaping control could be found with proper design of a storage function, by-passing the solution of the energy shaping equations. Such approach was more effective if the system was partially linearized by pre-compensation. The controller has a swing-up phase, applicable to a large set of initial conditions, that drives the system to a small neighborhood of the unstable equilibrium where a linear controller can regulate the system near the unstable equilibrium thanks to its robustness properties. A shortcoming of the present method is the necessity to design the controller gains. Enforcing control constraints such as limiting the maximum joint torques is also difficult. We are currently investigating a number of other nonlinear control approaches that would be applicable to systems with larger numbers of degrees of freedom.

References

1. Spong, M.W., Praly, L.: Control of Underactuated Mechanical Systems Using Switching And Saturation. In: Morse, A.S. (ed.) Control Using Logic-Based Switching. LNCIS, vol. 222, pp. 162–172. Springer, London (1997)
2. Spong, M.W.: Underactuated Mechanical Systems. In: Siciliano, B., Valavinis, K.P. (eds.) Control Problems in Robotics and Automation. LNCIS, vol. 230, pp. 135–150. Springer, London (1998)
3. Xinjilefu, Hayward, V., Michalska, H.: Stabilization of the Spatial Double Inverted Pendulum Using Stochastic Programming Seen as a Model of Standing Postural Control. In: Proc. 9th IEEE-RAS Int. Conf. on Humanoid Robots, pp. 367–372 (2009)
4. Murray, M., Li, Z., Sastry, S.S.: A Mathematical Introduction to Robotic Manipulation. CRC Press, Boca Raton (1994)
5. Marsden, J.E., Ratiu, T.S.: Introduction to Mechanics and Symmetry: A Basic Exposition of Classical Mechanical Systems, 2nd edn. Springer, Heidelberg (1999)
6. Oriolo, G., Nakamura, Y.: Control of Mechanical Systems with Second-order Nonholonomic Constraints: Underactuated Manipulators. In: Proc. 30th IEEE Conference on Decision and Control, pp. 2398–2403 (1991)

7. Wichlund, K.Y., Sørdalen, O.J., Egeland, O.: Control of Vehicles with Second-order Nonholonomic Constraints: Underactuated Vehicles. In: Proc. 3rd European Control Conference, pp. 3086–3091 (1995)
8. Ayoub, O.: Robotic Model of the Human Standing Posture, M.Eng. Thesis, Department of Electrical and Computer Engineering, McGill University (2005)
9. Sood, G.: Simulation and Control of a Hip Actuated Robotic Model for the Study of Human Standing Posture, M.Eng. Thesis, Department of Electrical and Computer Engineering, McGill University (2008)
10. Reyhanoglu, M., van der Schaft, A., McClamroch, N.H., Kolmanovsky, I.: Dynamics and Control of a Class of Underactuated Mechanical Systems. IEEE Transactions on Automatic Control 44(9), 1663–1671 (1999)
11. Spong, M.W.: Swing Up Control of the Acrobot. In: Proc. IEEE Int. Conf. on Robotics and Automation, pp. 2356–2361 (1994)

Closed-Loop Control of Plasma Osmolality

Kamel Zaarouri, Ahmad Haidar, and Benoit Boulet

Abstract. In this paper, a closed-loop system to regulate plasma osmolality in patients with central diabetes insipidus is investigated. Using model identification techniques, we obtained a 3rd-order LTI model of the renal/body fluid system from an 11th-order nonlinear system. A Smith Predictor and an $\mathscr{H}_\infty$ controller were developed. The effectiveness of the controller to reject a step disturbance in the plasma osmolality is examined. This paper shows the potential use of control theory in the context of central diabetes insipidus.

1 Introduction

Cells are the building blocks of our body, combining to form tissues and organs. All cells in the human body are surrounded by a liquid called tissue fluid that supplies them with nutrients and form the necessary environment that each cell needs to perform its specific function [1]. Plasma osmolality (OSMP) is a key environmental factor affecting cellular activity and is strictly maintained at constant level. Any alteration of the OSMP is detected by osmoreceptors in the hypothalamus. These receptors regulate the release of the antidiuretic hormone (ADH) based on the negative feedback principle in order to regulate OSMP back to its nominal value [2].

Acting directly on the kidneys, ADH is the primary determinant of free water excretion in the body. Central diabetes insipidus (CDI) is a syndrome that is characterized by decreased secretion of ADH. Patients affected with CDI have reduced ability to concentrate urine and consequently maintain healthy plasma osmolality levels.

Currently, OSMP in patients with CDI is regulated in open-loop by administering a synthetic ADH as a nasal spray. In this type of treatment, any momentary

Kamel Zaarouri · Ahmad Haidar · Benoit Boulet
Centre for Intelligent Machines McGill University 3480 University Street, Montreal, Quebec, Canada H3A 2A7
e-mail: {kamel.zaarouri,ahmad.haidar}@mail.mcgill.ca, benoit.boulet@mcgill.ca

J. Angeles et al. (Eds.): Brain, Body and Machine, AISC 83, pp. 217–225.
springerlink.com

disturbances to the normal osmolality levels will not be corrected until the next scheduled administration of the prescribed medication, causing frequent and significant OSMP variations. In this paper, we study a closed-loop regulation of the plasma osmolality level in patients with CDI. This regulatory system must contain three major components: An OSMP sensor, a mechanical pump capable of delivering the synthetic ADH and a control algorithm to regulate the pump's action.

Plasma osmolality can be estimated by [3]:

$$OSMP = 2(Plasma\left[Na^{+}\right]) + \frac{[Glucose]}{18} + \frac{[BUN]}{2.8}, \tag{1}$$

where $[Glucose]$ is the concentration of glucose in the blood stream expressed in mg/dl, $[BUN]$ is the concentration of blood urea nitrogen (BUN) expressed in mg/dl and $[Na^{+}]$ is the concentration of sodium expressed in mOsm/l.

Sodium and BUN implantable sensors have been readily available in the market for over 10 years with a measurement delay of few milliseconds. With implantable glucose sensors currently being developed to automatically measure interstitial glucose every few minutes [4], an OSMP sensor can be put together with a total sensing delay of that of the glucose. Furthermore, significant work has been done to develop piezoresistive embedded osmolality sensors that could measure accurately plasma osmolality [5]. The reliability of such devices continues to increase. Implantable pumps for drug delivery systems has been in service for over 25 years. Insulin pumps are a good example of such systems [6], [7]. Work is now being done on high-performance silicon implantable micropumps, with accurate pumping characteristics and intrinsic insensitivity to external condition [8].

To the best of our knowledge, no prior work has been done on any form of closed-loop regulation of OSMP in patients with CDI. In this paper, we first derive an LTI model for the renal/body fluid system, specifically the system describing the osmoregualtory renal function. Then, we study different control strategies to regulate ADH infusion and discuss their performance by simulation.

2 Mathematical Models

Many mathematical models of the renal/body fluid system have been previously developed. Guyton et al. [9] were pioneers in deriving a detailed model of the renal function from a system analysis point of view. Uttamsingh et al. [10] presented an overall representation of the renal/body fluid system with an emphasis on the undergoing control mechanisms. On the other hand, many models of the renal/body fluid system have been developed to highlight specific features of the control system, like Merletti and Weed [11] who investigated a relatively complete nonlinear model of fluid volume and osmolality control systems in the human body. Others, like G. Ciofani et al. [12], derived a simpler linear model to describe plasma osmolality control in the kidney.

The model of the renal system used in this paper is based on the one derived by Noraki Ikeda et al. [13] who built a biological system model capable of reproducing clinical findings and measurements. Based on this model, we were able to develop a system that describes the nonlinear dynamics between blood ADH concentration and plasma osmolality. We then proceeded in linearizing the model. Unfortunately, because of the complexity of the system, linearizing the system from input/output equations turned out to be very complex. Consequently, the system model was excited with a uniform Gaussian noise and I/O data were recorded and processed for system identification purposes. The LTI model was then used in deriving controllers for the closed-loop system.

The ARMAX estimator in Matlab's system identification toolbox was used and we were able to approximate with a fit of 97.4% the 11th order nonlinear system with an 3rd-order LTI system with transfer function:

$$G(s) = \frac{-0.0456s^2 + 0.0907s - 0.0451}{s^3 - 2.9867s^2 + 2.9736s - 0.9869}$$

Consider the following state-space representation of the model:

$$\dot{\mathbf{x}} = A\mathbf{x} + Bu$$
$$y = C\mathbf{x} + Du$$

Where u is the ADH concentration expressed as the ratio to the normal level, y the OSMP expressed in mOsm/l and **x** is the vector state of the system.

From the identified transfer function, the realization $G(s) = \left[\begin{array}{c|c} A & B \\ \hline C & D \end{array}\right]$ is obtained using the completely observable canonical form, where

$$A = \begin{bmatrix} 2.987 & 1 & 0 \\ -2.974 & 0 & 1 \\ 0.9869 & 0 & 0 \end{bmatrix}, B = \begin{bmatrix} -0.04565 \\ 0.09069 \\ -0.04506 \end{bmatrix},$$

$$C = \begin{bmatrix} 1 & 0 & 0 \end{bmatrix}, D = \begin{bmatrix} 0 \end{bmatrix}$$

A discretized state-space representation of the model is then computed using a sampling time of $Ts = 0.2$ minute.

3 Closed-Loop Control

If the osmolality sensor were implanted directly into the blood stream, the delay between any disturbance in the osmolality and the measurements would be in the order of few minutes, negligible when compared to our system's time constant. Consequently, any delay could be safely ignored. Unfortunately, biomedical sensors for safety reasons cannot be implanted directly in the blood stream and any measurement of the blood stream's biological analytes has to be done indirectly in the interstitial fluid (ISF). Therefore, any disturbances in the plasma composition will be kept undetectable in the ISF for about 25 min; When adding to that the delay of the

OSMP sensor, any measurement of the OSMP from the ISF has an inherent delay of about 30 minutes.

In this paper, two types of OSMP controllers are tested by simulation on the non-linear model: a proportional-integral-derivative (PID) controller in a Smith Predictor configuration, and an H_∞ controller.

3.1 PID Controller in a Smith Predictor Configuration

The PID controller was first put on the market in 1939 by A. Callender and A. Stevenson of Imperial Chemical Limited of Northwich and has remained the most widely used controller in process control until today. The PID controller is well known in the control area, famous for its simplicity and flexibility in trading off control performance with robustness.

However, a deadtime of 30 min in the system makes it hard to stabilize it with only a PID controller. Consequently, A PID controller in a Smith Predictor configuration is employed to compensate for the deadtime added by the sensors, increasing the system's stability margin.

To improve the design, let G_m represent the model of the systems dynamics, T_{dm} represent the model of the time delay, and e represent the error between the output of the model and the output of the plant. Thus, as shown in Figure 1 the model can be split into a pure delay and a delay-free LTI system. Although the fictitious signal v is unavailable, v_m, can be used as the feedback signal. In this setup the controller would act on the modeled delay-free system G_m and regain the lost stability margin. This arrangement controls the model well, but does not account for disturbances and model inaccuracies. To compensate for these errors, a second feedback loop is implemented using e. The controller K is typically a PI (Proportional plus Integral), PD (Proportional plus Derivative), or PID (Proportional plus Integral plus Derivative) controller, which can be tuned more closely because the effect of the time delay in the feedback loop has been minimized. In the ideal case where $G = G_m$ and $T_d = T_{dm}$, the closed loop transfer function reduces to:

$$\frac{Y(s)}{S(s)} = \frac{KGT_d}{1+KG}.$$

As can be seen, the dead time is no longer present in the denominator. The Smith Predictor controller K_{smith} is shown in Figure 2, which is an equivalent configuration to the one shown in Figure 1.

3.2 H-Infinity Controller

In order to utilize $\mathcal{H}_\infty$ control theory, the system has to be represented according to the standard augmented plant configuration shown in Figure 3. The augmented plant includes the weighting functions required to enforce the performance specifications.

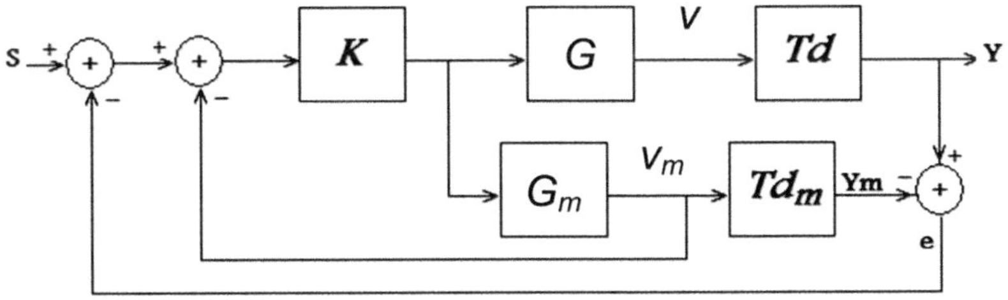

Fig. 1. The Smith Predictor configuration

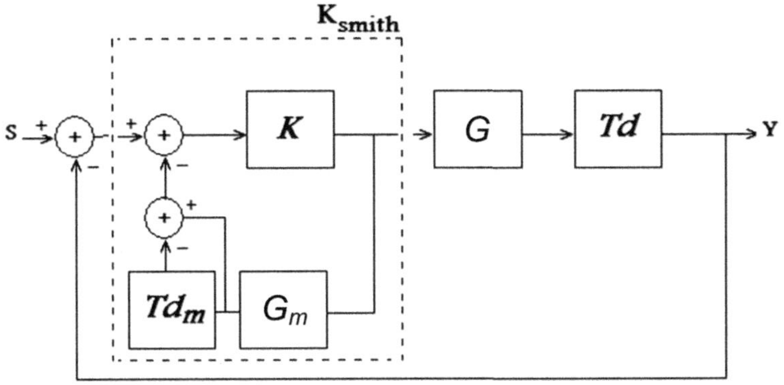

Fig. 2. Smith Predictor controller

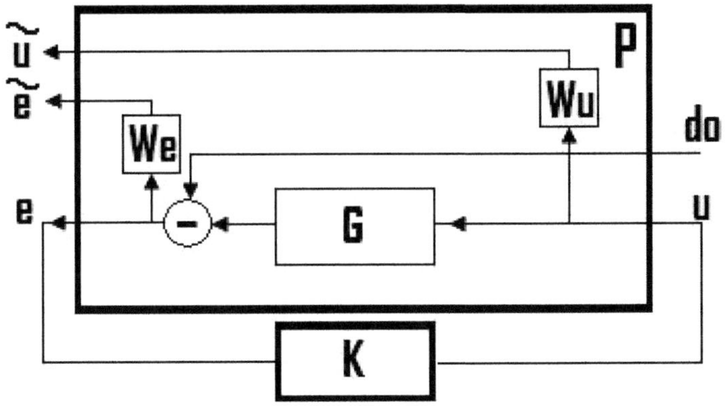

Fig. 3. The augmented plant P, showing the internal configuration and weighting functions.

Let,

$$\widetilde{e} := W_e e = (T_{d_o \widetilde{e}}) d_o$$
$$\widetilde{u} := W_u u = (T_{d_o \widetilde{u}}) d_o$$

where, W_e and W_u are scalar transfer functions. If a controller **K** has been found such that $\gamma < 1$, we can guarantee that:

$$||T_{d_o \widetilde{e}}||_\infty < 1$$
$$||T_{d_o \widetilde{u}}||_\infty < 1$$

And consequently,

$$|T_{d_o e}(jw)| < |W_e^{-1}(jw)| \quad \forall \omega$$
$$|T_{d_o u}(jw)| < |W_u^{-1}(jw)| \quad \forall \omega$$

implying

$$||e||_2 < ||We^{-1} do||_2$$
$$||u||_2 < ||Wu^{-1} do||_2$$

By choosing W_u appropriately, we can limit the amount of ADH required to control the osmolality. The choice of W_e will help to achieve the desired performance criteria. The weighting functions were chosen as follows:

$$We = 600 \frac{10^{1.5}s + 1}{10^{4.6}s + 1}$$
$$Wu = 10$$

Generalized plant P has two inputs, the exogenous input w that includes the reference signal and disturbances, and the control signal u (i.e., the manipulated variable). P has also two outputs, the controlled variables z and the error signal e.

Through mathematical manipulations, we obtain the closed-loop system:

$$z := \begin{bmatrix} \widetilde{u} \\ \widetilde{e} \end{bmatrix} = F_\ell(\mathbf{P}, \mathbf{K})\, d_o$$

where F_ℓ, known as the lower linear fractional transformation, is defined as:

$$F_\ell(\mathbf{P}, \mathbf{K}) = P_{11} + P_{12}\mathbf{K}(I - P_{22}\mathbf{K})^{-1} P_{21}$$
$$= \begin{bmatrix} 0 \\ -We \end{bmatrix} + \begin{bmatrix} Wu \\ -WeG \end{bmatrix} \mathbf{K}(I - (-G)\mathbf{K})^{-1}(-1)$$

The objective of $\mathscr{H}_\infty$ control is to find a controller **K** such that $||F_\ell(\mathbf{P}, \mathbf{K})||_\infty$ is minimized where:

$$||F_\ell(\mathbf{P},\mathbf{K})||_\infty = \sup_\omega \bar{\sigma}(F_\ell(\mathbf{P},\mathbf{K})(j\omega)) \leq \gamma \tag{2}$$

A γ level of 0.8996 was achieved with the following controller K:

$$K = \frac{-0.1383s^2 - 0.0091s - 0.0008}{s^3 + 1.9021s^2 + 0.1256s + 0.0111}$$

4 Simulation Results

Figure 4 shows the system response when the Smith Predictor and $\mathscr{H}_\infty$ controllers are used to regulate the OSMP after an output step disturbance of 20mOsm/l. The complete nonlinear system has been adopted in the simulation. Figure 5 shows the ADH level (expressed in ratio to nominal value) which was required to regulate the OSMP to normal level. Notice the 30 min delay in the ADH response. Faster sensors would lead to improvements in control action and, consequently, system performance.

Unlike the $\mathscr{H}_\infty$ controller, there is no inherent control over the output of the Smith Predictor; If kept unbounded, the controller could deliver abnormally high ADH levels. The output of the Smith Predictor was limited by the use of a saturation

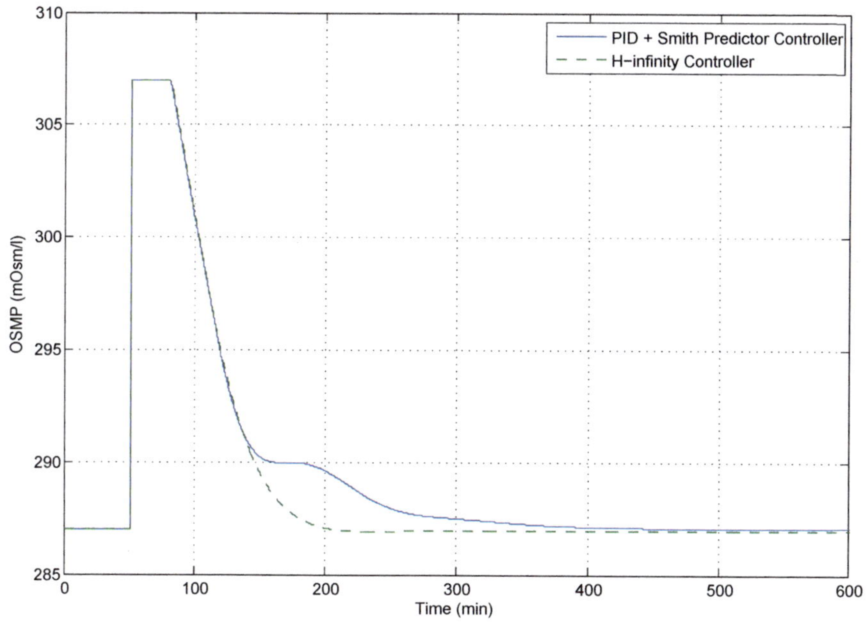

Fig. 4. System response to an output step disturbance in the plasma osmolality of 20 mOsm/l at t = 50 min.

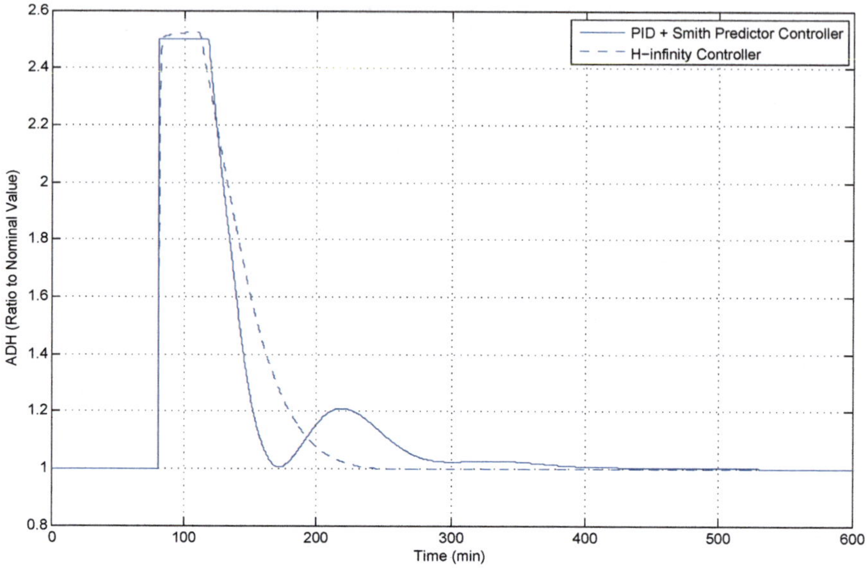

Fig. 5. ADH response to an output step disturbance in the plasma osmolality of 20 mOsm/l at t = 50 min.

function to a maximum ADH value of 2.5 (relative to nominal). Constraints on the control signal level would lead to performance degradation.

5 Conclusion

A closed-loop control of plasma osmolality in patients with central diabetes insipidus has been investigated. Two types of controllers were assessed: A PID controller in a Smith Predictor configuration and an $\mathscr{H}_\infty$ controller. An example to demonstrate the effectiveness of the controllers has been shown. The $\mathscr{H}_\infty$ controller was designed for the delay-free system and performed well when sensor delays where added. As a future work, the sensor's delay could be taken into account in the design stage of the $\mathscr{H}_\infty$ controller for performance comparison.

Acknowledgement. The authors would like to thank professor Elizabeth Jones. Her advice and experience are greatly appreciated.

References

1. Beckett, B.S.: Biology: a modern introduction. Oxford University Press, Oxford (1986)
2. Lote, C.J.: Principles of Renal Physiology, 4th edn. Kluwer Academic Publishers, Dordrecht (2000)

3. Koeppen, B.M., Stanton, B.A.: Renal Physiology, 3rd edn. Kluwer Academic Publishers, Dordrecht (2001)
4. Sensors for medical and science inc., smsi® glucose sensor (in development) (2009), http://www.s4ms.com/products_glucose.htm
5. Moulas, N.J., Gilchrist, K.H., Giovangrandi, L., Maluf, N.I., Kovacs, G.T.A.: Sensors and Actuators B, Chemical, vol. 83(1-3), p. 41 (2002) ISSN 0925-4005
6. Selam, J.L., Charles, M.A.: Diabetes Care, vol. 13, p. 955 (1990)
7. Kennedy, F.P.: Drugs, vol. 42, p. 213 (1991)
8. Maillefer, D., van Lintel, H., Rey-Mermet, G., Hirschi, R.: Proceedings of the MEMS 1999, Orlando, USA, pp. 17–21, 541–546 (1999)
9. Guyton, A.C., Coleman, T.G., Granger, H.J.: Ann. Rev. Physiol., vol. 34(1), p. 13 (1972)
10. Uttamsingh, R.J., Leaning, M.S., Bushman, J.A., Carson, E.R., Finkelstein, L.: Medical & biological engineering & computing, vol. 23(6), p. 525 (1985)
11. Merletti, R., Weed, H.R.: Proceedings of the 5th IFAC world congress, Paris, pp. 25.1–25.7 (1972)
12. Ciofani, G., Landi, A., Mazzei, D., Mazzoldi, A.: Proc. IEEE 44th Conference on Decision and Control and the European Control Conference, Seville, Spain, pp. 12–15 (2005)
13. Ikeda, N., Marumo, F., Shirtaka, M., Sato, T.: Annals of Biomedical Engineering, vol. 7, p. 135 (1979)

Cooperative Exploration, Localization, and Visual Map Construction

Ioannis M. Rekleitis, Robert Sim, and Gregory Dudek

Abstract. We examine the problem of learning a visual map of the environment based on discrete landmarks. While making this map we seek to maintain an accurate pose estimate for the mapping robots. Our approach is based on using a team of at least two (heterogeneous) mobile robots in a simple collaborative scheme. In many mapping contexts, a robot moves about the environment collecting data (images, in particular) which are later used to assemble a map; we view the map construction as both a knowledge acquisition and a training process. Without reference to the environment, as a robot collects training images, its position estimate accumulates errors, thus corrupting its estimate of the positions from which observations are taken. We address this problem by deploying a second robot to observe the first one as it explores, thereby establishing a *virtual tether*, and enabling an accurate estimate of the robot's position while it constructs the map. We refer to this process as *cooperative localization*. The images collected during this process are assembled into a representation that allows vision-based position estimation from a single image at a later time. In addition to developing a formalism and concept, we validate our approach experimentally and present quantitative results demonstrating the performance of the method in over 90 trials.

1 Introduction

Many robotic tasks require that the robot learn a representation, or map, of some property of the environment. Examples of such maps include measures of radiation hot-spots, magnetic declination, sonar representations and visual maps [16, 4, 6]. The specific problem we consider is mapping a property of interest over an

Ioannis M. Rekleitis · Gregory Dudek
Centre for Intelligent Machines and the School of Computer Science, McGill University
e-mail: {yiannis,dudek}@cim.mcgill.ca

Robert Sim
Microsoft Corporation
e-mail: rsim@microsoft.com

J. Angeles et al. (Eds.): Brain, Body and Machine, AISC 83, pp. 227–245.
springerlink.com

unknown environment. A significant issue faced by many map-building schemes is the management and estimation of positional (or pose) errors as the robot collects observations from the environment. That is, as a robot or a team of robots collects successive measurements from different positions and orientations (poses), the certainty of their pose estimates decreases with each new measurement. In some cases where the observations lie on a high-dimensional manifold, correlation between dimensions allows for globally consistent alignment of the observations via an expectation-maximization or iterative optimization approach to correcting the observation poses [20, 37]. That is, we can recover the spatial distribution of the measurements. However, it is often the case that either there is insufficient geometric constraint in the observations to produce confident pose estimates even *post hoc*, or that the computational cost of making the appropriate inferences is infeasible. Uncertainty modeling methods such as Kalman filtering can reduce the severity of the problem, but certainly do not eliminate it.

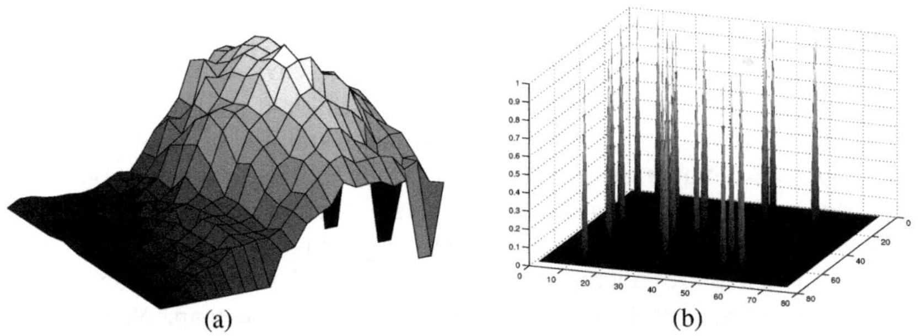

Fig. 1. Mapping: **(a)** Continuous function such as: Radiation, Visual appearance, Elevation, Magnetic field, Temperature, etc. **(b)** Discrete function such as: Mine detection, Lost objects, Holes, Electrical outlets, etc.

Our approach to the pose estimation problem for map building involves the use of two or more robots working in close cooperation. Several authors have also considered the use of marsupial robots or robot teams either in theory or practice [24, 13, 40].

This paper addresses the problem of establishing accurate pose estimates in the context of robotic mapping. The pose estimates can be used to collect accurately localized measurements in their own right, or as a precursor to a system that builds a map. The robot collecting measurements for the map operates in concert with a second robot that acts as an active observer. In our *cooperative localization* [31] scheme, this second robot tracks the motions of the first as it collects data and provides it with the information required to prevent odometric error from accumulating. We can view the robots as being "connected" by a *virtual tether* which is established between the two robots and which enables the task of mapping to be accomplished without significant error and independent of the ground surface conditions and the quality of the odometry estimate. In principle, more than one of these active observers could be used simultaneously, although this is not elaborated on in this paper. Beyond presenting the details of the approach and its implementation, this paper provides a quantitative evaluation validating the effectiveness of this methodology.

The remainder of this paper is structured as follows: Section 2 discusses the general framework in which our approach applies. Section 3 presents related work that addresses the problem of minimizing localization error during exploration. The *cooperative localization* strategy is introduced in Section 4. We then discuss a particular application of our approach to the task of visual landmark learning in Section 5 and experimental results are presented in Section 6. Finally, we discuss open questions and future directions in Section 7.

2 Motivation

The work presented here is motivated by the need to use a mobile robot in order to accurately map a spatially varying property of an unknown, possibly hazardous, environment. Such a property could be a continuous function, see Fig. 1a, over the accessible area such as radiation, temperature, magnetic field variation, elevation, or visual appearance, or the property could be a discrete function, see Fig. 1b, such as presence of mines, lost objects, holes/anomalies on the ground, or electrical outlets. In addition, the property could be a scalar or a vector-valued function. In most cases the sensor used to map arbitrary properties such as those noted above is not itself suitable for the accurate localization of the exploring robot – for example, a radiation meter cannot readily be used to accurately recover the pose of the exploring robot (except in very special cases). In practice, there are two different issues to consider. First the property of interest may vary too slowly to accurately assist in localization of the robot; second the values of the property of interest may be identical in many places, thus making the task of distinguishing two places impossible. Thus, the self-localization ability of a single robot using only the measurements of the function of interest may be poor in the absence of additional sensory apparatus. Furthermore, the terrain being explored may be uneven or otherwise problematic, resulting in wheel slippage, and rendering the odometry unreliable. Our approach employs cooperative localization [29] in order to recover the pose of the exploring robot with high accuracy, independent of the ground surface properties and the reliability of the odometry.

Another motivation for using more than one mobile robot is that several applications require the exploration or inspection of hazardous environments with an attendant risk to the robot doing the work. Such applications include but are not limited to: de-mining rural areas, inspecting nuclear facilities or marking/mapping of chemical spills. In order to improve robustness or reduce the potential cost in such a scenario we can deploy a team of heterogeneous robots consisting of a "base" robot which is equipped with the main computer, a communication module and the robot tracker sensor, and a team of lower-cost "exploring" robots that are equipped with only the mapping sensor (and the target for the robot tracker). In particular, our scheme obviates the need for accurate odometry on the exploring robots. The base robot is always located at a safe area keeping visual and radio contact with the exploring robots. If any of the exploring robots is destroyed the expense is limited, and the mission can continue with the surviving robots. The main advantage of this approach is that the (expensive) base robot is not endangered by moving to an unexplored hazardous environment. In the experiments presented in this paper there is one base robot equipped with the robot tracker sensor and one mapping robot equipped with the target and a measuring device (a camera).

3 Related Work

The problem that we have described is closely related to the problem of simultaneous localization and map-building, wherein the robot is tasked to explore its environment and construct a map [19]. The advantages of collaborative behavior have been examined extensively in the context of biological systems [38].

In the context of terrain coverage in particular, Balch and Arkin were among the first to quantitatively evaluate the utility of inter-robot communication [1]. Mataric was another pioneer in considering the utility of inter-robot communication and teamwork in space coverage [22]. Dudek, Jenkin, Milios and Wilkes proposed a multi-robot mapping strategy akin to that proposed here, but they only considered certain theoretical aspects of the approach as it applied to very large groups of robots. Several authors have also surveyed a range of possible approaches for collaborative robot interactions [3, 9, 8].

More recently, teams of mobile robots are used to reduce the localization error [17, 33]. In most cases the robots use each other to localize only if they meet by chance. Different estimation techniques have been employed to combine the information from the different robots: particle filters have gained popularity [11, 30] together with more traditional Kalman filter estimation [34, 10] or more recently maximum likelihood estimation [15].

A number of authors have considered pragmatic map-making in particular. Most existing approaches operate in the range-sensing domain, where it is relatively straightforward to transform observations from a given position to expected observations from nearby positions, thereby exploiting structural relationships in the data [18, 12, 2]. Such approaches usually differ in when the map is constructed. Off-line approaches post-process the data, usually by applying the *expectation maximization* (EM) paradigm [7] to the task by iteratively refining the map and the estimates of the observation points. On-line methods compute the maximum-likelihood map and robot pose as it explores, and are usually based on Kalman filtering or particle filtering and their extensions [33, 32, 23].

Several authors have investigated vision-based pose estimation and map construction [39, 14]. Se *et al* construct a map by extracting point features from images and localizing them in the world with a stereo camera [35]. Davison performs SLAM using a monocular camera by applying techniques from the structure-from-motion literature [5]. Other authors have considered generic features described implicitly by computing the principal components of sensor observations [26, 25].

Our work is similar to the earlier localization techniques in that it applies probabilistic methods to localization from feature observations. However, it is more similar to the latter techniques in that feature and camera geometry are not modeled explicitly but rather the (possibly complex) interaction of feature and sensor is learned as a function of pose.

4 Cooperative Localization

In previous work, Rekleitis, Dudek and Milios have demonstrated the utility of introducing a second robot to aid in the tracking of the exploratory robot's position [28]. In that work, the robots exchange roles from time to time during the exploration of a polygon-shaped world, thus serving to minimize the accumulation of odometry error. The authors refer to this procedure as *cooperative localization*. We have constructed a tracking device that can estimate the position and orientation of a mobile robot relative to a base robot equipped with the robot tracker sensor. The motion planning strategy is such that at any time one of the robots is stationary while the other robot is moving. The stationary robot acts as an artificial landmark in order for the moving robot to recover its pose with respect to the stationary one. Therefore, a detectable landmark is provided without any modification of the environment.

This paper builds on the results by Rekleitis *et al.* [27] by considering the task of exploring the visual domain. In the following section, we describe the method employed for tracking the position of the robot as it explores. Different types of sensors could be used depending on the required precision of the specific task.

4.1 Tracker-Based Pose Estimation

The Robot Tracker sensor returns three measurements, the triplet $T = [\rho\ \phi\ \theta]$ where ρ is the distance between the two robots, ϕ is the angle at which the observing robot

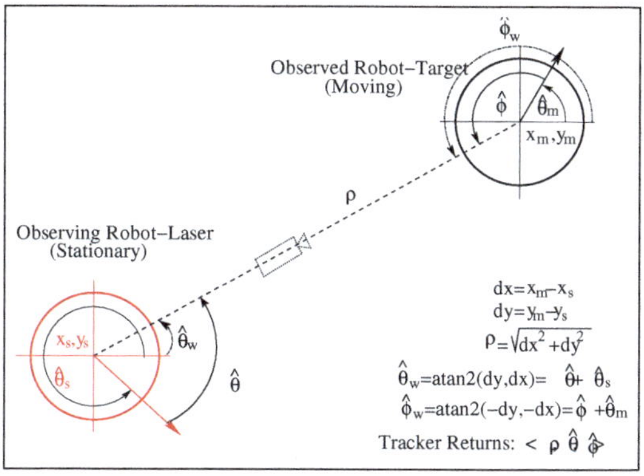

Fig. 2. Pose Estimation via Robot Tracker: Observation of the Moving Robot by the Stationary Robot.

sees the observed robot relative to the heading of the observing robot, and θ is the heading of the observed robot as measured by the observing robot relative to the heading of the observing robot; see Fig. 2. If the stationary robot is equipped with the Robot Tracker then the Pose ($\mathbf{X}_m$) of the moving robot is given by eq.(1), where $[x_s, y_s, \theta_s]^T$ is the pose of the stationary robot.

$$\mathbf{X}_m = \begin{pmatrix} x_m \\ y_m \\ \hat{\theta}_m \end{pmatrix} = \begin{pmatrix} x_s + \rho * \cos(\hat{\theta}_s + \hat{\theta}) \\ y_s + \rho * \sin(\hat{\theta}_s + \hat{\theta}) \\ \pi + \hat{\theta}_s + \hat{\theta} - \hat{\phi} \end{pmatrix} \qquad (1)$$

We have implemented two trackers based on this paradigm. The first operates in the visual domain using a helical target, whereas the second employs a laser range-finder and a geometric target. In the following sections we outline their implementations.

4.2 Implementation 1: Visual Robot Tracker

The first implementation of a visual robot tracker involved the mounting of a helical target pattern on the observed robot, while the observing robot was equipped with a camera.

The bottom part of the target pattern is a series of horizontal circles (in fact, these are cylinders and they project into a linear pattern in the image). This allows the robot to be easily discriminated from background objects: the ratio of spacing between the circles is extremely unlikely to occur in the background by chance. Thus, the presence of the robot is established by a set of lines (curves) with the appropriate length-to-width ratio, and the appropriate inter-line ratios. Figure 3a shows the gray-scale image and Fig. 3b presents the identifying stripes highlighted.

By mounting the observing camera above (or below) the striped pattern of the observed robot, the distance from one robot to the other can be inferred from the height of the stripe pattern in the image[1], due to perspective projection; scaling of the pattern could also be used. The difference in height between the observing camera and the target can be selected to provide the desired trade-off between range of operation and accuracy.

The second component of the target pattern is a helix that wraps once around the observed robot. The elevation of the center of the helix allows the relative orientation of the observed robot to be inferred; see Fig. 3.

In practice, the above process allows the observed robot's pose to be inferred with an accuracy of a few centimeters in position and 3 to 5 degrees in heading [29]. The range of the visual tracker depends on the height of the robots, in our experimental setup the effective range was between 1.5 m to 4.5 m. Experimental results showed the system to be sensitive to partial occlusions and especially to uneven floors. In

[1] After an initial calibration process a look-up table is constructed that relates y-coordinates in the image with distance.

Fig. 3. Robot Tracker: (a) The raw image of the moving robot as observed by the robot tracker. (b) The helical and cylindrical pattern detected in the image. (c) The laser-based tracker and target.

the next section we present a robot tracker sensor based on a laser range finder. The experiments presented in this paper utilized the laser-based robot tracker.

4.3 Implementation 2: Laser Robot Tracker

The second and more accurate implementation employs an *AccuRange*[2] laser range-finder mounted on the observing robot and a three-plane target mounted on the observed robot; see Fig. 3c. The AccuRange laser range-finder produces a range scan in the horizontal plane. The effective range of the Accurange scanner is up to 12 m with an angular resolution of 11.3 points per degree (4096 points for 360 degrees). The manufacturer specifications claim sub-millimeter accuracy per point; in practice we observed less than a centimeter variation regardless of the range. The target is a set of three vertical planes extending from the center of the target at three distinct angles $(100°, 120°, 140°)$ with length of 25cm and height of 30cm. From any direction around the observed robot at least two vertical planes are "visible". We employ the laser range-finder in order to detect the two planes. The intersection of the two planes defines a unique point in a fixed position with reference to the observed robot. Further on, the angle between the two planes combined with their orientations provides an estimate for the heading of the robot. When the approximate position of the observed robot is available (most of the time) then the laser points are filtered around that position and few lines have the correct size to select. At maximum range (12m) between 8 and 12 points are returned per target plane.

The precision of the laser range-finder subsystem is much higher than the precision of the visual tracker. The position estimation is accurate to half a centimeter and the heading estimation error is below one degree.

[2] AccuRange 4000 LineScanner laser range-finder from ACUITY RESEARCH Inc. (see http://www.acuityresearch.com/acculine.html)

5 Application: Landmark Learning and Landmark-Based Pose Estimation

In this section we demonstrate the effectiveness of our approach as it applies to the problem of constructing a visual map of an unknown environment. The visual map entails learning a set of visual landmarks which are useful for the task of estimating the pose of a single robot equipped with a monocular camera. We employ the tracker described in the previous section to properly register the landmark observations in the map, i.e. to provide "ground truth" positions while the robot explores the visual environment. We employ the visual mapping framework described in [36], which tracks the set of points output by a model of visual attention and attempts to construct representations of the landmarks as functions of the pose of the robot. The landmark representations do not recover 3D information about the landmark, nor do they depend on pre-defined camera calibration parameters. Instead, the representations learn the generating function $\mathbf{z} = \mathbf{f}(\mathbf{X}_m)$ that maps robot poses to landmark observations $\mathbf{z}$. This is accomplished by training a set of interpolation networks on a set of landmark training observations and ground truth poses. The remainder of this section will describe the visual mapping approach, and in subsequent sections we will present experimental results illustrating how the tracker can be employed to construct the map.

The learning method is depicted in Fig. 4 and operates as follows (refer to [36] for further details):

1. **Exploration:** One robot tracks the other as it collects images sampling a range of poses in the environment. The pose at which each image is taken is recorded as the estimate given by the tracker.

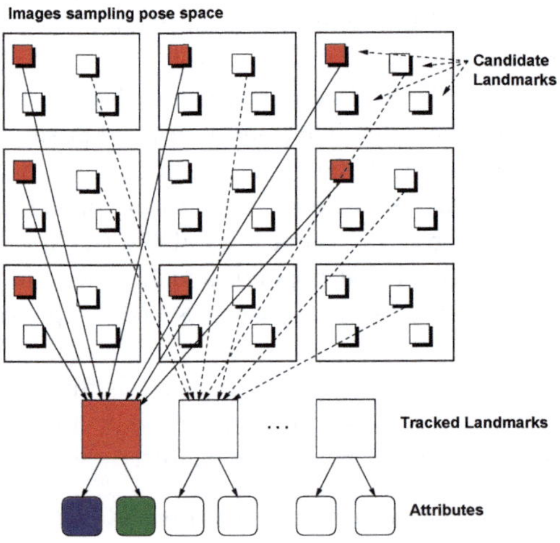

Fig. 4. The off-line training method. Images (large rectangles) are collected sampling the pose space. Landmarks are extracted from the images and matched across the samples. The *tracked landmarks* are parameterized as a function of pose and saved for future pose estimation.

2. **Detection:** *Landmark candidates* are extracted from each image using a model of visual attention. In this work, we identify candidates as local maxima of edge density in the image.
3. **Matching:** *Tracked landmarks* are extracted by tracking visually similar candidate landmarks over the configuration space. The measure of visual similarity we employ is the normalized correlation between candidate image regions.
4. **Parameterization:** The tracked landmarks are parameterized by computing the behavior of a set of computed landmark attributes (for example, position in the image, intensity distribution, edge distribution, etc), versus the "ground truth" pose provided by the tracker. The resulting models are then measured in terms of their *a priori utility* for pose estimation using cross-validation. The cross-validation covariance C, along with the inferred generating function $\mathbf{f}(\mathbf{X}_m)$ for each landmark will subsequently be employed for modeling the feature likelihood distribution $p(\mathbf{z}|\mathbf{X}_m)$.
5. The set of sufficiently useful landmark models is stored for future retrieval.

For the purposes of our experiments, the visual landmarks are initially selected from a subset of the training images using an attention operator that responds to local maxima of edge density in the image. The selected landmark candidates are then tracked over the remaining images along the robot's trajectory by maximizing correlation with the local appearance of the initially detected landmark. The set of matches to a given candidate constitute a *tracked landmark*, and is stored for parameterization and evaluation. Figure 5a depicts an example of a tracked landmark. Each thumbnail corresponds to the landmark as it was observed from the pose

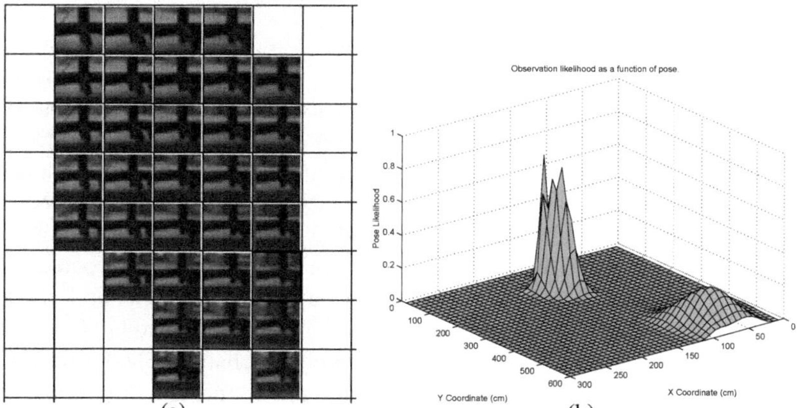

(a) (b)

Fig. 5. (a) A set of observations of an extracted scene feature. The grid represents an overhead view of the pose space of the camera, and feature observations are placed at the pose corresponding to where they were observed. Note that the observations capture variation in feature appearance. (b) The likelihood of an observation as a function of pose.

corresponding to its position in the grid (representing an overhead view of the pose space).

The parameterization of each landmark feature f_i is accomplished by employing a radial basis function regularization framework to model the observation generating function

$$\mathbf{z} = \mathbf{f}_i(\mathbf{X}_m), \tag{2}$$

where $\mathbf{z}$ is a low-dimensional vector-valued representation of the landmark attributes and $\mathbf{X}_m$ is the pose of the robot. In other words, $\mathbf{f}_i(\cdot)$ is the function that predicts the attributes of the landmark i as a function of the pose of the robot. Furthermore, the landmark is evaluated for its utility by computing the covariance C of a randomly sampled subset of leave-one-out cross-validation residuals over the training set. The cross-validation error provides an *a priori* estimate of the utility of the landmark and landmarks with large error can be discarded.

The parameterization of each landmark affords a maximum likelihood prediction of an observation, given an *a priori* pose estimate $\mathbf{X}_m$, as well as a measure of the uncertainty (C) of that prediction. As such, the landmark models are useful for the task of probabilistic robot localization. That is, we define a likelihood function $p(\mathbf{z}|\mathbf{X}_m)$ which allows us to measure the likelihood of an observation $\mathbf{z}$, assuming knowledge of the robot's pose $\mathbf{X}_m$:

$$p(\mathbf{z}_i|\mathbf{X}_m) = k \exp(-0.5(\mathbf{z}_i - \mathbf{f}_i(\mathbf{X}_m))^T C^{-1} (\mathbf{z}_i - \mathbf{f}_i(\mathbf{X}_m))$$

where k is a normalizing constant. This likelihood distribution can be employed in a *Bayesian framework* to infer the probability distribution of $\mathbf{X}_m$ given the observation $\mathbf{z}$:

$$p(\mathbf{X}_m|\mathbf{z}) = \frac{p(\mathbf{z}|\mathbf{X}_m)p(\mathbf{X}_m)}{p(\mathbf{z})} \tag{3}$$

where $p(\mathbf{X}_m)$ represents the prior information about $\mathbf{X}_m$ and $p(\mathbf{z})$ is a constant relative to the independent variable $\mathbf{X}_m$. Several such probability distributions can be generated—one for each observed landmark—and can be combined to obtain a full posterior pose distribution. Note that this framework is more generic than a Kalman filter in that it allows for a multi-modal representation of the pose likelihood.

The set of landmarks observed and computed over the environment during the mapping stage constitutes the visual map and can later be used for accurate single-robot pose estimation. When the robot requires a pose estimate without the aid of the robot tracker, it obtains a camera image and locates the learned landmarks in the image using the predictive model and the tracking mechanism. The differences in appearance and position between the prediction and the observation of each landmark are combined to compute the likelihood of the observation in the Bayesian framework. This process is illustrated in Fig. 6. The maximum posterior pose estimate is recovered by gradient ascent over the observation likelihood as a function of pose. An example likelihood function is plotted at a coarse scale in Fig. 5b. Note that the pose likelihood is a useful measure of confidence in the final estimate allowing for the rejection of outlier pose estimates on the basis of a user-defined threshold.

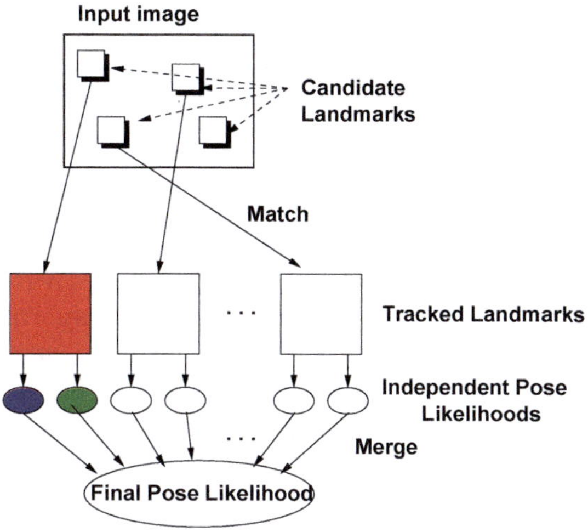

Fig. 6. Pose estimation based on learned visual landmarks. Landmarks (small squares) are extracted from the current camera observation and matched to the previously learned tracked landmarks. Each match generates a pose estimate, all the matches are filtered and combined to generate a final pose estimate.

The cost of computing the posterior distribution will be dependent on any prior pose information that is available. Uninformative priors will imply that every learned landmark will be considered for matching, and successfully matched landmarks will subsequently be evaluated over the search region of the pose space using their generative models. Assuming that landmark matching constitutes the greater computational cost of pose estimation (which is generally the case), the pose estimation process is linear in the number of landmarks in the map. This computational cost can be reduced by exploiting prior pose information to pre-select for matching only those landmarks that are *a priori* likely to be visible.

While this work computes only the mode of the posterior pose distribution, it should be noted that eq.(3) is well suited to alternative representations, such as computation of the posterior over a discretization of the pose space, or computation at the loci of a set of points in a particle filter. Furthermore, the parameterization of the landmark models is generative in nature, eq.(2), so that the representation can be employed in a more approximate approach, such as a Kalman filter or other hypothesis tracker. Such approaches are beyond the scope of this work, but are straightforward to apply in practice.

6 Experimental Results

In this section we present the results of deploying the tracking method for the task of landmark learning using a team of two robots. Two different experiments were conducted in our laboratory. In the first experiment we tested the ability of the team to move through two consecutive rooms while in the second experiment we maximized the area covered using all the available space.

6.1 Experiment 1

Our environment consisted of a laboratory partitioned into two "rooms" by room dividers, with an open doorway connecting them. The first two pictures in Fig. 7 are the robot's-eye-view of the two rooms, and the third picture presents the top view of the floor plan. At the outset, one robot remained stationary while the other used a seed-spreader exploration procedure [21] across the floor, taking image samples at 40cm intervals. When the robot had completed the first room, it moved beyond the door and the stationary robot followed it to the threshold, where it remained stationary while tracking the exploratory robot as it continued its exploration of the second room. In this experiment the robots first map one room, then move to the next room. The accumulated uncertainty is very small due to the sort path taken by the observing (base) robot. More significant is the odometric error that occurred when the observed (mapping) robot ran over a cable. The cooperative localization approach, though, corrected the pose estimate and the visual map was constructed accurately.

a) Odometry versus tracking: The trajectory of the exploratory robot was defined at the outset by a user. However, as the robot explored, accumulated error in odometry resulted in the robot straying from the desired path. The tracking estimate of the stationary robot was provided to the moving robot in order to correct this accumulated error. During the exploration the pose of the robot was corrected based on the observations of the robot tracker. During the experiment the pure odometry estimates were kept for comparison. Figure 8a plots the uncorrected odometric trajectory (plotted as 'x') and the actual trajectory of the robot, as measured by the tracker (plotted as 'o'). For the sake of clarity, the odometric error was reset to zero between the first and second rooms. Figure 8b displays the accumulated odometric error in the second room versus total distance traveled (after it was reset to zero).

b) Tracking versus vision-based pose estimation: Once image samples were obtained using the tracker estimates as ground truth position estimates, it was possible

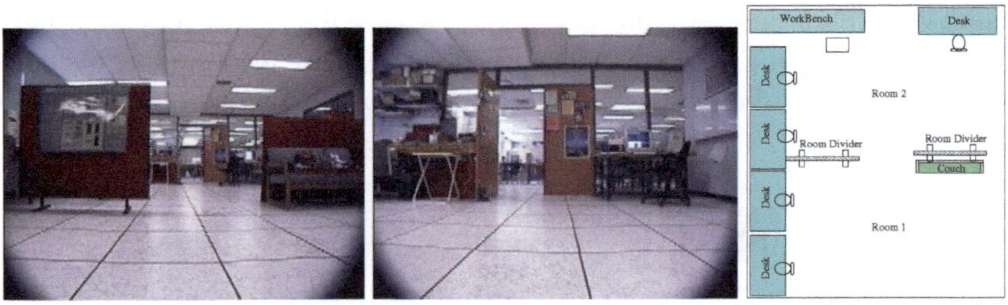

Fig. 7. Views of the two "rooms" as seen by the robot, and the floor plan of the two "rooms".

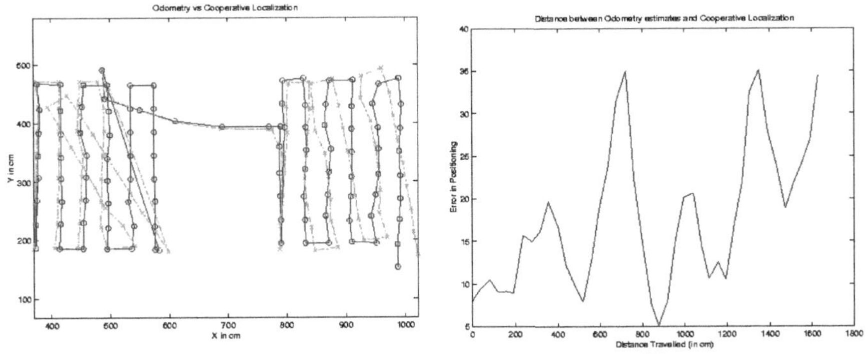

(a) (b)

Fig. 8. (a) Odometric (x) vs. Tracker-corrected (o) trajectories of the robot. (b) Odometric error versus distance traveled.

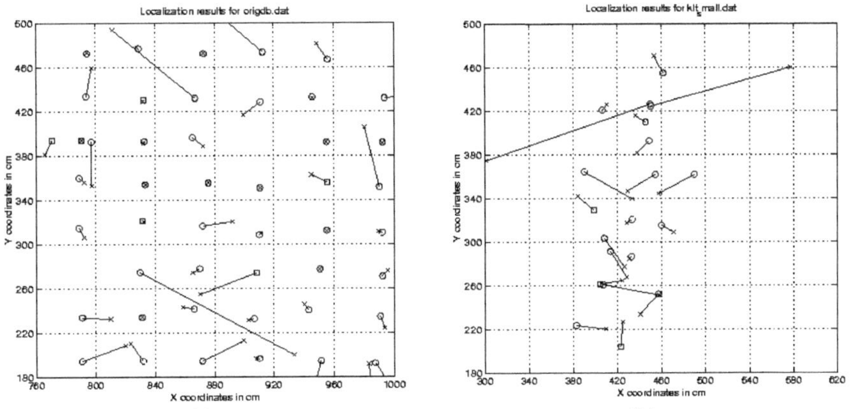

(a) (b)

Fig. 9. (a) Tracker estimates vs. Vision-based estimates for training images. (b) Tracker estimates vs. Image-based estimates for a set of 21 random positions.

to apply our landmark learning framework to the image samples in order to learn a mapping between appearance-based landmarks and the pose of the robot. Figure 9a shows the discrepancies between the pose estimates from the tracker (marked as circles) and the landmark-based vision pose estimator (marked as x) in Room 2. At each position, the two 2-D projections of the alternative pose estimates are joined by a line. While the tracker is clearly more accurate, the quality of the landmark-based pose estimates is sufficient for situations where only one robot is present. There are a few large outliers that can be easily eliminated using prior pose estimates.

Our final stage of this experiment involved navigating the robot to a series of random positions and acquiring image- and tracker-based pose estimates, which are plotted together in Fig. 9b. This final experiment illustrates the use of a multi-sensor estimator in removing outliers. Assuming that the tracker-based position is correct, the mean error in the image-based estimate was 33cm, a large part of which can be attributed to the two significant outliers from nearly the same position. The purpose of this experiment is to verify the accuracy of pure vision based localization. Clearly, combining probabilistically the vision based estimated and an odometry based prior estimate results in a more accurate pose estimation.

6.2 Experiment 2

A second experiment was performed where the two robots explored a single large room. At the outset, one robot remained stationary while the other used a seed-spreader exploration procedure [21] across the floor, taking image samples at 25cm intervals, and in four orthogonal viewing directions, two of which are illustrated in Fig. 10. In this experiment the goal was to map a larger open area and to obtain maps in different orientations.

As before, the trajectory of the exploratory robot was defined at the outset by a user. However, as the robot explored, accumulated error in odometry resulted in the robot straying from the desired path. The differential drive configuration of the exploratory robot, coupled with frequent rotations to capture the four viewing directions, led to a rapid, and somewhat systematic degradation in dead reckoning, as illustrated in Fig. 11a, where the uncorrected odometric trajectory is plotted as a dash-dotted line, and the actual trajectory of the robot, as observed by the tracker, is plotted as a solid line. The accumulated odometric error is plotted versus total distance traveled in Fig. 11b.

Once image samples were obtained using the tracker estimates as ground truth position estimates, it was possible to apply our landmark learning framework to the image samples in order to learn a mapping between appearance-based landmarks

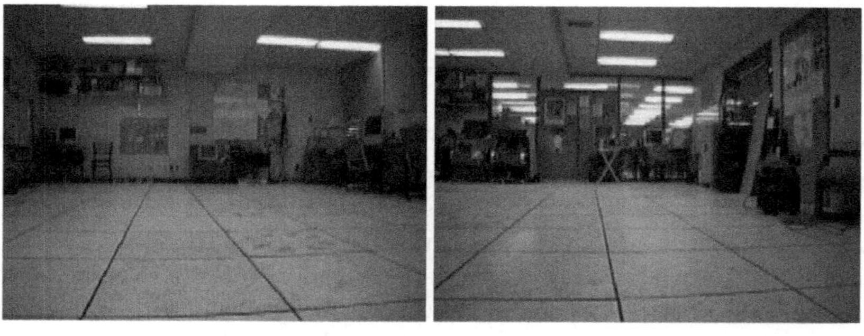

Fig. 10. Opposing views of the lab as seen by the exploring robot.

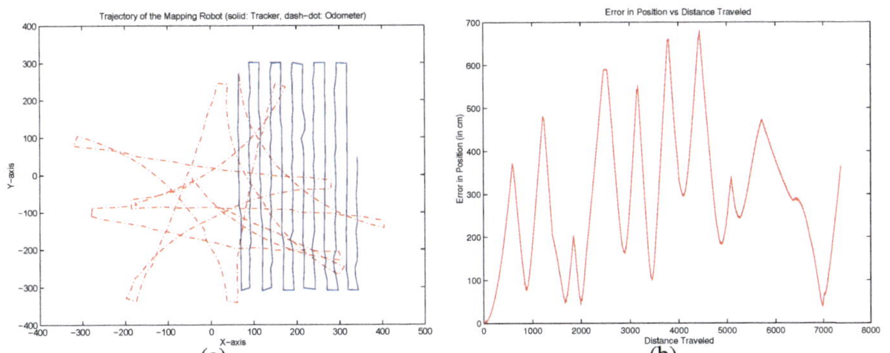

(a) (b)

Fig. 11. In this experiment the robot took pictures in four orientation; the higher number of rotations increased non-linearly the odometric error. (a) Odometric (denoted by dash-dotted line) vs. Tracker-corrected (denoted by a solid line) trajectories of the robot. (b) Odometric error versus distance traveled.

and the pose of the robot. Training was applied separately to each of the four viewing directions, developing a set of tracked landmark observations. Each thumbnail corresponds to an observation of the landmark. The relative position of the thumbnail in the image corresponds to the physical location from which it was acquired (that is, the positions of the thumbnails constitute an overhead view of locations visited by the robot). Note that in some locations there is no observation, as the tracker was not able to locate a visual match to the landmark.

Again, the final stage of our experiment involved navigating the robot to a series of 93 random positions and acquiring images along the four orthogonal viewing directions. Image- and tracker-based maximum likelihood pose estimates were then generated for one of the viewing directions, and outliers removed on the basis of a likelihood threshold. Of the 93 observations, 4 estimates were rejected. In general, these outliers corresponded to observations where the robot was very close to the wall it was facing. One would expect that an observation from a different viewing direction would return an estimate with higher confidence. We have omitted this application for the sake of brevity.

The remaining 89 image-based estimates of high confidence are plotted with their associated tracker-based estimates in Fig. 12a. Assuming that the tracker-based position is correct, the mean error in the image-based estimate was 17cm, (7.7cm in the x direction vs. 15cm in the y direction). The larger error in the y direction corresponds to the fact that the camera was pointed parallel to the positive y axis, and changes in observations due to forward motion are not as pronounced as changes due to side-to-side motion. The smallest absolute error was 0.49cm, which is insignificant compared to the "ground truth" error, and the largest error was 76cm. Note that most of the larger errors occur for large values of y. This is due to the fact that the camera was closest to the wall it was facing at these positions y, and as has been mentioned, tracking scene features over 25cm pose intervals became difficult.

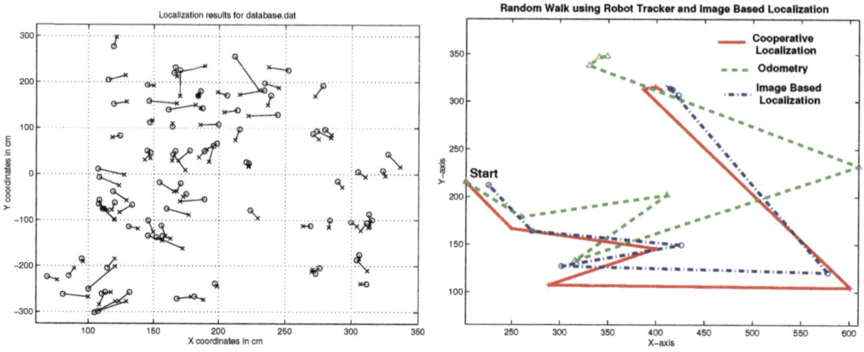

Fig. 12. (a) Tracker estimates vs. Image-based estimates for a set of 93 random positions. (b) The trajectory of the moving robot based on odometry estimates (triangles connected with a dashed line), the robot tracker cooperative localization ('+' connected with a solid line) and the image based localization ('o' connected with a dash-dotted line).

6.2.1 Random Walk

Figure 12b presents a random walk of the exploring robot through the mapped environment. The robot starts at a random location (marked as a "*"). Initially the odometry estimate is set to the value of the robot tracker estimate at that starting position, the pose estimate from the vision based system is approximately 30cm to the right of the robot tracker estimate. The robot took seven random steps and the three estimated trajectories are presented in Fig. 12b. First the odometer estimate (marked as triangles connected with a dashed line) is plotted; second, the robot tracker estimate (marked as "+" connected by a solid line), and third the visual pose estimator results (marked as "o" connected with a dash-dotted line). The robot tracker estimate provides a close approximation to ground truth at the end of the random walk the disparity between the robot tracker and the visual pose estimator is 17.5cm and between the robot tracker and the odometer is 68cm. The much higher disparity is a result of an increase in the accumulated error in orientation. The goal of this experiment was to evaluate the accuracy of the vision based localization; as noted earlier the fusion of odometry data and the visual pose estimation resulted in higher positioning accuracy.

7 Conclusions

We introduced a method for the automatic mapping of an arbitrary environment which utilizes *cooperative localization* in order to maintain a *virtual tether* between two robots as one explores the environment and the other tracks its pose. The implementation we presented relies on a mounted target whose pose can be estimated using a laser range-finder. The need for such an approach to maintaining a "ground truth" estimate of the exploring robot is validated by the magnitude of the accrued odometric error in our experimental results. Furthermore, we validate the utility of a set of learned landmarks for localization when the second robot cannot be deployed. This demonstrates conclusively that the virtual tether provides more accurate pose estimates, and hence a more accurate appearance-based map, than could be achieved with the robots operating independently.

Our approach does not eliminate positional error for the mapping robot but significantly reduces it. While the base robot does not move, the pose uncertainty of the mapping robot is equal to the uncertainty of the robot tracker sensor (a few centimeters) for an area as large as $225m^2$ for a robot tracker with range $12m$. Every time the base robot moves, its positional uncertainty increases by the positional uncertainty of the mapper plus the uncertainty of the robot tracker sensor. An analysis of the accumulation of the uncertainty is beyond the scope of this paper. In many environments, environmental features may be deemed accurate enough to be used to assist in localization. Such a SLAM formulation would further improve performance.

Our work demonstrates how co-operative localization can be employed as a mechanism for constructing a new map representation of the environment. One aspect of the problem that we did not consider is that of including feedback from the visual representation to further augment the robot's pose estimate as it explores. This approach would be an example of simultaneous localization and mapping (SLAM) using multiple sources of information. We will consider this problem in future work.

The particular map we produce, an appearance-based representation of the environment, allows a single robot to accurately estimate its position on subsequent visits to the same area. While such single-robot pose estimates are not as accurate as when two robots are used, their accuracy is substantially ameliorated by the fact that two robots were used in the initial mapping stage. The use of an appearance-based model obviates most dependences on the particular geometry of the local environment. In principle, if a pair of robots was used in this subsequent stage the accuracy of the estimates could be further improved, but the extent of this advantage remains to be determined.

It would appear that these advantages become even more profound if more than two robots are used for position estimation and mapping. In the particular algorithmic scheme the use of many more robots would be an issue, but it seems that several feasible solutions can be formulated; we hope to examine this problem in the future.

References

1. Balch, T., Arkin, R.C.: Communication in reactive multiagent robotic systems. Autonomous Robots 1(1), 27–52 (1994)
2. Burgard, W., Fox, D., Moors, M., Simmons, R., Thrun, S.: Collaborative multi-robot exploration. In: Proc. IEEE Int. Conf. on Robotics and Automation, pp. 476–481 (2000)
3. Cao, Y.U., Fukunaga, A.S., Kahng, A.B., Meng, F.: Cooperative mobile robotics: Antecedents and directions. In: Proc. IEEE/RSJ IROS, vol. 1, pp. 226–234 (1995)

4. Choset, H., Burdick, J.: Sensor based planning, part ii: Incremental construction of the generalized voronoi graph. In: Proc. IEEE Conf. on Robotics and Automation, pp. 1643–1648 (1995)
5. Davison, A.: Real-time simultaneous localisation and mapping with a single camera. In: Proc. IEEE Int. Conf. on Computer Vision (2003)
6. Dellaert, F., Burgard, W., Fox, D., Thrun, S.: Using the condensation algorithm for robust, vision-based mobile robot localization. In: IEEE Computer Society Conf. on Computer Vision and Pattern Recognition (1999)
7. Dempster, A.P., Laird, N.M., Rubin, D.B.: Maximum likelihood from incomplete data via the EM algorithm (with discussion). Journal of the Royal Statistical Society series B 39, 1–38 (1977)
8. Dudek, G., Jenkin, M.: Computational Principles of Mobile Robotics. Cambridge University Press, Cambridge (2000) ISBN: 0521568765
9. Dudek, G., Jenkin, M., Milios, E., Wilkes, D.: A taxonomy for multi-agent robotics. Autonomous Robots 3, 375–397 (1996)
10. Fenwick, J.W., Newman, P.M., Leonard, J.J.: Cooperative concurrent mapping and localization. In: Proc. IEEE Int. Conf. on Robotics and Automation, vol. 2, pp. 1810–1817 (2002)
11. Fox, D., Burgard, W., Kruppa, H., Thrun, S.: A probabilistic approach to collaborative multi-robot localization. Autonomous Robots 8(3), 325–344 (2000)
12. Fox, D., Burgard, W., Thrun, S.: Active markov localization for mobile robots. Robotics and Autonomous Systems (RAS) 25, 195–207 (1998)
13. Fox, D., Ko, J., Konolige, K., Limketkai, B., Schulz, D., Stewart, B.: Distributed multi-robot exploration and mapping. Proceedings of the IEEE 94(7), 1325–1339 (2006)
14. Hagen, S., Krose, B.: Towards global consistent pose estimation from images. In: Proc. IEEE/RSJ Int. Conf. on Intelligent Robots and Systems, pp. 466–471 (2002)
15. Howard, A., Mataric, M.J., Sukhatme, G.S.: Localization for mobile robot teams using maximum likelihood estimation. In: Proc. IEEE/RSJ Int. Conf. on Intelligent Robots and Systems, pp. 434–459 (2002)
16. Kuipers, B., Byun, Y.-T.: A robot exploration and mapping strategy based on a semantic hierachy of spatial representations. Robotics and Autonomous Systems 8, 46–63 (1991)
17. Kurazume, R., Hirose, S.: An experimental study of a cooperative positioning system. Autonomous Robots 8(1), 43–52 (2000)
18. Leonard, J.J., Durrant-Whyte, H.F.: Mobile robot localization by tracking geometric beacons. IEEE Transactions on Robotics and Automation 7(3), 376–382 (1991)
19. Leonard, J.J., Feder, H.J.S.: A computationally efficient method for large-scale concurrent mapping and localization. In: The 9th Int. Symposium on Robotics Research, pp. 169–176 (2000)
20. Lu, F., Milios, E.: Optimal global pose estimation for consistent sensor data registration. In: Int. Conf. in Robotics and Automation, vol. 1, pp. 93–100. IEEE, Los Alamitos (1995)
21. Lumelsky, V., Mukhopadhyay, S., Sun, K.: Sensor-based terrain acquisition: The 'sightseer' strategy. In: Proc. IEEE Conf. on Decision and Control Including The Symposium on Adaptive Processes, vol. 2, pp. 1157–1161 (1989)
22. M.: Environmental learning using a distributed representation. In: IEEE Int. Conf. on Robotics and Automation, vol. 1, pp. 402–406 (1990)
23. Montemerlo, M., Thrun, S., Koller, D., Wegbreit, B.: FastSLAM: A factored solution to the simultaneous localization and mapping problem. In: Proc. AAAI National Conf. on Artificial Intelligence (2002)

24. Murphy, R.R., Ausmus, M., Bugajska, M., Ellis, T., Johnson, T., Kelley, N., Kiefer, J., Pollock, L.: Marsupial-like mobile robot societies. In: Agents, pp. 364–365 (1999)
25. Nayar, S.K., Murase, H., Nene, S.A.: Learning, positioning, and tracking visual appearance. In: Proc. IEEE Conf. of Robotics and Automation, pp. 3237–3244 (1994)
26. Pourraz, F., Crowley, J.L.: Continuity properties of the appearance manifold for mobile robot position estimation. In: Proc. IEEE Computer Society Conf. on Pattern Recognition Workshop on Perception for Mobile Agents (1999)
27. Rekleitis, I.M.: Cooperative Localization and Multi-Robot Exploration. PhD thesis, School of Computer Science, McGill University (2003)
28. Rekleitis, I.M., Dudek, G., Milios, E.: Multi-robot collaboration for robust exploration. In: Proc. Int. Conf. in Robotics and Automation, pp. 3164–3169 (2000)
29. Rekleitis, I.M., Dudek, G., Milios, E.: Multi-robot collaboration for robust exploration. Annals of Mathematics and Artificial Intelligence 31(1-4), 7–40 (2001)
30. Rekleitis, I.M., Dudek, G., Milios, E.: Probabilistic cooperative localization and mapping in practice. In: Proc. IEEE Int. Conf. on Robotics and Automation, pp. 1907–1912 (2003)
31. Rekleitis, I.M., Dudek, G., Milios, E.E.: On multi agent exploration. In: Proc. Vision Interface, pp. 455–461 (1998)
32. Roumeliotis, S.I., Bekey, G.A.: Bayesian estimation and kalman filtering: A unified framework for mobile robot localization. In: Proc. IEEE Int. Conf. on Robotics and Automation, pp. 2985–2992 (2000)
33. Roumeliotis, S.I., Bekey, G.A.: Collective localization: A distributed kalman filter approach to localization of groups of mobile robots. In: Proc. IEEE Int. Conf. on Robotics and Automation, pp. 2958–2965 (2000)
34. Roumeliotis, S.I., Bekey, G.A.: Distributed multirobot localization. IEEE Transactions on Robotics and Automation 18(5), 781–795 (2002)
35. Se, S., Lowe, D.G., Little, J.: Mobile robot localization and mapping with uncertainty using scale-invariant visual landmarks. Int. Journal of Robotics Research 21(8), 735–758 (2002)
36. Sim, R., Dudek, G.: Learning generative models of scene features. Int. Journal of Computer Vision 60(1), 45–61 (2004)
37. Thrun, S., Fox, D., Burghard, W.: A probabilistic approach to concurrent mapping and localization for mobile robots. Autonomous Robots 5, 253–271 (1998)
38. Tinbergen, N.: The animal in its world; explorations of an ethologist, 1932-1972. Harvard University Press, Cambridge (1972)
39. Ulrich, I., Nourbakhsh, I.: Appearance-based place recognition for topological localization. In: Proc. IEEE Int. Conf. on Robotics and Automation, vol. 2, pp. 1023–1029 (2000)
40. Wurm, K.M., Dornhege, C., Eyerich, P., Stachniss, C., Nebel, B., Burgard, W.: Coordinated exploration with marsupial teams of robots using temporal symbolic planning. In: Proc. IEEE/RSJ Int.l Conf. on Intelligent Robots and Systems (2010)

Sliding-Mode Velocity and Yaw Control of a 4WD Skid-Steering Mobile Robot

Eric Lucet, Christophe Grand, and Philippe Bidaud

Abstract. The subject of this paper is the design and implementation of a robust dynamic feedback controller, based on the dynamic model of the four-wheel skid-steering RobuFAST A robot, undergoing high-speed turns. The control inputs are respectively the linear velocity and the yaw angle. The main objective of this paper is to formulate a sliding mode controller, robust enough to obviate the knowledge of the forces within the wheel-soil interaction, in the presence of sliding phenomena and ground-level fluctuations. Finally, experiments are conduced on a slippery ground to ascertain the efficiency of the control law.

1 Introduction

This paper considers the problem of a robust control of high-speed wheeled robots maneuvering on slippery grounds with varying properties.

The dynamic control of skid-steering robots was studied in particular in [1] using a dynamic feedback linearization paradigm for a model-based controller which minimizes lateral skidding by imposing the longitudinal position of the instantaneous center of rotation. Another algorithm reported in [2], offers a good robustness considering uncertainties on the robot dynamic parameters. In addition to the non-holonomic constraint of the controller designed by Caracciolo, the authors use an oscillator signal [3] for kinematic control.

We suggest a strategy based on the sliding-mode theory. The sliding-mode control law—or, more precisely, controller with variable structure generating a sliding regime—aims to obtain, by feedback, a dynamics widely independent from that

Eric Lucet
Robosoft, Technopole d'Izarbel, F-64210 Bidart, France
e-mail: eric.lucet@robosoft.fr

Christophe Grand · Philippe Bidaud
ISIR-UPMC, UMR 7222, F-75005 Paris, France
e-mail: christophe.grand@upmc.fr

J. Angeles et al. (Eds.): Brain, Body and Machine, AISC 83, pp. 247–258.
springerlink.com

of the process and its possible variations. Hence, the controller can be considered as belonging to the class of robust controllers. The sliding-mode control appears attractive for the handling of nonlinear and linear systems, multivariable and single-variable systems, as well as model or trajectory pursuit problems and problems of regulation.

Sliding-mode control allows a decoupling design procedure and good disturbance rejection. This control scheme is robust to the uncertainties in the dynamic parameters of the system, and is easily implementable. Indeed, robust control is widely used in the literature; particular [4] and [5] propose examples of dynamic sliding-mode controllers without taking into account the vehicle dynamics. In [6], and then [7], the authors consider the dynamics model of a unicycle system during the implementation of their control law by using the kinematic nonholonomic non-skidding constraint. The non compliance with nonholonomic constraints in real conditions is taken into account in [8]. However, the problem is formalized for the particular case of the partially linearized dynamics model of a unicycle robot.

Using the sliding-mode theory, we suggest here a new approach to control a fast skid-steering mobile robot with wheel torques as inputs, based on its dynamics model. The objective is to force the mobile robot to follow a desired path at relatively high speed, by controlling its yaw angle and its longitudinal velocity. The ground considered is nominally horizontal and relatively smooth as compared with the wheel dimensions. If most of the control laws consider that the conditions of movement without slippage are satisfied, this hypothesis is not valid at high speed, where wheel slippage becomes significant, thus reducing the robot stability. The implemented controller will have to be robust with respect to these phenomena in order to ensure an efficient path-following.

2 Application to a Skid-Steering Mobile Robot

Because it has proved to be robust enough to obviate the modeling of the forces in the wheel-soil interaction in the presence of slippage, a sliding-mode controller is applied to a skid-steering mobile robot. This scheme ensures the control of the hending velocity and the yaw angle.

2.1 System Modeling

Considering the integer $i \in [1;N]$ with N denoting the number of wheels of the skid-steering robot, let us define the two generalized torques τ_u and τ_θ, uniformly distributed throughout the torques τ_i of each wheel i according to the equations:

$$\tau_u = \sum_{i=1}^{N} \tau_i; \quad \tau_\theta = \sum_{i=1}^{N} \frac{-w_i}{R} \tau_i \tag{1}$$

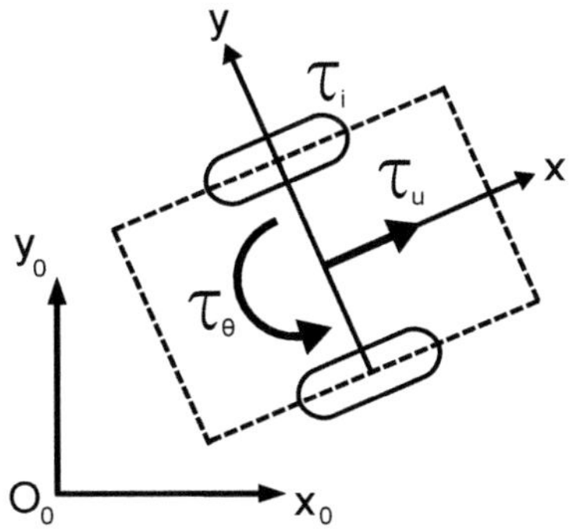

Fig. 1. Torques dispatching

2.2 *Control of the Yaw Angle*

2.2.1 Design of the Control Law

In the case of a skid-steering robot, let us express the yaw movement dynamics from the equations of the general dynamics:

$$I_z\dot{r} = \sum_{i=1}^{N}(-w_iF_{xi} + l_iF_{yi}) \tag{2}$$

Applying Newton's second law to the ith wheel, we have:

$$I_\omega\dot{\omega}_i = \tau_i - RF_{xi} \tag{3}$$

with R the wheel radius and I_w its centroidal moment of inertia, assumed to be the same for all the wheels.

Considering the torque definition τ_θ and equations (3) and (2), we have:

$$\dot{r} = \lambda\,\tau_\theta + \boldsymbol{\lambda}_\theta\dot{\omega} + \mathbf{D}_\theta\mathbf{F}_y \tag{4}$$

with:

$$\lambda = \frac{1}{I_z};\boldsymbol{\lambda}_\theta = \frac{I_\omega}{I_zR}[\cdots w_i\cdots];\dot{\omega} = [\cdots\dot{\omega}_i\cdots]^T;\mathbf{D}_\theta = \frac{1}{I_z}[\cdots l_i\cdots];\mathbf{F}_y = [\cdots F_{yi}\cdots]^T$$

The correction of the vehicle steering does not permit the system to converge to the desired trajectory. It is also necessary to correct the lateral error; otherwise, the system will aim towards a movement parallel to the reference path, not necessarily reaching it. This is why we are going to modify the desired yaw angle, as proposed in other works [9].

The robot has to follow the path, the reference point P being all the time the projection of its centre of mass G on this one. To take into account the lateral error, we add to θ_d a term limited between $-\frac{\pi}{2}$ and $\frac{\pi}{2}$ excluded, increasing with the lateral

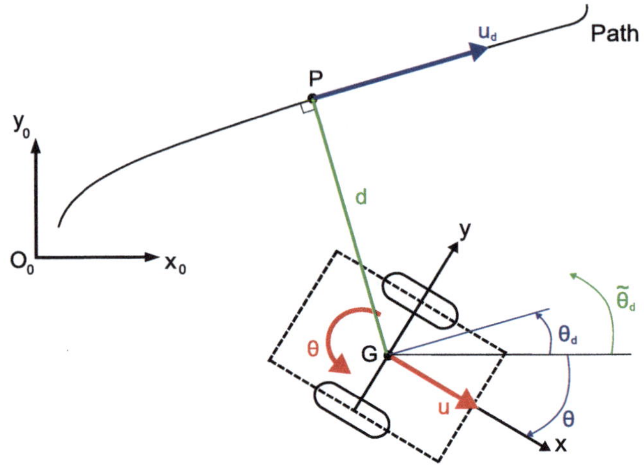

Fig. 2. Path following parameters

error d, the function defining this term being also odd to permit a similar behavior on both sides of the path. We thus define the modified desired steering angle $\tilde{\theta}_d$ such as:

$$\tilde{\theta}_d = \theta_d + \arctan\left(\frac{d}{d_0}\right) \tag{5}$$

with d_0 a positive gain of regulation of the intensity of the correction of the lateral distance d.

For the implementation of the controller, we proceed to a temporary change of the control variable by replacing the generalized torque τ_θ. To this end, we introduce $c_{d\theta}$, which represents the control law to be applied and $n(\theta, r, \dot{r})$ the uncertainty function on θ, r and $\dot{r}$ in the dynamics equations. We thus have the relation:

$$\dot{r} = c_{d\theta} - n(\theta, r, \dot{r}) \tag{6}$$

The control law is chosen as:

$$c_{d\theta} = \dot{r}_d + K_p^\theta \varepsilon_\theta + K_d^\theta \dot{\varepsilon}_\theta + \sigma_\theta \tag{7}$$

which includes four terms:

- $\dot{r}_d$, the second derivative of $\tilde{\theta}_d$, an anticipative term;
- $\varepsilon_\theta = \tilde{\theta}_d - \theta$, the yaw-angle error;
- K_p^θ and K_d^θ, two positive constants that permit to define the settling time and the overshoot of the closed-loop system;
- σ_θ, the sliding-mode control law.

2.2.2 Error State Equation

The second derivative of ε_θ is given below:

$$\begin{aligned}\ddot{\varepsilon}_\theta &= \dot{\tilde{r}}_d - \dot{r} = \dot{\tilde{r}}_d - c_{d\theta} + n \\ &= \dot{\tilde{r}}_d - \left(\dot{\tilde{r}}_d + K_p^\theta \varepsilon_\theta + K_d^\theta \dot{\varepsilon}_\theta + \sigma_\theta\right) + n \\ &= -K_p^\theta \varepsilon_\theta - K_d^\theta \dot{\varepsilon}_\theta + (n - \sigma_\theta)\end{aligned} \tag{8}$$

Let the error state vector be: $\mathbf{x} = \left[\,\varepsilon_\theta, \dot{\varepsilon}_\theta\,\right]^T$, the state equation then becoming

$$\dot{\mathbf{x}} = \mathbf{A}\mathbf{x} + \mathbf{B}\,(n - \sigma_\theta) \tag{9}$$

with:

$$\mathbf{A} = \begin{pmatrix} 0 & 1 \\ -K_p^\theta & -K_d^\theta \end{pmatrix} \; ; \; \mathbf{B} = \begin{pmatrix} 0 \\ 1 \end{pmatrix}$$

If $\sigma_\theta = 0$ (and so $n = 0$, without model error to correct), the system is linear, and we choose the values of K_p^θ and K_d^θ as $K_p^\theta = \omega_n^2$ and $K_d^\theta = 2\zeta\omega_n$ in order to define a second-order system. ω_n is the pulsation and ζ the damping ratio. To define numerical values, the 5% settling time T_r is introduced: $T_r = \frac{4.2}{\zeta\omega_n}$.

2.2.3 Stability Analysis

To approach the problem of the stability of the closed-loop system, the pursuit of the desired yaw angle $\tilde{\theta}_d$ can be studied by using the candidate Lyapunov function $V = \mathbf{x}^T\mathbf{P}\mathbf{x}$, with $\mathbf{P}$ a positive definite symmetric matrix. According to the Lyapunov theorem, [10], the state $\mathbf{x} = \mathbf{0}$ is stable if and only if:

$$V(0) = 0 \; ; \; \forall \mathbf{x} \neq \mathbf{0} \;\; V(\mathbf{x}) > 0 \; and \; \dot{V}(\mathbf{x}) < 0 \tag{10}$$

The first two foregoing relations being verified at once, it remains to establish the third. From eq.(9), we compute the derivative:

$$\begin{aligned}\dot{V}(\mathbf{x}) &= \dot{\mathbf{x}}^T\mathbf{P}\mathbf{x} + \mathbf{x}^T\mathbf{P}\dot{\mathbf{x}} \\ &= \left(\mathbf{x}^T\mathbf{A}^T + n\mathbf{B}^T - \sigma_\theta\mathbf{B}^T\right)\mathbf{P}\mathbf{x} + \mathbf{x}^T\mathbf{P}\left(\mathbf{A}\mathbf{x} + \mathbf{B}n - \mathbf{B}\sigma_\theta\right) \\ &= \mathbf{x}^T\left(\mathbf{A}^T\mathbf{P} + \mathbf{P}\mathbf{A}\right)\mathbf{x} + 2\mathbf{x}^T\mathbf{P}\mathbf{B}\,(n - \sigma_\theta)\end{aligned} \tag{11}$$

The last equality is obtained by considering that $s = \mathbf{B}^T\mathbf{P}\mathbf{x}$ is a scalar, so $\mathbf{B}^T\mathbf{P}\mathbf{x} = \mathbf{x}^T\mathbf{P}\mathbf{B}$. Then, the matrix $\mathbf{P}$ is computed to obtain the equation (12) below:

$$\mathbf{A}^T\mathbf{P} + \mathbf{P}\mathbf{A} = -\mathbf{Q}_l \tag{12}$$

with $\mathbf{Q}_l$ a positive defined symmetric matrix; it is the Lyapunov equation. In that case, the equation (11) is reformulated:

$$\dot{V} = -\mathbf{x}^T\mathbf{Q}_l\mathbf{x} + 2\mathbf{x}^T\mathbf{P}\mathbf{B}\,(n - \sigma_\theta)$$

It is necessary that $\dot{V}$ be negative for stability. The first term of the right-hand side of the above equation is negative, while the second term vanishes if $\mathbf{x}$ lies in the kernel of $\mathbf{B}^T\mathbf{P}$. Outside the kernel, the second term has to be as small as possible. Let us define $s = \mathbf{B}^T\mathbf{P}\mathbf{x}$. The equality $s = 0$ is the ideal case, represents the hyperplane defining the sliding surface. Keeping the sliding surface s equal to zero is then equivalent to the pursuit of the vector of the desired states, the error state vector $\mathbf{x}$ being zero. As this surface reaches the origin, the static error ε_θ will be equal to zero.

We suggest the sliding-mode control law σ_θ:

$$\sigma_\theta = \mu \frac{s}{|s|} \tag{13}$$

where we use the norm of s, and μ is a positive scalar. This choice leads to:

$$s(n - \sigma_\theta) = sn - \mu \frac{s^2}{|s|} = sn - \mu |s| \leq |s| (|n| - \mu)$$

Thus, the conditions of convergence are: $|n| \leq n_{Max} < \infty$ and a choice of $\mu > n_{Max}$ which guarantee the Lyapunov theorem hypotheses. Stability is guaranteed if we adopt the control law (13).

Finally, we have the control law:

$$c_{d\theta} = \dot{\bar{r}}_d + K_p^\theta \varepsilon_\theta + K_d^\theta \dot{\varepsilon}_\theta + \mu \frac{s}{|s|} \tag{14}$$

with $s = \mathbf{B}^T\mathbf{P}\mathbf{x}$.

2.2.4 Solution the Lyapunov Equation

To solve the equation (12), the matrix $\mathbf{Q}_l$ is chosen as:

$$\mathbf{Q}_l = \begin{pmatrix} a & 0 \\ 0 & b \end{pmatrix}$$

with $a > 0$ and $b > 0$. With matrix $\mathbf{A}$ determined, matrix $\mathbf{P}$ is:

$$\mathbf{P} = \begin{pmatrix} \frac{1.05b}{\zeta^2 T_r} + \frac{5a\zeta^2 T_r}{21} + \frac{aT_r}{16.8} & \frac{a\zeta^2 T_r{}^2}{35.28} \\ \frac{a\zeta^2 T_r{}^2}{35.28} & \frac{bT_r}{16.8} + \frac{a\zeta^2 T_r^3}{296.352} \end{pmatrix} \tag{15}$$

We determine the influence of the $\mathbf{Q}_l$ matrix components. As previously defined, the equation of the sliding hyperplane is:

$$s = \mathbf{B}^T\mathbf{P}\mathbf{x} = p_{21}\varepsilon_\theta + p_{22}\dot{\varepsilon}_\theta$$

with p_{21} and p_{22} being two entries of the positive-definite and symmetric matrix $\mathbf{P}$ occurring in the expression of the candidate Lyapunov function. Here, this hyperplane is a straight line.

At the neighborhood of this straight line, we have: $s = p_{21}\varepsilon_\theta + p_{22}\dot{\varepsilon}_\theta = 0$, whose integral is:

$$\varepsilon_\theta(t) = \varepsilon_\theta(\tau)\exp\left[\left(-p_{21}/p_{22}\right)(t-\tau)\right]$$

with $\varepsilon_\theta(\tau)$ a real constant which depends on the initial conditions at $t = \tau$, when the system arrives at the neighborhood of the sliding straight line.

Consequently, we derive from this solution that to correct the error, it is necessary to increase the value of p_{21} and to decrease the value of p_{22}. According to expression (15) for $\mathbf{P}$, we know these two parameters. Hence, to eliminate quickly the position error, it is necessary to increase the value of a and to decrease that of b. As far as the sliding straight line is concerned, this modification of the various coefficients increases the straight line slope.

2.3 Control of the Longitudinal Velocity

We use the dynamics model according to the longitudinal axis from the equations of the general dynamics:

$$M(\dot{u} - rv) = \sum_{i=1}^{N} F_{xi} \tag{16}$$

From the definition of the torque τ_u and equations (3) and (16), we solve for the longitudinal acceleration:

$$\dot{u} = \gamma\tau_u + \Lambda_u \sum_{i=1}^{N} \dot{\omega}_i + rv \tag{17}$$

with:

$$\gamma = \frac{1}{MR}; \quad \Lambda_u = \frac{-I_\omega}{MR}$$

As stated previously, c_u is the control law and $m(u,\dot{u})$ is a function of uncertainties on u and $\dot{u}$ in the equations of the system dynamics. We have the following relationship:

$$\dot{u} = c_u - m(u,\dot{u}) \tag{18}$$

with the control law defined as:

$$c_u = \dot{u}_d + K_p^u \varepsilon_u + \sigma_u \tag{19}$$

and:

- $\dot{u}_d$, an anticipative term;
- $\varepsilon_u = u_d - u$, the velocity error;
- K_p^u, a positive constant that permits to define the settling time of the closed-loop system;
- σ_u, the sliding-mode control law.

Using the Lyapunov candidate function $V = (1/2)\,\varepsilon_u^2$, it can be immediately verified that the stability of the system is guaranteed by the choice of the sliding-mode control law$\sigma_u = \rho \frac{\varepsilon_u}{|\varepsilon_u|}$, with ρ a positive scalar, large enough to compensate the uncertainties on the longitudinal dynamics of the system.

2.4 *Expression of the Global Control Law*

In practice, uncertainty about the dynamics of the system to control leads to uncertainty in the sliding hyperplane $s = 0$. As a consequence $s \neq 0$ and the sliding control law s, which has a behavior similar to the signum function, induces oscillations while trying to reach the sliding surface $s = 0$ with a theoretically zero time. These high-frequency oscillations around the sliding surface, called chattering, increase the energy consumption and can damage the actuators. In order to reduce them, we can replace the signum function by an arctan function or, as chosen here, by adding a parameter with a small value ν in the denominator. So, we use the function $\frac{s}{|s|+\nu}$.

Finally, the following torques are applied to each of the N wheels:

$$\tau_i = \frac{1}{N}\left(\tau_u - \frac{R}{w_i}\tau_\theta\right) \tag{20}$$

with τ_u and τ_θ re-computed with a change of variable, from the inverse of the robot dynamics model—equations (17) and (4) with the accelerations $\dot{u}$ and $\dot{\theta}$ replaced respectively by the control laws (19) and (14)—namely,

$$\tau_u = \frac{1}{\gamma}\left(\dot{u}_d + K_p^u \varepsilon_u + \rho \frac{\varepsilon_u}{|\varepsilon_u| + \nu_u} - \Lambda_u \sum_{i=1}^{N} \dot{\omega}_i - r\nu\right) \tag{21}$$

$$\tau_\theta = \frac{1}{\lambda}\left(\dot{r}_d + K_p^\theta \varepsilon_\theta + K_d^\theta \dot{\varepsilon}_\theta + \mu \frac{\mathbf{B}^T\mathbf{P}\mathbf{x}}{|\mathbf{B}^T\mathbf{P}\mathbf{x}| + \nu_\theta} - \Lambda_\theta \dot{\omega} - \mathbf{D}_\theta \mathbf{F}_y\right) \tag{22}$$

To estimate the value of the lateral forces $\mathbf{F}_y$, Pacejka theory [11] could be used, by taking into account the slip angle. Because of the robustness of the sliding-mode control, however we can consider that $\mathbf{F}_y$ is a perturbation to be rejected, and we do not include it in the control law. A slip-angle measure being in practice not very efficient, this solution is better.

3 Application to the RobuFAST a Robot

3.1 *Experiments*

3.1.1 Control Law Implementation

The sliding-mode control law is tested with the RobuFAST A robot on a slippery flat ground. This experimental platform, shown in Fig.3, consists of an electric off-road vehicle, whose maximum reachable speed is 8 m/s. Designed for all-terrain

Table 1. Experimental robot inertial parameters

Total mass	$M = 350\ kg$
Yaw inertia	$I_z = 270\ kg.m^2$
Wheelbase	$l = 1.2\ m$
Rear half-wheelbase	$w = 0.58\ m$

mobility, the robot can climb slopes of up to 45° and has the properties displayed in Table 1. The front and rear directions of the vehicle are blocked to allow the skid-steering mode operation.

The implementation of the control law in real-life conditions requires some measures and observations. In particular the norm of the velocity vector, measured by the GPS, must be decomposed into its lateral and longitudinal components. This decomposition is made possible by the addition of an observer of the slippage angle [12], the knowledge of this angle and the velocity vector norm allowing us to make a projection on the axes of the robot frame.

The controller is implemented in two steps: first, a proportional derivative controller is settled for path following; then, the sliding-mode controller is added and its gains tuned.

This sliding-mode controller being a torque controller, a difficulty is that the robuFast A robot inputs are its wheel velocities. It is thus necessary to convert the amplitude of the torques generated by the controller.

Referring to eq.(3) of the wheel dynamics, we can consider that the addition of a force differential in a wheel is equivalent to a differential in its angular acceleration, i.e.,

$$\Delta\dot{\omega} \equiv \frac{R}{I_\omega}\Delta F$$

The losses in the movement transmission, due to wheel-soil contact, are disturbances to be compensated by the robust controller. This method is justified in particular in a patent [13].

The value of I_ω is obtained by the sum $I_\omega = I_{eq} + I'_\omega$. For a Marvilor BL 72 motor of a mass of 2.06kg and a reduction gear with $K = 16$, we have $I_\omega = 0.364 kg.m^2$, where I'_ω is the wheel inertia and $I_{eq} = K^2 I_m$ is the inertia equivalent to the motor and reduction gear unit, with K the reduction gear speed reducing ratio and I_m the motor inertia.

3.1.2 Experiment Results

The robot moves at a velocity of 3 m/s along a sinusoidal road. A derivative-proportional controller is applied to the robot, then the sliding-mode controller is implemented.

The position is plotted on figure 4 in m, the gains being tuned for optimum path-following: $K_p^u = 0.05\ s^{-1}$, $K_d^\theta = 0.02\ s^{-1}$, $K_p^\theta = K_d^{\theta 2}/4\zeta^2$, $\zeta = 0.70$, $T_r = 0.5\ s$, $v_u = 0.01\ ms^{-1}$, $v_\theta = 0.02$, $a = 0.10$, $b = 0.1$, $\mu = 0.1$ and $\rho = 0.01\ ms^{-2}$.

Fig. 3. RobuFAST A on site with its DGPS fixed base (upper left hand side)

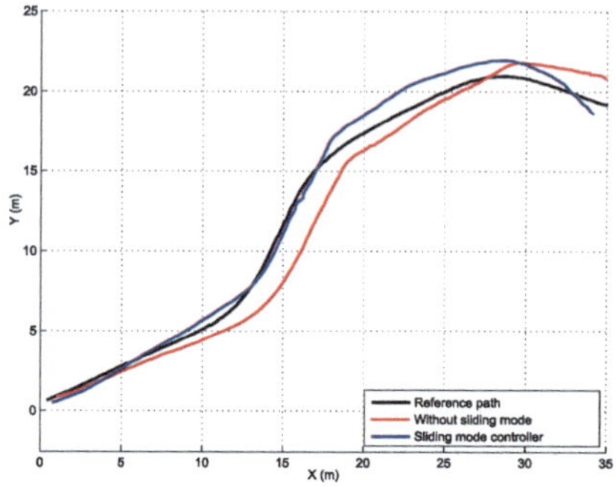

Fig. 4. Position (m)

Fig. 5, which indicates the curvature of the reference path as a function of the time, yields the robot position for the analysis of the evolution of its state variables in the time, the lateral error of Fig. 6, the yaw error of Fig. 7, the longitudinal velocity error of Fig. 8, and the torque inputs of Fig. 9 and Fig. 10.

Without sliding-mode, we see on the position curve that the vehicle follows the desired path with a small position offset. After a bend, the robot takes time to converge to the path. The succession of bends maintains this position error. With sliding-mode, the position curve converges much better to the desired path with however a small lateral error of about the length of a vehicle track, between the second and the third bend.

The delay of the actuators is due to the inertia of the vehicle slipping on a wet ground. There is almost no yaw-angle error during this period, as we can see at around a time of 14 s on the sliding mode controller curve Fig. 7. There is thus only a lateral error to be corrected, which explains a slower convergence. The convergence time can be tuned with parameter d_0 of eq.(5), but too high a value will bring about the risk of yaw instability, which could occur during bend.

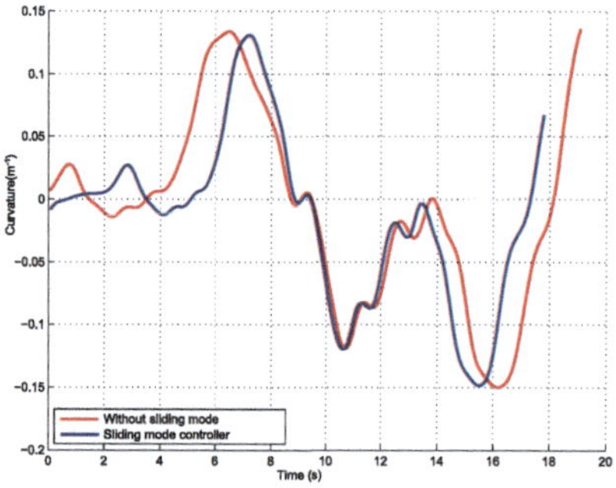

Fig. 5. Curvature of the path to follow during the time (1/m)

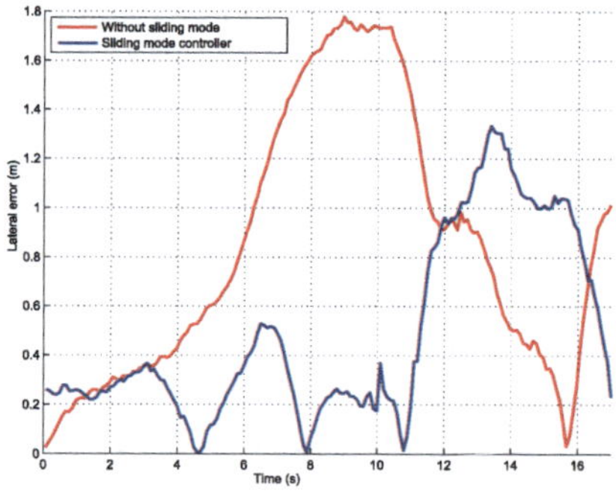

Fig. 6. Lateral error (m)

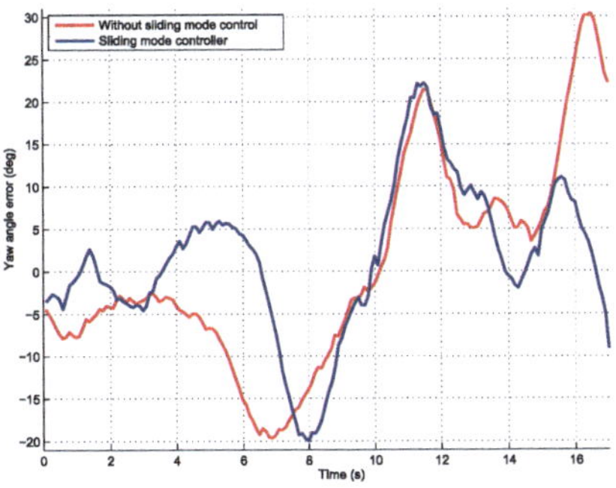

Fig. 7. Yaw angle error (deg)

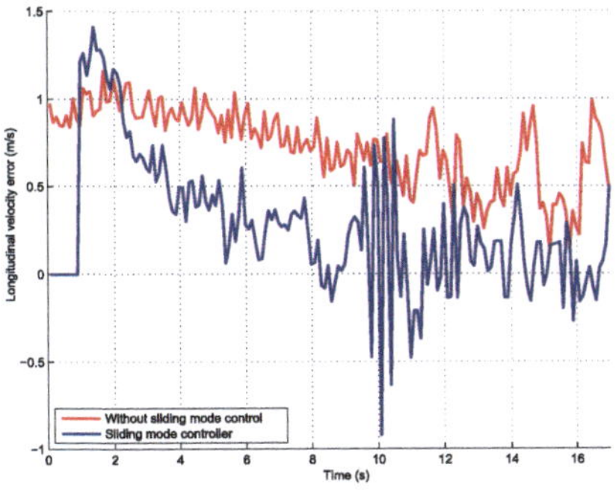

Fig. 8. Longitudinal velocity error (m/s)

On the lateral error curves displayed on Fig. 6, we notice a good following until the second bend (11 s) with an error that remains under 0.6 m oscillating around 0.2 m, whereas the error reaches 1.8 m without sliding-mode.

There is no significant difference between the yaw error with and without the sliding-mode control law; however, after the last bend (15 s) the robot has some difficulties to reach the path without the sliding-mode controller. The longitudinal velocity error is globally well reduced with the sliding-mode controller, as we can see in Fig. 8. We observe the chattering phenomenon during the second bend (11 s), with the strong oscillations of the velocity curve.

To correct the velocity errors, higher torque values are applied with the sliding-mode controller (torque curve for the velocity regulation of the Fig. 10), stabilized around 30 Nm. On the torque curve for the yaw regulation, we observe few peaks for three bends (7 s, 11 s et 15 s), with higher values without sliding-mode because of larger errors to correct in order to reach the path.

Finally, during several trial days, we noticed an increase of the energy consumption of 20 % to 30 % with the sliding-mode controller, the robot batteries emptying faster because of a higher frequency request of the actuators. This last point is a

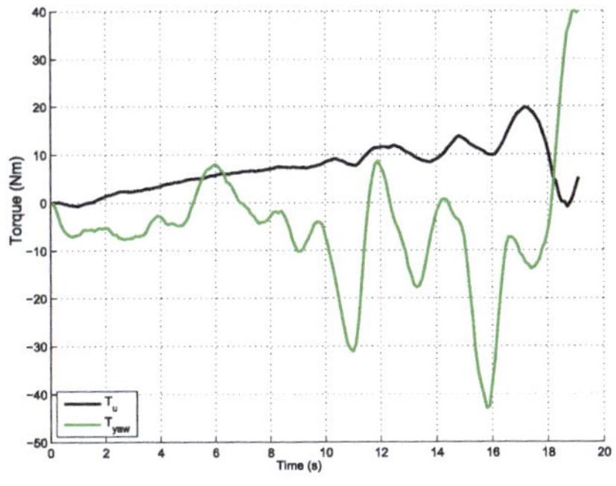

Fig. 9. Torques without the sliding-mode controller (Nm)

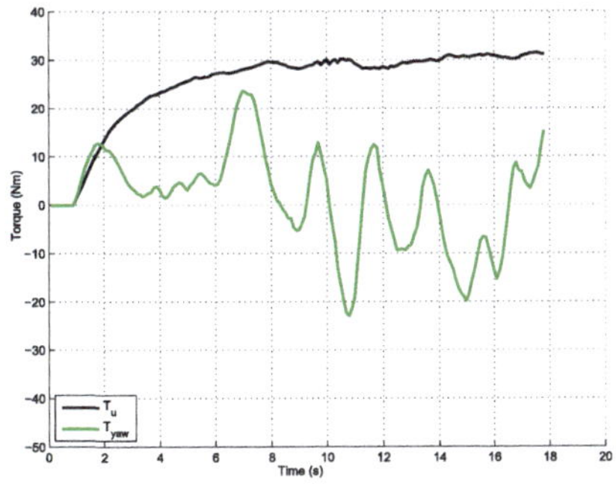

Fig. 10. Torques with the sliding-mode controller (Nm)

constraint, the supply of energy being an important problem in mobile robotics. This control law could be used at intervals, when it turns out to be necessary.

4 Conclusions

The sliding-mode controller introduced here was tested in real-life conditions with a torque-controlled four-wheel skid-steering robot. It was proven to be robust on sinusoidal paths, and despite wheel slippage. The chattering noticed during the experiments led to a higher energy consumption. In order to reduce it, we defined the gain variables according to some criteria such as the velocity or the path curvature. If the energy consumption becomes a concern, a higher-order sliding-mode control law should be considered.

References

1. Caracciolo, L., De Luca, A., Iannitti, S.: Trajectory tracking control of a four-wheel differentially driven mobile robot. In: Proceedings of the IEEE International Conference on Robotics & Automation, Detroit, Michigan, pp. 2632–2638 (May 1999)
2. Kozlowski, K., Pazderski, D.: Modeling and control of a 4-wheel skid-steering mobile robot. International journal of applied mathematics and computer science 14, 477–496 (2004)
3. Dixon, W.E., Dawson, D.M., Zergeroglu, E., Behal, A.: Nonlinear Control of Wheeled Mobile Robots, London (2001)
4. Jorge, A., Chacal, B., Sira-Ramirez, H.: On the sliding mode control of wheeled mobile robots. In: IEEE International Conference on Systems, Man, and Cybernetics, Humans, Information and Technology, vol. 2, pp. 1938–1943 (1994)
5. Aguilar, L.E., Hamel, T., Souères, P.: Robust path following control for wheeled robots via sliding mode techniques. In: IROS (1997)
6. Yang, J.-M., Kim, J.-H.: Sliding mode control for trajectory tracking of nonholonomic wheeled mobile robots. In: IEEE (1999)
7. Corradini, M.L., Orlando, G.: Control of mobile robots with uncertainties in the dynamical model: a discrete time sliding mode approach with experimental results. In: Elsevier Science Ltd. (ed.) Control Engineering Practice, Pergamon, vol. 10, pp. 23–34 (2002)
8. Hamerlain, F., Achour, K., Floquet, T., Perruquetti, W.: Higher order sliding mode control of wheeled mobile robots in the presence of sliding effects. In: 44th IEEE Conference on Decision and Control, and 2005 European Control Conference. CDC-ECC 2005, December 12-15, pp. 1959–1963 (2005)
9. Lhomme-Desages, D., Grand, C., Guinot, J.C.: Trajectory control of a four-wheel skid-steering vehicle over soft terrain using a physical interaction model. In: Proceedings of ICRA 2007: IEEE/Int. Conf. on Robotics and Automation, Roma, Italy, April 2007, pp. 1164–1169 (2007)
10. Sastry, S.S.: Nonlinear systems: Analysis, Stability and Control. Springer, Heidelberg (1999)
11. Pacejka, H.B.: Tyre and vehicle dynamics (2002)
12. Cariou, C., Lenain, R., Thuilot, B., Berducat, M.: Automatic guidance of a four-wheel-steering mobile robot for accurate field operations. J. Field Robot. 26, 504–518 (2009)
13. Yoshikawa, T.: Open-loop torque control on joint position controlled robots (October 2008)

On the Design and Validation of an Intelligent Powered Wheelchair: Lessons from the SmartWheeler Project

Joelle Pineau, Amin Atrash, Robert Kaplow, and Julien Villemure

Abstract. New-generation, intelligent, powered wheelchairs promise to increase the mobility and freedom of individuals with serious chronic mobility impairments. And while rapid progress continues to be made in terms of the engineering capabilities of robotic wheelchairs, many projects fall short of the target in terms of ease of use, conviviality, and robustness. This paper describes the SmartWheeler, a multi-functional intelligent wheelchair, which leverages state-of-the-art probabilistic techniques for both autonomous navigation and user interaction modeling, to provide a novel robust solution to the problem of assistive mobility. We also discuss the use of standardized evaluation in the development and testing of such technology.

1 Introduction

Many people suffering from chronic mobility impairments use a powered wheelchair to help them move around their environment. However, there are many factors which may make the use of such wheelchairs difficult or impractical, including fatigue, muscle spasms, or sensory impairments, to name just a few. It has been reported that up to 40% of patients found daily steering and maneuvering tasks to be difficult or impossible [1]. Considering that there are 4.3 million users of power mobility in the US alone [2], the potential for intelligent solutions to assistive mobility is immense.

Several intelligent wheelchair platforms have been developed over the years. A thorough review of research in this area is available for interested readers [3]. Technical innovation on these projects usually targets one or more of three key aspects: autonomous navigation, safe interaction with the environment, and interaction between the user and the robotic wheelchair. In general, researchers have been most successful at advancing technology in the first two of these aspects, arguably

Joelle Pineau
McGill University, Montreal QC Canada
e-mail: jpineau@cs.mcgill.ca

J. Angeles et al. (Eds.): Brain, Body and Machine, AISC 83, pp. 259–268.
springerlink.com

because physical environments are easier to handle than humans. Yet one of the greatest challenges for the smooth integration of assistive mobility technology is in providing a usable interface between the technology and the person.

This paper provides an overview of the SmartWheeler platform, a multi-functional intelligent wheelchair designed to assist individuals with mobility impairments in their daily locomotion. The robotic components are integrated onboard a commercially available powered wheelchair. The main innovation of this project is in incorporating state-of-the-art artificial intelligence techniques to tackle both autonomous navigation of the wheelchair, and robust human-robot interaction.

While the long-term objective of our work is to increase the autonomy and safety of individuals in the context of their daily activities, the development of the platform to date has focused on satisfying a specific corpus of tasks, as defined by the Wheelchair Skills Test [4]. The use of such a well-defined set of tasks has many advantages for the objective evaluation of the robot platform; in particular, it allows a direct comparison with non-intelligent powered wheelchairs using an objective performance measure.

This paper provides an overview of the physical robot platform. We briefly describe the autonomous navigation system, and focus at more length on the human-robot interaction manager. Finally we describe the evaluation protocol and preliminary results. Formal experiments with the target population are still underway, and will be reported in later publications.

2 Wheelchair Platform Design

The current SmartWheeler platform, developed between 2006 and 2009 at the Center for Intelligent Machines (McGill University), is shown in Figure 1. The robotic components are integrated onboard a commercially available Sunrise Quickie Freestyle wheelchair. The robot senses obstacles in its environment through two (one forward-facing, one rear-facing) SICK laser range-finders. Robot positioning information is acquired through custom-made wheel odometers. Both of these components are essential for the autonomous navigation of the robot. Different sensors could be used (e.g. sensors, infrared), but the precision and reliability of the information is of the utmost importance for achieving robust navigation.

Communication between the intelligent wheelchair and the user occurs through a Lilliput 8" touch-sensitive LCD and a two-way voice interface. The voice input (microphone) modality is used primarily for acquiring user commands, whereas the display is used primarily to provide feedback to the user about the wheelchair's understanding of the dialogue.

The robot platform also includes an onboard computer, which interfaces with the wheelchair's motor control board to provide navigational commands. Wireless communication is also available onboard, though not required for the main functionalities of the platform ,as described below.

The robotic components listed above could easily be ported to standard commercial platforms other than the Sunrise Quickie Freestyle. In practice, the primary

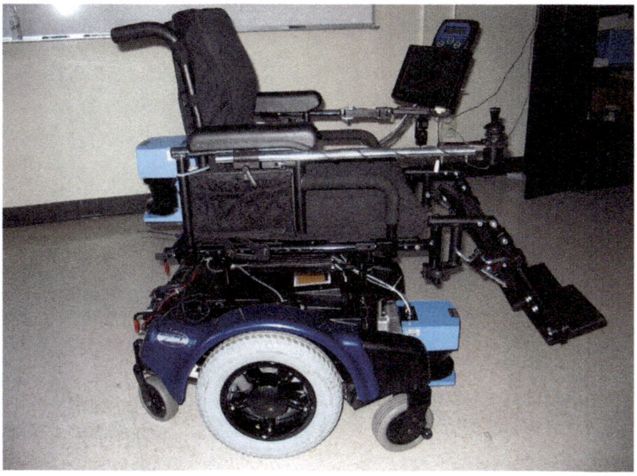

Fig. 1. SmartWheeler robot platform.

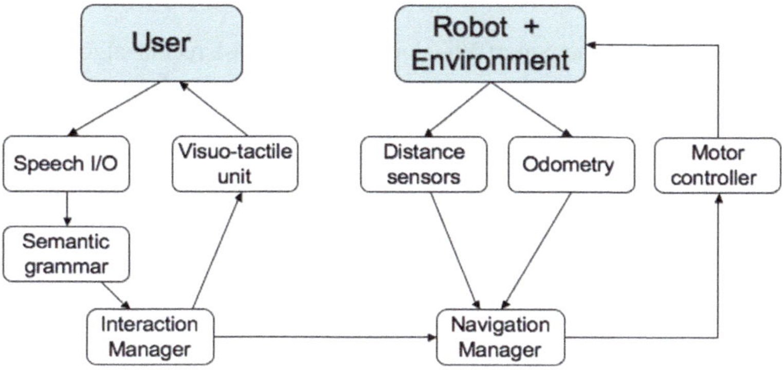

Fig. 2. SmartWheeler Architecture

challenge of such an integration is in interfacing the onboard computer with the proprietary wheelchair controller.

Figure 2 displays an overview of the software architecture operating onboard the robot. The primary modules of interest are the Navigation Manager and the Interaction Manager. Sections 3 and 4, respectively, provide a detailed discussion of the main models and algorithms used in these two components.

3 Autonomous Navigation

Core components of the autonomous robot navigation system include automatic map-building, localization within the environment, avoidance of obstacles (both stationary and moving), and automatic path planning. In recent years, statistical techniques have been used extensively to provide robust mathematical frameworks for

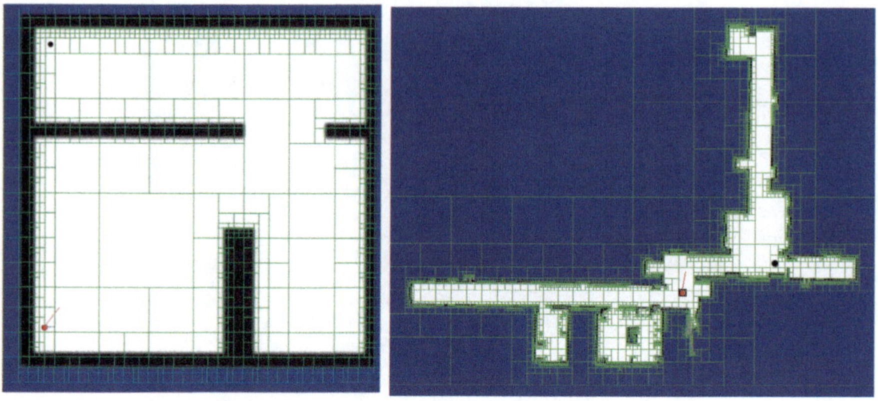

Fig. 3. Variable resolution decomposition of a simple robot navigation domain (left) and the first floor of the McConnell Engineering building (right). Light green lines indicate the state space representation considered for POMDP-based path planning.

these operations [5]. The SmartWheeler leverages most recent algorithmic techniques for performing these basic operations. The overall software architecture is written within the publicly available Player application [6]. Specific robot navigation routines, including mapping, localization, and low-level path planning and localization are implemented using the Carmen robot navigation toolkit [7].

A number of sophisticated statistical techniques–such as particle filtering—are included within the Carmen toolkit to perform state estimation based on the information acquired through the laser range-finders and odometry measurements. This allows the robot to build robust estimates of its environment, and its location within that space. The path planning methods included, however, are somewhat less robust, in that they assume the pose of the robot is known and do not take state uncertainty into account when making control decisions.

As part of the SmartWheeler project, we developed and implemented probabilistic models of decision-making, which are more robust for navigating in uncertain domains. The primary mathematical framework for this component is the Partially Observable Markov Decision Process (POMDP), which provides a stochastic model for sequential decision-making under uncertainty [8, 9]. One of the advantages of the POMDP paradigm is its ability to optimize plans contingent on partial state observability. In recent years, the development of efficient algorithms for POMDPs has allowed the framework to handle increasingly large domains [10]. Yet applicability for robot navigation remains elusive, due to the curse of dimensionality.

We recently developed a method for automatically generating a POMDP representation of an environment using variable resolution decomposition techniques. We can then apply state-of-the-art POMDP solution methods to optimize the action-selection policy of this model, such as to provide a near-optimal control strategy [11]. Using this method, the spatial discretization of the planning problem is generated automatically from any given metric map of the environment (sensor-built or hand-coded). As shown in Figure 3, this is done by taking advantage of the regular structure of indoor environments, such as identifying open spaces and abstracting them into a small number of states, while preserving significant resolution in areas near objects or walls. This results in a compact, yet accurate model of the environment. By applying POMDP planning methods using this representation, we can achieve robust planning under position uncertainty. Mapping uncertainty may also be included using more recently developed methods [12].

4 User Interaction

The Interaction Manager provides the mechanism for communication between the user and the wheelchair. Various communication modalities have been explored on-board smart wheelchairs, to supplement the traditional joystick input. Voice recognition is a popular choice as it requires little user training, as well as offering high information bandwidth at a low cost. Yet achieving reliable performance with a voice-driven wheelchair can be difficult due to high recognition error rates.

The goal of our system is to provide robustness to noise in the recognition system, as well as allow the user to speak naturally to the wheelchair, without needing to memorize a specific set of commands. To handle this, the SmartWheeler architecture includes a combination of intelligent systems techniques.

The speech recognition component is handled by Dragon Naturally Speaking, one of the more reliable commercially available software packages for speech to text transcription. We then process the output of the speech recognizer using a semantic grammar [13], which removes unnecessary lexical and syntactic information from the transcribed text. The final, and most important step, is the actual dialogue management. This module is responsible for the production of appropriate responses by the robot. There is no available software package for this step, and in fact the development of such systems is a highly active area of research, motivated largely by complex automated calling systems now in place in many organizations.

In the context of our intelligent wheelchair project, it was imperative to have a dialogue management system which offers both robustness and flexibility, such as to maximize ease-of-use. To achieve robustness, we turned once again to the POMDP paradigm. In this context, the POMDP model is used by the Interaction Manager to select appropriate responses subject to the observations acquired through the speech interface. Here again, the ability for the POMDP model to reason about uncertainty in the state (in this case the user's spoken word and intention) is crucial to the robustness of the system. In particular, the POMDP model can suggest clarification questions whenever the input received is incomplete or ambiguous. Figure 4 shows a short dialogue between the intelligent wheelchair and a test subject.

While the POMDP offers the appropriate mathematical tools for achieving robust interaction, it can be quite inflexible. In particular, standard POMDP planning algorithms require a (mathematically) accurate predictive model of the dynamics of the conversation. In general, it is challenging to define such a model *a priori*,

User command	Speech recognition output	Robot action
"Turn right."	`turn right`	Turns right 90 deg.
"Stop."	`this`	*Please repeat your command.*
"Turn left a little."	`turn left little`	Turn left 22.5 deg.
"Drive forward."	`drive forward`	Drive forward (max. 100m).
"Follow right wall."	`right wall`	Drive along the right wall (distance to wall = 0.65m).
"Turn around."	`turn around`	Turn to the left 180 deg.
"Follow left wall."	`left all`	*Please repeat your command.*
"Follow left wall."	`fall left wall`	Drive along the left wall (distance to wall = 0.65m).
"Stop."	`stop`	Stop

Fig. 4. Sample dialogue between a test subject and the intelligent wheelchair. The first column shows the user's actual spoken words. The second column reports the output of the automated speech recognition module. The third column shows the action chosen by the POMDP-based Interaction Manager (entries in *italic* represent clarification queries; all other action choices are transmitted to the Navigation Manager.)

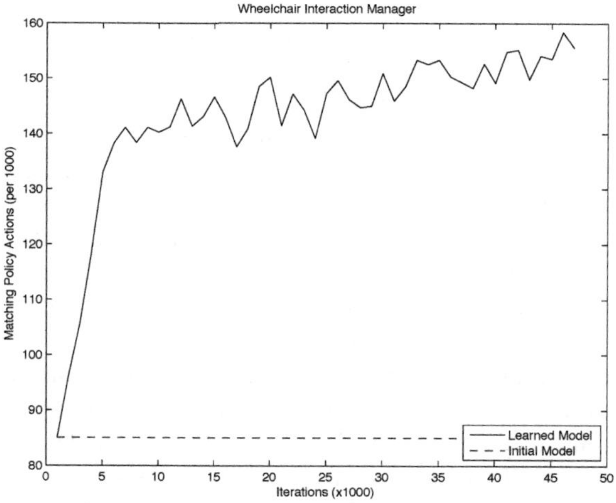

Fig. 5. Improvement in the quality of actions selected by the Interaction Manager as a function of the observed training data.

as human behavior is difficult to quantify, and varies substantially between individuals. To address this issue, we leveraged machine learning techniques to build the model directly from observed data [14]. Figure 5 shows the effect of learning on the performance of the Interaction Manager; these results were compiled using a simulated user.

5 Validation Methodology

As outlined by Simpson in his review paper, there have been substantial efforts directed towards the development of smart wheelchairs, yet much less attention has been devoted to the evaluation of these prototypes [3]. The evaluation methodology for the SmartWheeler was developed in close consultation with rehabilitation researchers and clinicians, with the goal of having an evaluation protocol that met a number of criteria:

- it could be undertaken by users with disabilities,
- it did not require extensive user training or customization,
- it allowed comparison between standard and intelligent powered wheelchairs,
- it was sufficiently repeatable to allow aggregation of user results,
- it admitted an objective performance measure.

We settled on using the Wheelchair Skills Test, developed by researchers at Dalhousie University [4]. This testing procedure consists of a set corpus of motion tasks, which the user must accomplish safely and to completion. It was originally designed for objectively evaluating manual wheelchair skills, then extended to powered wheelchairs. It has not previously been used to evaluate intelligent wheelchairs, yet applies readily without modification. In the case of the SmartWheeler, the main

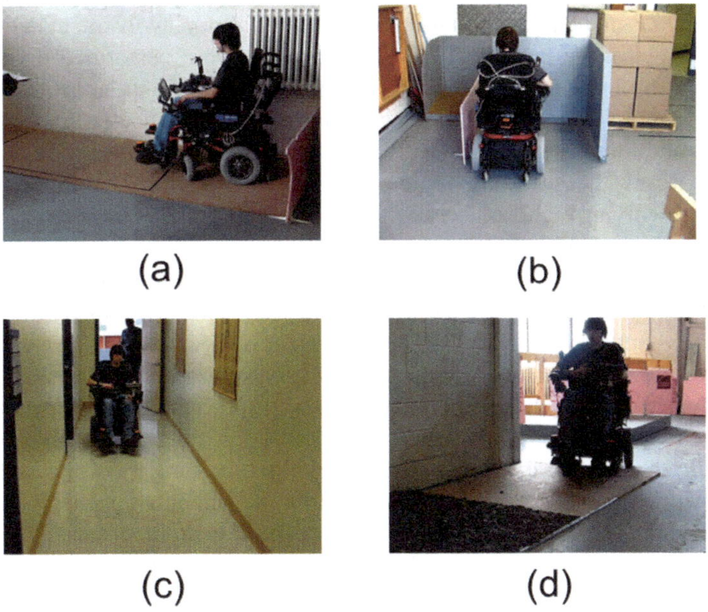

Fig. 6. Various stations of the Wheelchair Skills Test. The wheelchair must (a) travel along a sloped platform; (b) be aligned to the left wall; (c) move forward through a door; (d) travel through increased rolling resistance (in this case, gravel).

difference is that the user controls the chair through vocal (rather than manual) commands, and the internal controller is responsible for the low-level navigation.

The test includes tasks with a wide range of difficulty. We focus on the subset of skills which are relevant for robot-assisted mobility. Figure 6 shows an experimenter undergoing testing for some of the skills included in the test. Based on the selected set of tasks, we specified the vocabulary and grammar necessary for the speech interface. Neither of these are so restrictive as to require the user to use specific phrases. The initial vocabulary contained approximately 60 keywords which triggered appropriate grammatical rules and served as the input set for the Interaction Manager. The vocabulary itself, and the probability distributions describing the expected frequency of specific words for certain tasks were extended in the course of the user experiments using simple statistical techniques. Individual customization was not applied, given that each user only ran the test a single time. Rather, the learning was used to improve the model for the full population.

6 Results and Discussion

A preliminary evaluation of the Interaction Manager involving seven healthy subjects, all of them university students without involvement in the project, was performed early on [15]. These results provided an initial set of data for the model customization. A second round of experiments involving eight healthy subjects, all of them clinicians in local rehabilitation centers but without involvement in the project, was performed more recently. These experiments were performed on a different robotic platform developed at Ecole Polytechnique de Montréal [16]; this second platform features substantial differences from the SmartWheeler in terms of hardware and autonomous navigation software, however the user interaction modules are the same. Results of these experiments are recorded in Figure 7. A third round of experiments involving eight subjects with mobility impairments is currently underway (also using the second robot).

As shown in Figure 7, the Interaction Manager provides a robust architecture for handling communication with the user. Users were able to complete the test using between 114 and 219 commands. The word error rate for some subjects (subjects 4 and 8) was quite high. However, the appropriate use of queries allowed the system to reach a performance level comparable to that of other users, as shown by the low incidence of incorrect actions.

Overall, the test subjects were satisfied by the functionality of the interface and appreciated the visual feedback capabilities. Some subjects felt they needed more time to become familiar with the system to exploit it more successfully. Training time for all subjects was on the order of 30 minutes. The system was judged to be sufficiently usable and robust to move forward with experiments involving the target population.

The current experimental protocol is quite constrained, both in time and type of tasks evaluated. This is useful to allow formal testing with a number of subjects. But it has important limitations in terms of evaluating the long-term usability of the

Subject id	Number of commands	Word error rate	Number of queries	Number of correct actions	Number of incorrect actions
1	136	8.8%	10	121	5 (3.7%)
2	159	13.8%	18	136	5 (3.1%)
3	165	13.5%	11	152	2 (1.2%)
4	201	23.6%	37	155	9 (4.5%)
5	114	6.2%	13	97	4 (3.5%)
6	219	2.3%	10	208	1 (0.5%)
7	210	13.1%	25	175	10 (4.8%)
8	141	19.3%	26	111	4 (2.8%)

Fig. 7. Performance of the Interaction Manager for the Wheelchair Skills Test. The second column shows the number of vocal commands issued by the user throughout the test. The third column reports the raw speech recognition error rate. The fourth column shows the number of clarification queries issued by the robot in cases where the command was misunderstood or ambiguous. The fifth column presents the number of correct actions carried by the robot, as identified by human labeling of video sequences. Finally, the last column reports the number of times the robot selected an incorrect actions; users were instructed to recover from such situations by issuing a *Stop* command, or starting a new command.

system. However, a number of challenges remain before the intelligent wheelchair is ready for deployment in natural living environments. From a practical engineering perspective, it is necessary to extend the intelligent wheelchair's knowledge of the world. This means acquiring maps of the environment such as to allow navigation over a much larger footprint. It also means extending the vocabulary and grammar to accommodate a larger variety of commands. We foresee a number of important technological challenges in carrying out these extensions. First, there is the issue of computational scalability of the models and algorithms. Some of the inference techniques used by both the Navigation and Interaction component will require more efficient approximation to handle larger dimensional domains. However the most important challenge will likely be to develop models and algorithms which allow life-long learning, so that an intelligent wheelchair can automatically adapt to new environments, new habits, and new activities, along with its passenger.

Acknowledgements. The project involved close collaboration with researchers at the Ecole Polytechnique de Montreal (Paul Cohen, Sousso Kelouwani, Hai Nguyen, Patrice Boucher), at the Universite de Montreal (Robert Forget, Wormser Honore, Louise Demers), at the Centre de readaptation Lucie Bruneau (Claude Dufour) and at the Constance-Lethbridge Rehabilitation Centre (Paula Stone, Daniel Rock, Jean-Paul Dussault). Financial support was provided through the Canadian Foundation for Innovation, the Natural Sciences and Engineering Council of Canada, and the Fonds quebecois de la recherche sur la nature et les technologies. Additional funding was provided by the Fondation du Centre de readaptation

Lucie Bruneau, the Fondation Constance Lethbridge, and Robovic. Materials were donated by Sunrise Medical Canada and LiPPERT.

References

1. Fehr, L., Langbein, E., Skaar, S.B.: Adequacy of power wheelchair control interfaces for persons with severe disabilities: A clinical survey. J. of Rehabilitation Research & Development 37, 353–360 (2000)
2. Simpson, R.C., LoPresti, E.F., Cooper, R.A.: How many people would benefit from a smart wheelchair? J. of Rehabilitation Research & Development 45, 53–72 (2008)
3. Simpson, R.C.: Smart wheelchairs: A literature review. J. of Rehabilitation Research & Development 42, 423–436 (2005)
4. Routhier, F., Vincent, C., Desrosiers, J., Nadeau, S., Guerette, C.: Development of an obstacle course assessment of wheelchair user performance (OCAWUP). Technology & Disability 16, 19–31 (2004)
5. Thrun, S., Burgard, W., Fox, D.: Probabilistic Robotics. MIT Press, Cambridge (2005)
6. Collett, T.H.J., MacDonald, B.A., Gerkey, B.P.: Player 2.0: Toward a practical robot programming framework. In: Proceedings of the Australasian Conference on Robotics and Automation (2005)
7. Montemerlo, M., Roy, N., Thrun, S.: Perspectives on standarization in mobile robot programming: The Carnegie Mellon Navigation (CARMEN) Toolkit. In: International Conference on Robotics and Systems, pp. 2436–2441 (2003)
8. Kaelbling, L.P., Littman, M.L., Cassandra, A.R.: Planning and acting in partially observable stochastic domains. Artificial Intelligence 101(1-2), 99–134 (1998)
9. Sondik, E.J.: The optimal control of partially observable Markov processes. PhD thesis, Stanford University (1971)
10. Pineau, J., Gordon, G., Thrun, S.: Anytime point-based approximations for large POMDPs. Journal of Artificial Intelligence Research 27, 335–380 (2006)
11. Kaplow, R., Atrash, A., Pineau, J.: Variable resolution decomposition for robotic navigation under a POMDP framework. In: Proceedings of the International Conference on Robotics and Automation (ICRA) (2010)
12. Guez, A., Pineau, J.: Multi-tasking SLAM. In: Proceedings of the International Conference on Robotics and Automation, ICRA (2010)
13. White, M.: Open ccg: The open nlp ccg library. Software (2001), `http://openccg.sourceforge.net`
14. Atrash, A., Pineau, J.: A bayesian reinforcement learning approach for customizing human-robot interfaces. In: International Conference on Intelligent User Interfaces (IUI) (2009)
15. Atrash, A., Kaplow, R., Villemure, J., West, R., Yamani, H., Pineau, J.: Towards the deployment of an intelligent wheelchair in a standardized rehabilitation environment. Interaction Studies (2009)
16. Honoré, W., Atrash, A., Boucher, P., Kaplow, R., Kalouwani, S., Nguyen, H., Villemure, J., West, R., Routhier, F., Stone, P., Dufour, C., Dussault, J.-P., Rock, D., Cohen, P., Demers, L., Forget, R., Pineau, J.: Human-oriented design and initial validation of an intelligent powered wheelchair. In: RESNA Annual Conference (2010)

Devon Island as a Proving Ground for Planetary Rovers

Timothy D. Barfoot, Paul T. Furgale, Braden E. Stenning, Patrick J.F. Carle, John P. Enright, and Pascal Lee

Abstract. The future of space exploration will be increasingly surface-based and extended-duration. Planetary rovers, both manned and autonomous, will play vital roles in transporting instruments, astronauts, and equipment across rugged and unfamiliar surfaces. To enable this vision, it is advisable to deploy prototype rover vehicles in analog environments on Earth, in order to learn how best to use these tools. Devon Island, in the Canadian High Arctic, has been used as a proving ground for planetary rovers, due to its vast scale, variety of topography/geology, challenging lighting, lack of vegetation, existing infrastructure at the well-established Haughton-Mars Project Research Station, and wealth of interesting scientific mission objectives. In this paper we review the suitability of using Devon Island for the continued testing of planetary rovers; several examples of previously conducted tests are provided. We conclude that despite the typical logistical challenges associated with remote field work, Devon Island should be considered a strong candidate for ongoing rover field deployments.

1 Introduction

Several past space exploration achievements have demonstrated the benefit of surface mobility: the Russian Lunakhod Rovers (1970-71), the NASA Apollo Lunar

Timothy D. Barfoot · Paul T. Furgale · Braden E. Stenning · Patrick J.F. Carle
University of Toronto Institute for Aerospace Studies, Toronto, Canada
e-mail: {tim.barfoot,paul.furgale}@utoronto.ca,
{braden.stenning,patrick.carle}@utoronto.ca

John P. Enright
Ryerson University, Toronto, Canada
e-mail: jenright@ryerson.ca

Pascal Lee
Mars Institute, Vancouver, Canada
e-mail: pascal.lee@marsinstitute.net

J. Angeles et al. (Eds.): Brain, Body and Machine, AISC 83, pp. 269–281.
springerlink.com

Roving Vehicles (1971-1972), the NASA Mars Pathfinder Sojourner Rover (1997), and the NASA Mars Exploration Rovers (2004-present). All but Sojourner saw traverses on the order of tens of kilometers in length. Future missions, including the NASA Mars Science Lab (2011) and the ESA ExoMars Mission (~2016), will also enjoy the freedom of surface roving, with similar large-scale traverses planned. Indeed, it is the very nature of exploration that will ensure rovers, both manned and autonomous, continue to play enabling roles on future surface missions.

Mars and the Moon represent two of the most important surface exploration targets in the near- to mid-term (i.e., 10-30 years), as indicated by the Global Exploration Strategy (2007). Future missions will almost certainly be focussed on more challenging regions of the lunar and Martian surfaces than in the past. On the Moon, we will seek the permanently shadowed regions in the rugged South Pole Aitken Basin in search of water ice; traverses of several hundred kilometers have been considered. On Mars, we will search for signs of past and present life at sites exhibiting signs of potential hydrothermal activity, and eventually attempt to return a sample to Earth [11]. These and other new challenges will require rovers that can travel further, faster, and through more difficult terrain than their predecessors. Increased autonomy will play a major role in achieving this aim. As part of the path to spaceflight, field testing will need to be conducted in relevant operational environments. One site that has seen a number of robotic field deployments is the Haughton impact crater and surrounding area on Devon Island in the Canadian High Arctic.

Devon Island presents unique qualities for planetary analogue studies because it offers an unusually wide variety of geological features and microbiological attributes of strong planetary analogue value or potential [9]. It has been used for rover testing in the past [2, 3, 5, 6, 7, 12] because it presents real challenges to field exploration that are analogous in fundamental ways to those expected in planetary exploration. Being an impact basin in a polar desert environment, the lack of vegetation and variety of terrain make it well suited for rover field tests. Moreover, the long-term presence of the Haughton-Mars Project Research Station has drastically offset the logistical burden associated with carrying out field tests at this remote venue.

The rest of this paper makes an attempt to review the suitability of this Devon Island site for planetary rover testing. The key issues discussed are scale, terrain variety, lighting, lack of vegetation, infrastructure/logistics, and potential for scientific discovery. Our aim is not to contrast Devon with other potential analog sites, but rather to evaluate it on its own merits. We approach the review from a mobility perspective, for this is the main function of the rover (both manned and autonomous). The two main facets of mobility are (i) guidance, navigation, and control (GN&C), which includes aspects of autonomous operations, and (ii) locomotion; our focus will be placed primarily on testing the former, but we will also discuss the latter. We begin with a summary of the Haughton impact crater/surroundings and past rover deployments, followed by our review, and finally a conclusion.

2 Site Description

In this section we provide a brief description of the Haughton impact crater and surroundings on Devon Island, followed by a summary of past rover deployments.

2.1 Haughton Impact Crater and Surroundings

The Haughton impact crater is located at latitude 75° North on Devon Island[1], within the Arctic Archipelago, as shown in Figure 1(a). The crater itself is 20 kilometers in diameter, resulting from a massive impact 23 million years ago. Haughton crater has been well preserved owing to the fact that it exists in a polar desert, with relatively slow erosional processes. Craters are considered prime sites for planetary exploration because the impact churns up deep geological layers, exposing them on the surface; basement material from 1700 meters has been found in the Haughton crater. Within the crater, a large variety of features may be found including rock outcrops,

(a) Devon Island in Canadian High Arctic.

(b) Haughton Crater on Devon Island.

(c) View from inside Haughton Crater.

(d) Haughton-Mars Project Research Station.

Fig. 1. Haughton Impact Crater on Devon Island, Nunavut.

[1] Notably, Devon Island borders the fabled Northwest Passage; explorer Sir John Franklin and his crew wintered on Beechey Island, just off the southwest shore of Devon in 1845-46.

(a) CMU's Hyperion (2001). *Photo credit: Carnegie Mellon University*

(b) NASA Ames' K10s (2007). *Photo credit: M. Deans*

(c) UofT's Pushcart Rover (2008).

(d) UofT's ROC6 (2009).

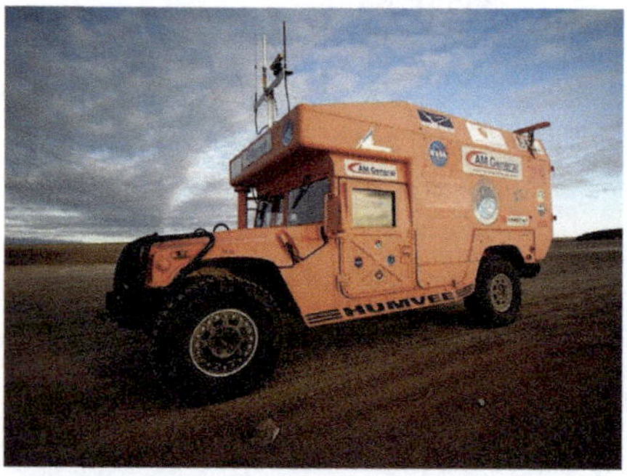

(e) HMP's Mars-1 (2003-present).

(f) HMP's Moon-1 (2010-present).

Fig. 2. Recent rover deployments on Devon Island including actual visit dates.

impact melt breccias, gullies, and rivers (see Figure 1(c)). Outside the crater, a number of other interesting features exist including ejecta blocks, canyons, lakes, mesas, plains, and polygonal terrain. The Haughton-Mars Project (HMP) Research Station is positioned near the northwest rim of the crater at $75°26'00''$ N, $89°51'47''$ W. HMP is comprised of several permanent buildings and weather havens, and boasts a fleet of all-terrain vehicles (ATVs), generators, and advanced communication systems [8]. Site access is via Twin Otter aircraft.

2.2 *Past Rover Deployments*

There have been several research-level rover vehicles tested on Devon Island. Figure 2 shows Carnegie Mellon University's (CMU) autonomous Hyperion sun-synchronous rover [12, 13], NASA Ames' dual autonomous K10 rovers [6, 7], the University of Toronto's (UofT) Pushcart Rover [2], UofT's autonomous ROC6 [5], and HMP's manned Mars-1. HMP's manned Moon-1 rover will join the fleet in the summer of 2010 and the NASA Ames K10 rovers will also return for a second campaign[2]. To date, no flight-design rovers have been deployed on Devon

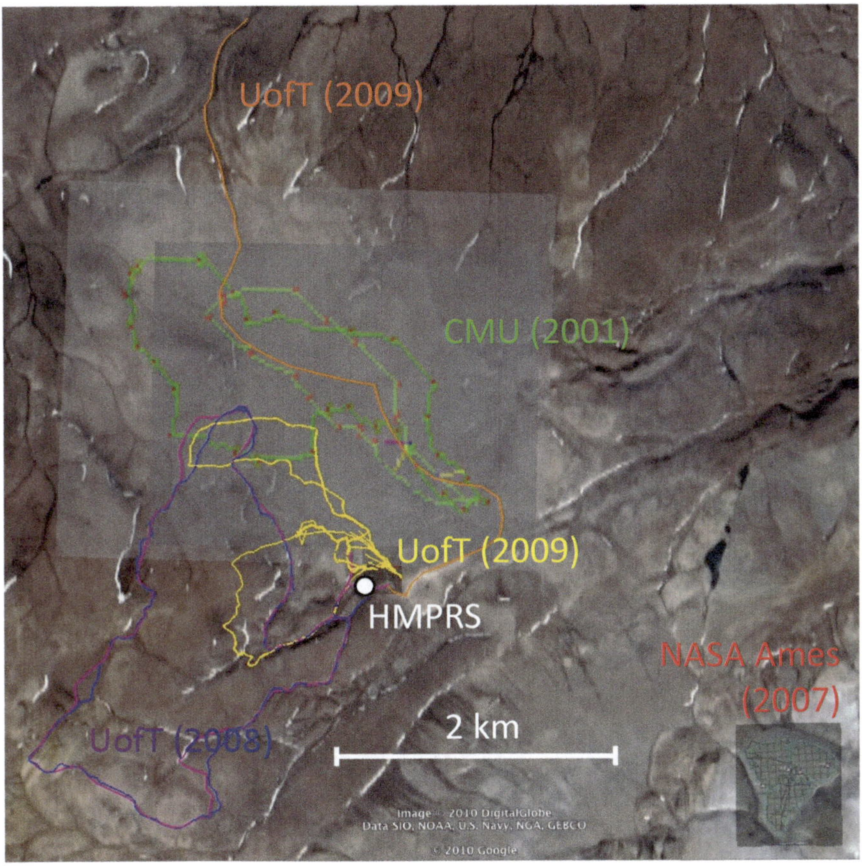

Fig. 3. Recent long-distance rover traverses on Devon Island. Note that NASA Ames traverse was a compact systematic grid in bottom right. *Photo credit: Google Earth*

[2] Personal communication with Dr. Terry Fong.

Island; testing has focussed on the operational and GN&C aspects of rover missions. Matthews [10] provides a photo montage of earlier prototypes by additional organizations.

3 Review

In this section we discuss a number of issues that affect the quality of a rover field deployment: scale, terrain variety, lighting, lack of vegetation, infrastructure/logistics, and scientific merit. The discussion is primarily focussed on guidance, navigation, and control testing, including autonomous operations.

3.1 Scale

Scale is a major issue. Past rover flight missions saw traverses on the order of tens of kilometers. Field deployments used to validate rover technologies should ideally see traverses of this scale. Past Devon rover deployments were able to accomplish mission-scale traverses: CMU Hyperion (~15 km in 2001), NASA Ames K10s (~42 km in 2007), UofT Pushcart (~20 km in 2008), and the UofT ROC6 (~44 km in 2009). Figure 3 shows a compilation of these traverses. The length of traverse is essentially only limited by testing time.

A second major issue related to scale is sensor range. Most rover guidance, navigation, and control architectures employ vision sensors such as panoramic cameras, stereo cameras, and lidar (light distance and ranging). The images captured by these sensors are typically used in appearance-based vision algorithms for localization, motion estimation, terrain assessment, and path-planning. Figure 4 shows the scale of what is visible for both camera and lidar sensors at a typical site on Devon. A panoramic camera's range is limited only by topographic occlusions; the figure shows that terrain out to 5 km may be viewed in this case. A lidar typically has a shorter range (~1 km in practice) and we see from the figure that the Devon terrain allows the sensor to exercise its maximum range in this case.

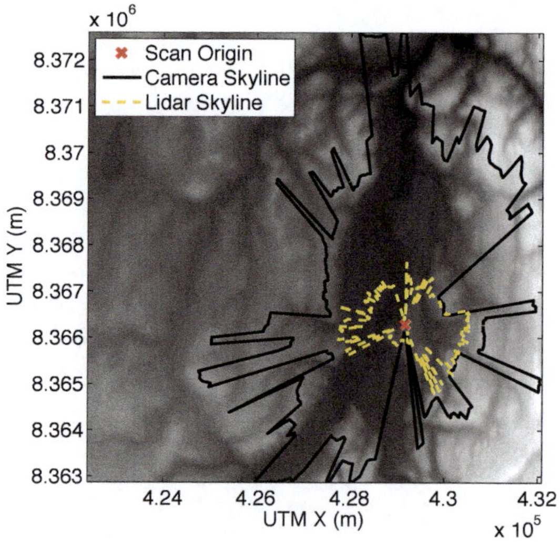

Fig. 4. Sensor ranges: lidar (~1 km), camera (~5 km)

(a) Polygon Trough (b) Canyon (c) Mound/Rocks (d) Valley (e) Vast Plain (f) Rocky Slope

Fig. 5. Variety of rover testing terrain; UofT ROC6 rover shown.

3.2 *Terrain Variety*

Throughout their long lives, NASA's Mars Exploration Rovers have encountered a wide variety of terrain types from terraced crater walls to flat plains. Even with high-resolution imagery gathered from orbit, it can be difficult to predict the nature of the rover-scale terrain on the ground.

Large-scale topography (10s to 100s of meters) is very important for testing certain rover GN&C methods as well as chassis capabilities. Long traverses need to encounter a variety of large terrain features to provide realism in terms of the three-dimensional motion of the platform and occlusions to vision sensors [5]. Topography is also important for testing the performance of long-range localization, which relies on topographic features and horizon silhouettes to determine position [3].

A variety of different rock distributions is also desirable to test rover chassis and GN&C methods such as terrain assessment and path planning. Fist-sized rocks and up can constitute hazards that need to be avoided. Varying the rock distribution can vary the difficulty of the traverse immensely.

Figure 5 shows a sample of the large variety in available testing terrain near HMP. These photos were all gathered on the yellow UofT (2009) path in Figure 3, except for Figure 5(a), which was gathered at a polygon terrain site 10 km north of HMP. The Haughton crater itself (see Figure 1(c)) is also within access distance of HMP and offers a wide variety of features including rock outcrops, steep impact melt breccias, and gullies.

3.3 Lighting

For rover testing, the Devon Island field season is typically early July to mid August, as this is when the ground is mostly free of snow. At latitude 75° north, the sun remains above the horizon 24 hours a day in this period. We see in Figure 6[3], that the sun's elevation hovers between zero and 30 degrees throughout the day, meaning the shadows are typically very long. From the azimuth we see that the sun moves continuously around the horizon, meaning the shadows move rapidly.

One of the main challenges for rover GN&C is the effect of lighting on visual navigation. Shadows can be a major source of error in appearance-based methods. It is therefore important to perform tests under a variety of different lighting situations. High-latitude venues tend to work well in this respect, exactly because the sun is low on the horizon and the shadows are long and moving quickly. Figure 7 shows an example of a vision algorithm failure from a UofT field test on Devon Island in 2009 [5]. A stereo image was gathered at midday; a second image from four hours later could not be matched to the original (i.e., in order to re-localize the rover) due to the dramatic changes in shadowing. A third image was taken the next morning and was successfully matched.

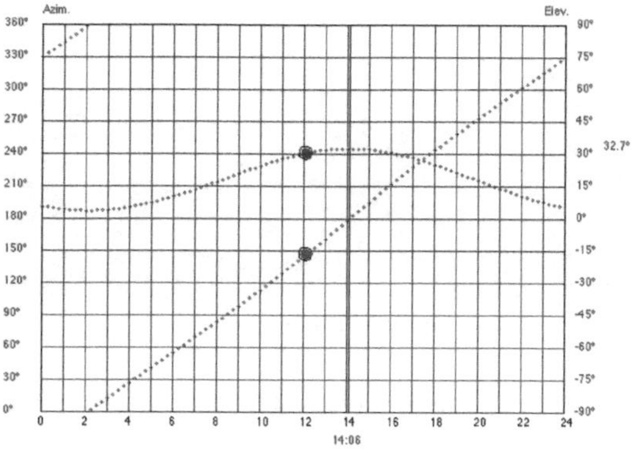

Fig. 6. Sun Azimuth and Elevation on July 31 at HMP.

Figure 8 shows a quantitative test of matching distinctive image keypoints (i.e., SURF features) from a stationary stereo camera over a nine hour test . We see that the algorithm is able to match many features well for over six hours in this case, but

[3] Created using web applet at `http://www.jgiesen.de/azimuth/`.

(a) Sunny initial image, mid-day.

(b) 4 hours later, match to (a) fails due shadows shifting in bright sunlight.

(c) Overcast next morning, match to (a) succeeds due to similar shadowing.

Fig. 7. Example of a vision algorithm failing to recognize a location due to shadows changing.

eventually fails to produce an adequate number of matches. The early 'dip' at 02:30 is likely due to the presence of temporary cloud cover, underscoring the importance of lighting on these algorithms.

It should also be pointed out that these difficult lighting conditions are representative of several mission scenarios for both Mars and the Moon. Polar missions for both targets have been considered and present major challenges not just for vision algorithms but also solar power generation. Carnegie Mellon University's Hyperion experiment conducted on Devon Island in 2001 [13] was aimed specifically at testing a mission planner that could ensure a rover would traverse a well-lit path despite dynamic shadowing. They showed that it was possible to remain synchronous with the sun over a 24 hour period. Devon has also been used for celestial navigation experiments [4], wherein a sun sensor was used to determine a rover's absolute heading; the near-polar environment makes this test relevant to high-latitude missions.

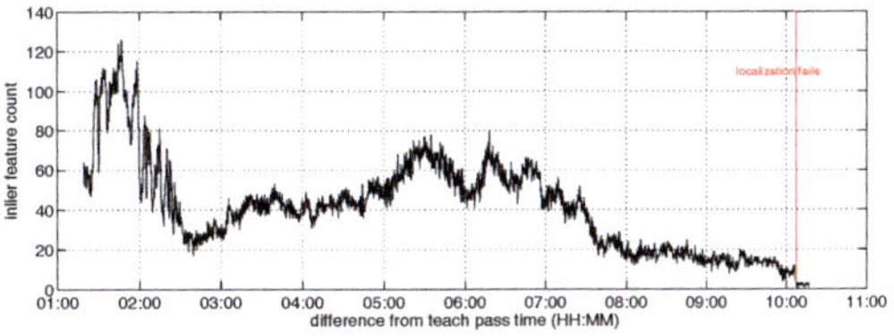

Fig. 8. Vision algorithm matching over a nine hour period. Distinct keypoints taken from a stationary stereo camera are matched to an initial image. Matched feature count remains high for over six hours, but eventually fails due to dramatic changes in appearance due to shadows shifting.

3.4 Lack of Vegetation

Perhaps the most appealing aspect of the Devon site, from a planetary rover testing perspective, is the almost complete lack of vegetation. There is some low-lying vegetation in localized areas, but most areas are completely clear of noticeable specimens. Figure 9 depicts the 'Lake Sediments' area, which is one of the most heavily vegetated areas near HMP (and not appropriate to rover testing due to environmental sensitivity). Figure 5 shows the more typical vegetation-free terrain. To understand why vegetation is problematic for rover testing we must again discuss the GN&C architecture, which has a number of sub-components. We will discuss a few of these sub-components, each impacted by vegetation in different ways.

Fig. 9. Worst-case vegetation from a rover-testing point of view near HMP. These areas are few and can/should be avoided.

3.4.1 Short-Range Terrain Assessment and Obstacle Avoidance

This GN&C component uses sensors onboard a rover (e.g., stereo camera and/or lidar) to (i) build up a local model of the terrain (i.e., within 10 m), (ii) classifies patches of the terrains as driveable or not, then (iii) plans a collision-free path to a short-range goal. The presence of vegetation makes the classification step more difficult than it needs to be. Due to the limited processing available on planetary rovers, a simple geometry-based terrain classification is typically employed (e.g., simple plane fitting). With this type of system, tall grasses, for example, can cause an area of terrain to look undriveable even though it is. A more sophisticated terrain classification method could be used (to handle vegetation), but then it would not be planetary-relevant and would likely need to be tailored to the specific vegetation encountered. Devon does not require vegetation-specific terrain assessment.

3.4.2 Long-Range Terrain Assessment and Path Planning

This GN&C method is similar to the previous one but operates on a much longer scale (i.e., hundreds of meters). A sensor such as a long-range lidar is used to build a very detailed model of the terrain in a wide region around the rover (see Figure 4). This model has scientific value, but also engineering value as it may be used to plan paths for a rover. The presence of trees and tall bushes introduce far more undriveable areas than would be encountered in a planetary situation. Moreover, they severely limit the size of area that can be scanned (e.g., using a camera or long-range lidar). This is very problematic when vegetation is taller than the rover being tested. Thus, the ability to scan several hundred meters to assess terrain on a large scale is limited when vegetation is present. Devon does not suffer from this problem.

3.4.3 Localization

There are a few different types of localization techniques. Short-range techniques such as visual odometry are employed to estimate the relative motion of a rover platform using onboard sensors (e.g., stereo cameras) [5]. Long-range techniques such as horizon matching are used to estimate global position by matching camera views to orbital digital elevation maps [3]. There are three negative aspects imposed by vegetation for localization. First, although it is possible to use localization methods in the presence of vegetation, there is an underlying assumption in many of these techniques that the terrain is not changing over time. Vegetation can change over time (e.g., wind, growth). Second, certain types of vegetation (e.g., grass, leaves) are very self-similar and thus make it difficult to find distinct visual features for tracking. In other words, the texture of the appearance is too fine. Third, other types of vegetation (e.g., far away trees silhouetted against the sky) provide features that are too easy to track (for relative localization) or too difficult to match (for global localization, due to occlusions). In summary, the presence of vegetation severely limits the quality of the test results for localization methods. Devon does not suffer from any of these problems and has proven to be a very good test venue for vision-based localization techniques.

3.5 Infrastructure and Logistics

Fig. 10. ROC6 loaded onto a Twin Otter.

Although Devon Island is extremely remote, access is not difficult in the summer months. Commercial flights may be taken from Ottawa, Canada, to Resolute Bay on Cornwalis Island and then a chartered Twin Otter to Devon Island (see Figure 10). Once on Devon Island, the Haughton-Mars Project Research Station [8] is well-equipped to serve the needs of a rover testing team. Food, water, electricity, fuel, office space, internet access, all-terrain vehicles, and guides are all available, allowing maximum time to be spent conducting experiments. Moreover, because experiments can be conducted right out of HMP basecamp, the work day is effectively 24 hours long. These attributes make it possible to conduct a very large number of experiments in a 2-3 week stay. An issue worth noting is that government and Inuit permits are required to access Devon Island and that equipment typically needs to be shipped well in advance of arrival. Also, bugs are minimal on Devon and weather is variable in July to August.

3.6 Scientific Merit

The main reason that HMP Research Station exists is the ongoing geology, geomorphology, and biology this unique site hosts. Each field season an international group of scientists visits to collect samples and take measurements. There are countless opportunities for rover researchers to team up with scientists in order to conduct [2]. missions that do real science using robotic tools. Figure 11 shows one example of an ice-prospecting mission conducted on polygonal terrain on Devon It also worth noting that Devon is a fragile environment containing valuable scientific information and wildlife. Keeping the environmental footprint to a minimum is essential; before any rover mission is planned, the impact on Devon must be weighed against the scientific merit of the experiment.

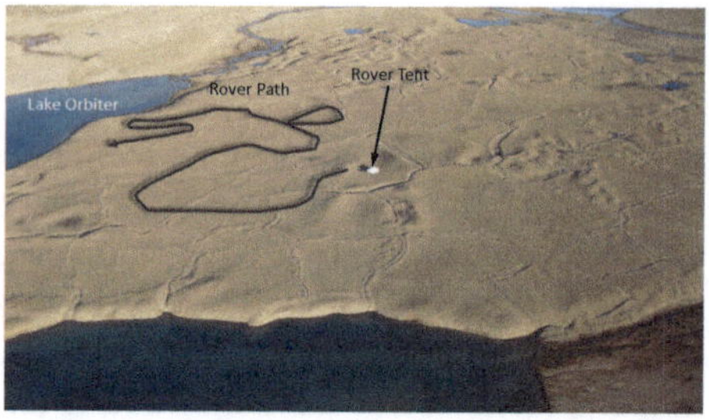

Fig. 11. A simulated rover mission to prospect for ground-ice was conducted at this polygonal terrain near Lake Orbiter on Devon Island in 2009. The UofT ROC6 used ground-penetrating radar along this 734 m path to look for ice deposits (2009).

4 Conclusion

We have reviewed the merits of using Devon Island, and specifically the Haughton crater and surroundings, as a proving ground for planetary rovers. Based on the scale, variety of terrain, lighting conditions, lack of vegetation, existing infrastructure, and wealth of interesting scientific objectives, we believe Devon Island is a strong candidate for continued rover field deployments, despite its remote locale.

Acknowledgements. The authors would like to thank the dedicated teams at the Haughton-Mars Project, the Polar Continental Shelf Project, and Technical Field Support Services for logistics and field support in accessing Devon Island. Funding for the UofT projects described in this article was provided by the Canadian Space Agency under its Canadian Analogue Research Network Grant Program, the Northern Scientific Training Program, the Natural Sciences and Engineering Research Council of Canada, and the Canada Foundation for Innovation. We also thank Professor David Wettergreen of Carnegie Mellon University and Dr. Terry Fong of NASA Ames for providing information related to their Devon field campaigns for inclusion in this article.

References

[1] ASI, BNSC, CNES, CNSA, CSA, CSIRO, DLR, ESA, ISRO, JAXA, KARI, NASA, NSAU, and Roscosmos, The Global Exploration Strategy: The Framework for Coordination, Technical report (2007)

[2] Barfoot, T.D., Furgale, P.T., Osinski, G.R., Ghafoor, N., Williams, K.: Field Testing of Robotic Technologies to Support Ground-Ice Prospecting in Martian Polygonal Terrain. Planetary and Space Science, special issue on Exploring other worlds by exploring our own: The role of terrestrial analogue studies in planetary exploration 58(4), 671–681 (2010)

[3] Carle, P., Furgale, P.T., Barfoot, T.D.: Long-Range Rover Localization by Matching Lidar Scans to Orbital Elevation Maps. Journal of Field Robotics 27(3), 344–370 (2010)
[4] Enright, J., Furgale, P., Barfoot, T.D.: Sun Sensing for Planetary Rover Navigation. In: Proc. of the IEEE Aerospace Conference, Big Sky, MT (2009)
[5] Furgale, P.T., Barfoot, T.D.: Visual Teach and Repeat for Long-Range Rover Autonomy. Journal of Field Robotics, special issue on Visual mapping and navigation outdoors (2010)
[6] Fong, T., Deans, M., Bualat, M., Flueckiger, L., Allan, M., Utz, H., Lee, S., To, V., Lee, P.: Analog Lunar Robotic Site Survey at Haughton Crater. In: Proc. of the Workshop on Enabling Exploration: The Lunar Outpost and Beyond, Abs. 3058, Lunar Exploration Analysis Group, Houston, TX (2007)
[7] Fong, T., Allan, M., Bouyssounouse, X., Bualat, M., Deans, M., Edwards, L., Fluckiger, L., Keely, L., Lee, S., Lees, D., To, V., Utz, H.: Robotics Site Survey at Haughton Crater. In: Proc. of the 9th Int. Symp. on Artificial Intelligence, Robotics and Automation in Space (iSAIRAS), Los Angeles, CA (2008)
[8] Lee, P., Braham, S., Boucher, M., Schutt, J., Glass, B., Gross, A., Hine, B., McKay, C., Hoffman, S., Jones, J., Berinstain, A., Comptois, J.-M., Hodgson, E., Wilkinson, N.: Haughton-Mars Project: 10 Years of Science Operations and Exploration Systems Development at a Moon/Mars Analog Site on Devon Island, High Arctic. In: Proceedings of the 38th Lunar and Planetary Science Conference, League City, Texas, pp. 2426–2427 (2007)
[9] Lee, P., Bunch, T.E., Cabrol, N., Cockell, C.S., Grieve, R.A.F., Rice, J.W., Mckay, C. P., Chutt, J.W., Zent, A.P.: Haughton-Mars 97 - I: Overview of Observations at the Haughton Impact Crater, a Unique Mars Analog Site in the Canadian High Arctic. In: Proceedings of the 29th Lunar and Planetary Science Conference, Houston, Texas, pp. 1973–1974 (1998)
[10] Matthews, J.: Development of the Tumbleweed Rover, Technical report, Jet Propulsion Laboratory (2003)
[11] Schenker, P.S., Huntsberger, T.L., Pirjanian, P., Baumgartner, E.T., Tunstel, E.: Planetary Rover Developments Supporting Mars Exploration, Sample Return and Future Human-Robotic Colonization. Autonomous Robots 14(2), 103–126 (2003), doi:10.1023/A:1022271301244
[12] Wettergreen, D., Dias, M., Shamah, B., Teza, J., Tompkins, P., Urmson, C., Wagner, M., Whittaker, W.: First Experiment in Sun-Synchronous Exploration. In: Proceedings of the IEEE International Conference on Robotics and Automation (ICRA), Washington, DC, pp. 3501–3507 (2002)
[13] Wettergreen, D., Thompkins, P., Urmson, C., Wagner, M., Whittaker, W.: Sun-Synchronous Robotic Exploration: Technical Description and Field Experimentation. International Journal of Robotics Research 24(1), 3–30 (2005)

Leader-Follower Cucker-Smale Type Flocking Synthesized via Mean Field Stochastic Control Theory

Mojtaba Nourian, Peter E. Caines, Roland P. Malhamé, and Minyi Huang

Abstract. In this paper we study a controlled Leader-Follower (L-F) flocking model (where the state of each agent consists of both its position and its controlled velocity) by use of the Mean Field (MF) Stochastic Control framework. We formulate the large population stochastic L-F flocking problem as a dynamic game problem. In this model, the agents have similar dynamics and are coupled via their nonlinear individual cost functions which are based on the uncontrolled Cucker and Smale (C-S) flocking algorithm. The cost of each leader is based on a trade-off between moving its velocity toward a certain reference velocity and a C-S type weighted average of all the leaders' velocities. Followers react by tracking the C-S type weighted average of the velocities of all the leaders and followers. For this nonlinear dynamic game problem we derive two sets of coupled deterministic equations for both leaders and followers approximating the stochastic model in large population. Subject to the existence of unique solutions to these systems of equations we show that: (i) the set of MF control laws for the leaders possesses an ε_N-Nash equilibrium property with respect to all other leaders, (ii) the set of MF control laws for the followers is almost surely ε_N-Nash equilibrium with respect to all the other agents (leaders and followers), and (iii) $\varepsilon_N \to 0$ as the system's population size, N, goes to infinity. Furthermore, we analyze the MF system of equations for the leaders and followers with the linear coupling cost functions to retrieve similar MF equation systems in Linear-Quadratic-Gaussian (LQG) dynamic game problems.

Mojtaba Nourian · Peter E. Caines
Centre for Intelligent Machines (CIM), Department of Electrical and Computer Engineering, McGill University, Montreal, QC H3A 2A7, Canada; also affiliated with GERAD, Montreal
e-mail: mnourian, peterc@cim.mcgill.ca

Roland P. Malhamé
GERAD, Montreal and Department of Electrical Engineering, École Polytechnique de Montréal, Montreal, QC H3C 3A7, Canada
e-mail: roland.malhame@polymtl.ca

Minyi Huang
School of Mathematics and Statistics, Carleton University, Ottawa, ON K1S 5B6, Canada
e-mail: mhuang@math.carleton.ca

J. Angeles et al. (Eds.): Brain, Body and Machine, AISC 83, pp. 283–298.
springerlink.com

1 Introduction

Collective motion such as the flocking of birds, schooling of fish and swarming of bacteria, is one of the most widespread phenomena in nature (see Figs. 1 and 2). For the past fifty years, scientists from different disciplines have studied such emergent behaviour to understand the general mechanisms of cooperative phenomena and their potential applications (see [1] among many other papers).

The study of collective motion in nature is of interest not only to model, analyze and interpret these widespread phenomena, but also because ideas from these behaviours can be used by engineers to develop efficient algorithms for a wide range of applications (see [2] among other papers).

Fig. 1. Collective motion and emergent behaviour in animals (Top: sardines schooling. Bottom: birds Flocking). Top Right: taken by David Doubilet. Bottom Left: taken by Manuel Presti.

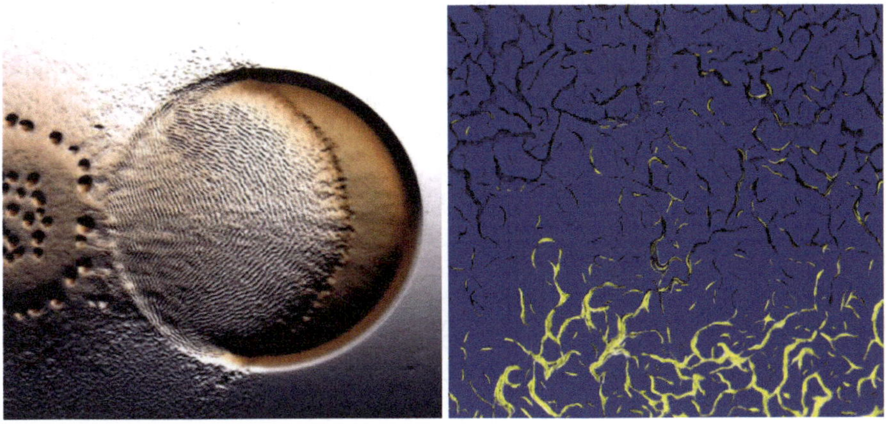

Fig. 2. Left: swarm of M. Xanthus bacteria (left) moving over their prey (right), John Kirby, University of Iowa Carver College of Medicine. Right: two strains of Proteus mirabilis bacteria swarm towards each other, K. Gibbs, J. Milner-Brage, E. P. Greenberg, Science 2008.

There are two main classes of models for flocking and swarming behaviour: (i) Individual Based Models (IBMs) in the form of coupled ordinary (stochastic) differential (difference) equations (see for example [3, 4]), (ii) Continuum models in the form of Partial Differential Equations (PDEs) to model the collective motion in the case of systems with large populations (see for example [5, 6, 7]). The continuum models can be derived from the IBMs in the large population limit by use of the kinetic theory of gases, hydrodynamic and mean field theory [5, 7, 8]. A comprehensive survey on these models can be found in [5].

In [4] Cucker and Smale introduce an interesting flocking model for a group of agents. This model is motivated by the collective motion of a group of birds such that each bird updates its velocity as a weighted velocity of other birds. The weights in this model are functions of the relative distance between the birds.

On the other hand, decision making and collective behaviour almost always involve some form of leader-follower behaviour [9]. This behaviour is observed in humans and many other species in nature [10], and is studied in variety of disciplines such as game theory [11], collective decision [12], flocking [13] and biology [10], among others. Such behaviour in nature is often attributed to the fact that there exist some individuals in the group which have more information than others, for instance the location of resources or migratory routes [10].

For instance, a hierarchical leadership among flock members of pigeons has been experimented in [14] where the authors suggest that "hierarchical organization of group flight may be more efficient than an egalitarian one, at least for those flock sizes that permit regular pairwise interactions among group members, during which leader-follower relationships are consistently manifested" [14].

In this paper we study a controlled Leader-Follower (L-F) flocking model (where the state of each agent consists of both its position and its controlled velocity) by use of the Mean Field (MF) (or Nash Certainty Equivalence) Stochastic Control framework.

Since 2003, inspired by mean field approximations in statistical mechanics, the MF framework has been studied for large population dynamic noncooperative games in a series of papers by Huang, Caines and Malhamé (see [15, 16, 17, 18] and the references therein).

The common situation of the studied models is that the dynamics and cost of any given agent are influenced by certain averages of the mass multi-agent behavior. For instance, consider the fundamental scalar Linear-Quadratic-Gaussian (LQG) dynamic game problem considered in [18]. In this model, the dynamics of the agents in a system with population N are given by

$$dz_i = (a_i z_i + b u_i)dt + \sigma_i dw_i, \quad 1 \le i \le N, \ t \ge 0,$$

where $z_i, u_i \in \mathbb{R}$ are the state and control input, respectively, and $\{w_i, 1 \le i \le N\}$ denotes a set of independent standard Wiener processes. The cost function for each individual agent is given by

$$J_i^{(N)}(u_i, u_{-i}) \triangleq E \int_0^\infty e^{-\rho t} \left(\left[z_i - \gamma(\frac{1}{N} \sum_{k \ne i}^{N} z_k + \eta) \right]^2 + r u_i^2 \right) dt, \quad 1 \le i \le N,$$

where $r, \rho > 0$ and $\gamma \in (0,1)$.

The specification of a consistency relationship between each individual decision and the overall effect of the population on that agent, in the population limit, is the key idea of the MF framework. For our example, in the population limit (continuum of systems), we have the fundamental MF equation system

$$\begin{aligned}
\rho s_a &= \frac{ds_a}{dt} + a s_a - \frac{b^2}{r} \Pi_a s_a - z^*, \\
\frac{d\bar{z}_a}{dt} &= (a - \frac{b^2}{r} \Pi_a) \bar{z}_a - \frac{b^2}{r} s_a, \\
\bar{z} &= \int_{\mathscr{A}} \bar{z}_a dF(a), \\
z^* &= \gamma(\bar{z} + \eta),
\end{aligned}$$

where $\Pi_a > 0$ is the solution of the algebraic Riccati equation and $F(\cdot)$ is the probability distribution function of the populations' dynamical parameters $\{a_i, i \in \mathscr{A}\}$.

Each agent's MF control law, $u_i(\cdot) = -b/r(\pi_{a_i} z_i(\cdot) + s_i(\cdot))$, $1 \le i \le N$, consists of the feedback of its own local stochastic state and the feedback of the precomputed deterministic mass effect ($z^*(\cdot)$ in the MF equation system of our example). Therefore, each agent ignores in the process the fine details of the behaviour of any other individual agent [18].

One of the key features of the MF framework is the Nash optimality property of its control laws which makes each individual behaviour optimal with respect to all other agents in a game theoretic sense [18, 19]. A set of control laws $\{u_i^o(\cdot)\}_{i=1}^N$ such that $u_i^o(\cdot)$ minimizes $J_i^{(N)}(u_i, u_{-i}^o)$ for $1 \leq i \leq N$, is called a Nash equlibrium.

The standard equations of consensus theory have been shown to be derivable from the basic LQG MF equations in [20, 21, 22]. Furthermore, the MF methodology has close connections with the McKean-Vlasov equations of statistical mechanics [23, 24]. Recently, the authors in [25] used a MF approach for the synchronization of coupled oscillators.

The estimation based L-F dynamic games model for large population systems was introduced in [26, 27, 28]. In this model agents have linear stochastic dynamics and are coupled via their quadratic cost functions.

The cost function of each leader is based on a trade-off between moving toward a certain reference trajectory which is unknown to the followers and staying near the centroid of both leaders and followers,

$$J_l^L \triangleq \limsup_{T\to\infty} \frac{1}{T} \int_0^T \left(\|z_l^L - \left(\lambda_1 h + \lambda_2 \Phi^L + \lambda_3 \Phi^F\right)\|_Q^2 + \|u_l^L\|_R^2\right) dt,\ l \in \mathscr{L},$$

where $\mathscr{L}$ is the set of the leaders; $\lambda_1, \lambda_2, \lambda_3 \in [0,1]$ such that $\lambda_1 + \lambda_2 + \lambda_3 = 1$; $h \in C_n^b$ is a reference trajectory; $\Phi^L(\cdot) = 1/N_L \sum_{i\in\mathscr{L}} z_i^L(\cdot)$, is the centroid of the leaders, and $\Phi^F(\cdot) = 1/N_F \sum_{i\in\mathscr{F}} z_i^F(\cdot)$, is the centroid of the followers. The integers N_L and N_F denote the population size of the leaders and followers, respectively. $\|x\|_Q \triangleq (x^T Q x)^{1/2}$ for any appropriate dimension vector x and matrix $Q \geq 0$.

Followers react by tracking a convex combination of their own centroid and the centroid of the leaders [28]. The LRA cost function for an individual follower is given by

$$J_f^F \triangleq \limsup_{T\to\infty} \frac{1}{T} \int_0^T \left(\|z_f^F - \left(\eta \Phi^L + (1-\eta)\Phi^F\right)\|_Q^2 + \|u_f^F\|_R^2\right) dt, f \in \mathscr{F},$$

where $\mathscr{F}$ is the set of the followers and η is a scalar in $[0,1]$.

Moreover, we studied the case where the followers do not know $h(\cdot)$ (the reference trajectory of the leaders) and they need to identify the member of a given finite class of models which is generating $h(\cdot)$ by use of the Maximum likelihood estimator [26, 27, 28]. We showed that, under suitable identifiability conditions, the followers find the true reference trajectory of leaders in finite time by observing a suitable random fraction of the leaders' states [26, 27, 28].

This paper is an extension of the models in [26, 27, 28] to the nonlinear Cucker-Smale (C-S) type cost coupling functions. We formulate the large population stochastic L-F flocking problem as a dynamic game problem. In this model, the agents have similar dynamics and are coupled via their nonlinear individual cost functions which are based on the uncontrolled C-S flocking algorithm. Hence this model may be regarded as a controlled game theoretic formulation of the uncontrolled C-S flocking model in which each agent, instead of responding to an ad-hoc algorithm, obtains its control law from a game theoretic Nash equilibrium depending upon its individual cost function and those of all other agents.

The cost of each leader is based on a trade-off between moving its velocity toward a certain reference velocity and a C-S type weighted average of all the leaders' velocities. Followers react by tracking a C-S type weighted average of the velocities of all agents (leaders and followers).

For these controlled flocking dynamic game problems, similar to [25, 29], we derive two sets of coupled deterministic equations approximating the stochastic model in the large population limit. These sets of equations consist of coupled Hamilton-Jacobi-Bellmann (HJB) and Fokker-Planck-Kolmogrov (FPK) equations in the control optimized form, and an infinite population cost coupling function.

Subject to the existence of unique solutions to these systems of equations we show that: (i) the set of MF control laws for the leaders possesses an ε_N-Nash equilibrium property with respect to all other leaders, (ii) the set of MF control laws for the followers is almost surely ε_N-optimal with respect to all the other agents, and (iii) $\varepsilon_N \to 0$ as the system's population population size, N, goes to infinity. Finally, we analyze the MF system of equations for the leaders and followers with the linear coupling cost functions to retrieve similar MF equation systems in LQG dynamic game problems.

The following notation will be used throughout the paper. We use the integer valued subscript as the label for a certain agent of the population and superscripts L and F for a leader and follower agent, respectively. $\|\cdot\|$ denotes the 2-norm of vectors and $\|\cdot\|_\infty$ denotes the infinity or sup norm. $\|x\|_Q \triangleq (x^T Q x)^{1/2}$ for any appropriate dimension vector x and matrix $Q \geq 0$. Let C_n be the family of all n-dimensional continuous functions on $\mathbb{R}^+$ and $C_n^b \triangleq \{f \in C_n, \|f\|_\infty \triangleq \sup_{t\geq 0}\|f(t)\| < \infty\}$. The gradient of a scalar function f is denoted by ∇f where ∇ denotes the vector differential operator. $\nabla_v\cdot$ and $\nabla_x\cdot$ are the divergence operators with respect to v and x, respectively. The Laplacian of a function f is denoted by the symbol Δf.

2 Background and Problem Formulation

2.1 The Uncontrolled C-S Model

The fundamental uncontrolled C-S flocking model for a system of population N is given by the nonlinear dynamical system:

$$\begin{cases} dx_i(t) = v_i(t)dt, \\ dv_i(t) = \frac{1}{N}\sum_{j=1}^{N} a(\|x_i(t)-x_j(t)\|)(v_j(t)-v_i(t))dt, \end{cases} \qquad 1 \leq i \leq N, \tag{1}$$

where $t > 0$, $x_i \in \mathbb{R}^n$ and $v_i \in \mathbb{R}^n$ are, respectively, position and velocity vectors of the i-th agent, $1 \leq i \leq N$, and $x_i(0), v_i(0)$, $1 \leq i \leq N$, are given. The communication rates $a(\|x_i - x_j\|)(\cdot)$ are given by

$$a(\|x_i - x_j\|)(t) \triangleq \frac{K}{(\sigma^2 + \|x_i(t) - x_j(t)\|^2)^\beta}, \tag{2}$$

for some fixed K, $\sigma > 0$ and $\beta \geq 0$.

It is shown in [4] that the agents' velocities converge to a common value and the distance between agents remain bounded regardless of the initial configurations when $\beta < 1/2$, and subject to conditions on the initial configurations when $\beta \geq 1/2$. This result was improved in [7] in the case of $\beta = 1/2$ for any initial configurations.

The C-S kinetic (continuum) model, derived from the individual based uncontrolled C-S algorithm (1), is:

$$f_t^{cs} + v \cdot \nabla_x f^{cs} = -\nabla_v \cdot (m^{cs} f^{cs}),$$
$$m^{cs}(x,v,t) \triangleq \int_{\mathbb{R}^{2n}} a(\|x - x'\|)(v' - v) f^{cs}(x',v') dx' dv',$$

where $f^{cs}(x,v,t)$ denote the population density function of agents positioned at (x,t) with velocity v (see (15)-(16) in [5]).

2.2 Stochastic L-F Dynamic Game Flocking Model

Let $\mathscr{N}$ be the set of all leader and follower agents, $\mathscr{L}$ be the set of leaders (L-agents) and $\mathscr{F}$ be the set of followers (F-agents). We assume that $\mathscr{N} = \mathscr{L} \cup \mathscr{F}$, $\mathscr{L} \cap \mathscr{F} = \emptyset$ and the sets of leaders and followers are not empty. The integers N, N_L and N_F are reserved to denote the population size of the system, the number of the leaders and the number of the followers in the system, respectively.

The dynamics for the leaders are given by:

$$\begin{cases} dx_i^L = v_i^L dt, \\ dv_i^L = (A^L v_i^L + B^L u_i^L)dt + C^L dw_i, \end{cases} \quad 1 \leq i \leq N_L \tag{3}$$

where $x_i^L \in \mathbb{R}^n$ is the position of the i-th leader, $v_i^L \in \mathbb{R}^n$ is its velocity, $u_i^L \in \mathbb{R}^m$ is its control input, and $\{w_i, 1 \leq i \leq N_L\}$ denotes a set of N_L independent p-dimensional standard Wiener processes. The set of initial data $\{(x_i^L(0), v_i^L(0)) : 1 \leq i \leq N_L\}$ are assumed to be mutually independent and also independent of $\{w_i, 1 \leq i \leq N_L\}$ with finite second moments. The matrices A^L, B^L and C^L have compatible dimensions.

The Long Run Average (LRA) cost function for an individual leader i is given by

$$J_i^L \triangleq \limsup_{T\to\infty} \frac{1}{T} \int_0^T \left(\|\phi_i^L\|_Q^2 + \|u_i^L\|_R^2\right) dt, \quad 1 \leq i \leq N_L, \tag{4}$$

where the matrices Q and R are symmetric positive definite with compatible dimensions. In (4)

$$\phi_i^L \triangleq \lambda(r - v_i^L) + (1-\lambda)\frac{1}{N_L}\sum_{j\in\mathscr{L}}^{N_L} a(\|x_i^L - x_j^L\|)\,(v_j^L - v_i^L),$$

where λ is a scalar in $[0,1]$; $r(\cdot) \in C_n^b$ is a reference velocity signal, and the functions $a(\|x_i^L - x_j^L\|)(\cdot)$ are defined in (2). The cost of each leader is based on a trade-off between moving its velocity toward a certain reference velocity and a C-S type weighted average of all the leaders' velocities.

The dynamics for the followers are given by

$$\begin{cases} dx_i^F = v_i^F\, dt, \\ dv_i^F = (A^F v_i^F + B^F u_i^F)dt + C^F d\nu_i, \end{cases} \qquad 1 \le i \le N_F \tag{5}$$

where $x_i^F \in \mathbb{R}^n$ is the position of the i-th follower, $v_i^F \in \mathbb{R}^n$ is its velocity, $u_i^F \in \mathbb{R}^m$ is its control input, and $\{\nu_i, 1 \le i \le N_F\}$ denotes a set of N_F independent p-dimensional standard Wiener processes independent of leaders' initial values and $\{w_i, 1 \le i \le N_L\}$. The set of initial data $\{(x_i^F(0), v_i^F(0)) : 1 \le i \le N_F\}$ are assumed to be mutually independent and also independent of all initial states and Wiener processes, with finite second moments. The matrices A^F, B^F and C^F have compatible dimensions.

Followers react by tracking the C-S type weighted average of the velocities of all the leaders and followers. The LRA cost function for an individual follower i is given by

$$J_i^F \triangleq \limsup_{T\to\infty} \frac{1}{T}\int_0^T \left(\|\phi_i^F\|_Q^2 + \|u_i^F\|_R^2\right) dt, \quad 1 \le i \le N_F, \tag{6}$$

where

$$\phi_i^F \triangleq \frac{1}{N}\sum_{j=1}^{N} a(\|x_i^F - x_j\|)(v_j - v_i^F),$$

and the functions $a(\|x_i^L - x_j\|)(\cdot)$ are defined in (2). Each follower is coupled to all the leaders and followers through ϕ_i^F, $1 \le i \le N_F$, in which , for simplicity of notation, we dropped the superscripts L and F for the state of all the other leaders and followers, respectively.

We assume that the portion of the leaders in the system is fixed in large population:

(A1) There exists $\alpha \in (0,1)$ such that $\lim_{N\to\infty} N_L/N = \alpha$.

For each leader i, let $z_i^L \triangleq [x_i^L, v_i^L]^T$ and rewrite (3) as

$$dz_i^L = (F^L z_i^L + G^L u_i^L)dt + D^L dw_i, \quad 1 \le i \le N, \tag{7}$$

where

$$F^L = \begin{pmatrix} 0 & I_{n\times n} \\ 0 & A^L_{n\times n}\end{pmatrix},\ G^L = \begin{pmatrix} 0_{n\times m} \\ B^L_{n\times m}\end{pmatrix},\ D^L = \begin{pmatrix} 0_{n\times p} \\ C^L_{n\times p}\end{pmatrix}.$$

In a similar way, for each follower i, let $z_i^F \triangleq [x_i^F, v_i^F]^T$ and rewrite (5) as

$$dz_i^F = (F^F z_i^F + G^F u_i^F)dt + D^F dv_i, \quad 1 \le i \le N, \tag{8}$$

where

$$F^F = \begin{pmatrix} 0 & I_{n\times n} \\ 0 & A^F_{n\times n} \end{pmatrix}, \; G^F = \begin{pmatrix} 0_{n\times m} \\ B^F_{n\times m} \end{pmatrix}, \; D^F = \begin{pmatrix} 0_{n\times p} \\ C^F_{n\times p} \end{pmatrix}.$$

3 Mean Field Stochastic Control Theory

3.1 Preliminary Nonlinear Optimal Control Problem

In this section first we consider a single agent with linear stochastic dynamics

$$dz = (Fz + Gu)dt + Ddw, \tag{9}$$

where z_t, u_t are the state and the control input, respectively, and $z(0)$ is given. The nonlinear cost function for this agent is defined as

$$J(u) \triangleq \limsup_{T\to\infty} \frac{1}{T}\int_0^T \left(m(z) + \|u\|_R^2\right) dt, \tag{10}$$

where $m(z)$ is a positive function.

Let

$$\rho^u \triangleq \limsup_{T\to\infty} \frac{1}{T}\int_0^T \int_{\mathbb{R}^{2n}} \left(m(z^u) + \|u\|_R^2\right) f_\infty^u(z) dz dt < \infty,$$

where $u \in \mathscr{U}^\infty$, the set of state dependent steady state controls u for which is assumed to exist a smooth invariant measure f_∞^u for the state process.

Inspired by [25], let the cost perturbation function be defined as:

$$h_T^u(z,t) = E_{f_\infty^u}\left[\left(\int_t^T \left\{m + \|u\|_R^2 - \rho^u\right\} ds\right) |z(t) = z\right],$$

where $u \in \mathscr{U}^\infty$.

In case $|\lim_{T\to\infty} h_T^u(z,t)| < \infty$, for all $t \ge 0$ and all z, set $\lim_{T\to\infty} h_T^u(z,t) = h^u(z,t)$, then we can obtain

$$(Fz + Gu)\cdot \nabla_z h^u + (u)^T Ru + m + \sigma \triangle h^u = \rho^u,$$

where $\sigma = 1/2\,\mathrm{tr}(DD^T)$ and $u \in \mathscr{U}^\infty$.

Hence, we define the HJB optimality equation:

$$H(z, \nabla_z h) = \rho^o, \tag{11}$$

$$H(z, \nabla_z h) \triangleq \min_{u\in\mathscr{U}^\infty} \left\{(Fz + Gu)\cdot\nabla_z h + u^T Ru + m + \sigma\triangle h\right\},$$

$$u^o = \mathrm{argmin}_{u\in\mathscr{U}^\infty} \left\{(Fz + Gu)\cdot\nabla_z h + u^T Ru + m + \sigma\triangle h\right\}, \tag{12}$$

and the FPK equation:

$$f_t^u + \nabla_z \cdot ((Fz + Gu) f^u) - \sigma \triangle f^u = 0,$$

where

$$\rho^o = \inf_{u \in \mathscr{U}^\infty} \limsup_{T \to \infty} \frac{1}{T} \int_0^T \int_{\mathbb{R}^{2n}} \left(m(z^u) + \|u\|_R^2 \right) f_\infty^u(z) dz dt < \infty,$$

and f_∞^u is the steady-state solution of the FPK equation.

The solution of (12) can be expressed as $u^o(z) \triangleq -1/2 R^{-1} G^T \nabla_z h(z)$, which by substitution into (11) gives the nonlinear HJB equation:

$$\left(Fz - \frac{1}{4} G R^{-1} G^T \nabla_z h \right) \cdot \nabla_z h + m + \sigma \triangle h = \rho^o. \tag{13}$$

3.2 *The Nonlinear MF (NLMF) System of Equations*

3.2.1 The NLMF System of Equations for Leaders

Applying the approach above to our flocking model (4)-(7) we obtain the controlled *NLMF system of equations* for the leaders:

$$(F^L z - \frac{1}{4} G^L R^{-1} (G^L)^T \nabla_z h^L) \cdot \nabla_z h^L + m^L + \sigma^L \triangle h^L = \rho^{L,o}, \tag{14}$$

$$f_t^L + \nabla_z \cdot \left((F^L z - \frac{1}{2} G^L R^{-1} (G^L)^T \nabla_z h^L) f^L \right) - \sigma^L \triangle f^L = 0, \tag{15}$$

$$m^L(x,v) = \|\lambda (r - v) + (1 - \lambda) \int_{\mathbb{R}^{2n}} a(\|x - y\|)(\omega - v) f_\infty^L(y, \omega) dy d\omega\|_Q^2, \tag{16}$$

in the leader's infinite population limit as $N_L \to \infty$, where $\sigma^L = 1/2 \, \mathrm{tr}(D^L (D^L)^T)$. In (14)-(16) let $f^L(x,v,0) = f_0(x,v)$ be the given initial density for $x, v \in \mathbb{R}^n$, and suitable initial and boundary conditions for the MF-HJB equation (14).

The set of equations in (14)-(15) expresses the controlled mean field stochastic evolution of the infinite leader's population system, *i.e.* $N_L \to \infty$. More precisely, the NLMF system of equations consists of:

- The **Leader's MF-HJB** equation (14) which describes the solution to the minimization of the LRA cost function (10) under the optimal control $u^{L,o}(z) \triangleq -1/2 R^{-1} (G^L)^T \nabla_z h^L(z)$, with respect to the nonlinear cost coupling $m^L(x,v)$.
- The **Leader's MF-FPK** equation (15) which describes the evolution of the leader's population density function with the optimal control $u^{L,o}$ obtained from the solution of the Leader's MF-HJB equation in (14).
- The **Leader's MF-CC** (Cost Coupling) (16) which is the cost coupling aggregate effect in the leader's infinite population limit. The MF-CC, $m^L(x,v)$, is intended to approximate $\|\phi^L\|_Q^2$ defined in (4) in the leader's infinite population limit by use of the steady-state population density function, $f_\infty^L(x,v)$, obtained from the solution of the Leader's MF-FPK equation in (15).

- The leader's optimal cost:

$$\rho^{L,o} = \limsup_{T\to\infty} \frac{1}{T} \int_0^T \int_{\mathbb{R}^{2n}} \left(m^L(x,v) + \|u^{L,o}\|_R^2\right) f_\infty^L(x,v)dxdvdt.$$

It is important to note that the Leader's MF-HJB and Leader's MF-FPK equations are coupled through (i) $f_\infty^L(x,v)$ in the Leader's MF-CC, (ii) $h^L(x,v)$ in the Leader's MF-FPK equation, and (iii) $f_\infty^L(x,v)$ and $u^{L,o}$ in $\rho^{L,o}$.

Subject to existence conditions, the equilibrium solution of the NLMF system of equations for leaders (14)-(16) yields a consistent relation between the Leader's MF-CC, $m^L(x,v)$, and the leader's population density function, $f^L(x,v,t)$, generated by the Leader's MF-FPK equation (15) by use of u^o obtained from the solution of the Leader's MF-HJB equation (14).

3.2.2 The NLMF System of Equations for Followers

Using a similar approach we obtain the controlled *NLMF system of equations* for the followers:

$$(F^F z - \frac{1}{4} G^F R^{-1} (G^F)^T \nabla_z h^F) \cdot \nabla_z h^F + m^F + \sigma^F \triangle h^F = \rho^{F,o}, \tag{17}$$

$$f_t^F + \nabla_z \cdot \left((F^F z - \frac{1}{2} G^F R^{-1} (G^F)^T \nabla_z h^F) f^F \right) - \sigma^F \triangle f^F = 0, \tag{18}$$

$$\begin{aligned} m^F(x,v) = \|\alpha & \int_{\mathbb{R}^{2n}} a(\|x-y\|)(\omega - v) f_\infty^L(y,\omega) dy d\omega \\ & + (1-\alpha) \int_{\mathbb{R}^{2n}} a(\|x-y\|)(\omega - v) f_\infty^F(y,\omega) dy d\omega\|_Q^2, \end{aligned} \tag{19}$$

in the leader's infinite population limit as $N_L \to \infty$, where $\sigma^F = 1/2\,\mathrm{tr}(D^F (D^F)^T)$ and α is defined in (**A1**).

The NLMF system of equations (17)-(19) consists of: (i) the **Follower's MF-HJB** equation (17), (ii) the **Follower's MF-FPK** equation (18), and (iii) the **Follower's MF-CC** (Cost Coupling) (19) which is the cost coupling aggregate effect in the system's infinite population limit.

The MF-CC, $m^F(x,v)$, is intended to approximate $\|\phi^F\|_Q^2$ defined in (6) in the system's infinite population limit by use of both the steady-state leader's population density function, $f_\infty^L(x,v)$, obtained from the NLMF system of equation for leaders (14)-(15) and the steady-state follower's population density function, $f_\infty^F(x,v)$. The follower's optimal cost $\rho^{F,o}$ is defined in a similar way to $\rho^{L,o}$ with respect to $f_\infty^F(x,v)$.

It is important to note that the NLMF system of equations for leaders (14)-(16) and that for followers (17)-(19) are coupled through the leader's population density function, $f^L(x,v,t)$, in the Follower's MF-CC, $m^F(x,v)$.

4 Optimality Properties

In this section we assume that there exist unique solutions to the NLMF systems of equations (14)-(16) and (17)-(19). In a finite population, $N < \infty$, we assume that the finite sets of the leaders and followers use the infinite population MF control inputs:

$$u_i^{L,o} \triangleq -\frac{1}{2}R^{-1}(G^L)^T\nabla_z h^L,\ 1 \leq i \leq N_L \tag{20}$$

$$u_i^{F,o} \triangleq -\frac{1}{2}R^{-1}(G^F)^T\nabla_z h^F,\ 1 \leq i \leq N_F, \tag{21}$$

where $h^L(x,v)$ and $h^F(x,v)$ are the solutions of the Leader's and Follower's MF-HJB equations ((14) or (17)), respectively. Hence, in the leader and follower finite population cases, the closed loop dynamics of the agents are, respectively:

$$dz_i^{L,o} = (F^L z_i^{L,o} - \frac{1}{2}G^L R^{-1}(G^L)^T\nabla_z h^L)dt + D^L dw_i, \quad 1 \leq i \leq N_L, \tag{22}$$

$$dz_i^{F,o} = (F^F z_i^{F,o} - \frac{1}{2}G^F R^{-1}(G^F)^T\nabla_z h^F)dt + D^F dv_i, \quad 1 \leq i \leq N_F. \tag{23}$$

In this case each agent has sufficient a priori information to solve the NLMF system of equation offline and obtain its decentralized MF control law ((20) or (21)) which depends only on the precomputable solutions, h^L or h^F, obtained from the leader's or follower's NLMF system of equations.

Let the global admissible control set for the leader-follower agents be $\mathscr{U} \triangleq \{u|u(t)$ is adapted to the $\sigma-$algebra $\sigma(z_l^L(s), z_f^F(s), s \leq t, l \in \mathscr{L}, f \in \mathscr{F}), \forall t \geq 0\}$. In a population of N agents denote $u_{-i} \triangleq (u_1, \cdots, u_{i-1}, u_{i+1}, \cdots, u_N)$.

Definition 1. *[19] A set of controls $\{u_k^o \in \mathscr{U}, 1 \leq k \leq N\}$ for N agents is called an almost sure ε_N-Nash equilibrium with respect to the costs J_k, $1 \leq k \leq N$, if there exists $\varepsilon_N \geq 0$ such that for any fixed i, $1 \leq i \leq N$, we have*

$$J_i(u_i^o; u_{-i}^o) - \varepsilon_N \leq \inf_{u_i \in \mathscr{U}} J_i(u_i; u_{-i}^o) \leq J_i(u_i^o; u_{-i}^o). \quad a.s.$$

Theorem 1. *[30] Assume (**A1**) hold. For the system of leader-follower agents in finite population, (4), (6) and (22), we have the following:*

(i) The set of the leaders' NLMF control laws $\{u_i^{L,o}, 1 \leq i \leq N_L\}$ in (20) obtained from (14)-(16) is an almost sure ε_{N_L}-Nash equilibrium with respect to all other leaders, that is to say,

$$J_i^L(u_i^{L,o}; u_{-i}^{L,o}) - \varepsilon_{N_L} \leq \inf_{u \in \mathscr{U}} J_i^L(u; u_{-i}^{L,o}) \leq J_i(u_i^{L,o}; u_{-i}^{L,o}). \quad 1 \leq i \leq N_L, \quad a.s.$$

(ii) The set of the followers' NLMF control laws $\{u_i^{F,o}, 1 \leq i \leq N_F\}$ in (21) obtained from (17)-(19) is almost surely ε_N-Nash equilibrium with respect to all the other leaders and followers, that is to say, let $u^{L,o} \triangleq (u_1^{L,o}, \cdots, u_{N_L}^{L,o})$ then for any follower we have

$$J_i^F(u_i^{F,o};u^{L,o},u_{-i}^{F,o}) - \varepsilon_N \leq \inf_{u\in\mathscr{U}} J_i^F(u;u^{L,o},u_{-i}^{F,o}) \leq J_i^F(u_i^{F,o};u^{L,o},u_{-i}^{F,o}),\ 1\leq i\leq N_F,$$

almost surely.

(iii) $\lim_{N_L\to\infty} \varepsilon_{N_L} = 0$, *and* $\lim_{N\to\infty} \varepsilon_N = 0$, *almost surely.*

5 Analyses of the Linear Cost Coupling Case ($\beta = 0$)

In this section let $A^L = A^F = 0$ and $B^L = B^F = I$ in (3)-(5), $\beta = 0$ in (4)-(6) (linear coupling cost case), $K = 1$ in (4)-(6), and $r(\cdot) = \eta$ a constant in (4). For simplicity of analysis, we consider the scalar leader and follower system case, in (3)-(5) we let $C^L = C^F = \varsigma$, a positive constant, and in (4)-(6) we let $R = Q = I$.

Therefore, the NLMF system of equations for the leaders and followers take, respectively, the forms:

$$v\cdot\nabla_x h^L - \frac{1}{4}\|\nabla_v h^L\|^2 + m^L + \frac{1}{2}\varsigma^2\Delta h^L = \rho^{L,o}, \tag{24}$$

$$f_t^L + v\cdot\nabla_x f^L = \nabla_v\cdot((\nabla_v h^L)f^L) + \frac{1}{2}\varsigma^2\Delta f^L, \tag{25}$$

$$m^L(v) = \|\lambda(\eta - v) + (1-\lambda)\int_{\mathbb{R}}(\omega - v)f_\infty^L(\omega)d\omega\|^2, \tag{26}$$

and

$$v\cdot\nabla_x h^F - \frac{1}{4}\|\nabla_v h^F\|^2 + m^F + \frac{1}{2}\varsigma^2\Delta h^F = \rho^{F,o}, \tag{27}$$

$$f_t^F + v\cdot\nabla_x f^F = \nabla_v\cdot((\nabla_v h^F)f^F) + \frac{1}{2}\varsigma^2\Delta f^F, \tag{28}$$

$$m^F(v) = \|\alpha\int_{\mathbb{R}}(\omega - v)f_\infty^L(\omega)d\omega + (1-\alpha)\int_{\mathbb{R}}(\omega - v)f_\infty^F(\omega)d\omega\|^2, \tag{29}$$

where the optimal control with respect to $m^L(v)$ and $m^F(v)$ are

$$u^{L,o}(v) \triangleq -\frac{1}{2}\nabla_v h^L(v), \quad u^{F,o}(v) \triangleq -\frac{1}{2}\nabla_v h^F(v), \tag{30}$$

for the leaders and followers, respectively. It is important to note that the solutions of the MF-HJB equations (24)-(27), h^L and h^F, are independent of variable x since $m^L(v)$ and $m^F(v)$ in (26)-(29) are independent of x. Therefore, both terms $v.\nabla_x h^L$ and $v.\nabla_x h^F$ are zero in (24) and (27), respectively.

Let

$$h^{L,o}(v) \triangleq \lambda(\eta - v)^2 + (1-\lambda)\int_{\mathbb{R}}(\omega - v)^2 f_\infty^L(\omega)d\omega, \tag{31}$$

$$h^{F,o}(v) \triangleq \alpha\int_{\mathbb{R}}(\omega - v)^2 f_\infty^L(\omega)d\omega + (1-\alpha)\int_{\mathbb{R}}(\omega - v)^2 f_\infty^F(\omega)d\omega. \tag{32}$$

Then $h^{L,o}(v)$ and $h^{F,o}(v)$ are the solutions of the MF-HJB equations (24) and (27), respectively, where $\rho^{L,o} = \rho^{F,o} = \varsigma^2$.

Let $\mu_\infty^L \triangleq \int_{\mathbb{R}} \omega f_\infty^L(\omega) d\omega$ and $\mu_\infty^F \triangleq \int_{\mathbb{R}} \omega f_\infty^F(\omega) d\omega$ then by (30) we have

$$u^{L,o}(v) = -\frac{1}{2}\nabla_v h^L(v) = -v + \lambda\eta + (1-\lambda)\mu_\infty^L,$$
$$u^{F,o}(v) = -\frac{1}{2}\nabla_v h^F(v) = -v + \alpha\mu_\infty^L + (1-\alpha)\mu_\infty^F.$$

Therefore, the closed loop dynamics of the leaders and followers for a finite population take the form:

$$dv_i^L = (-v_i^L + \lambda\eta + (1-\lambda)\mu_\infty^L)dt + \varsigma dw_i, \quad 1 \le i \le N_L, \tag{33}$$
$$dv_i^F = (-v_i^F + \alpha\mu_\infty^L + (1-\alpha)\mu_\infty^F)dt + \varsigma d\nu_i, \quad 1 \le i \le N_F. \tag{34}$$

From (26)-(29) we get

$$m^{L,o}(v) = \|\lambda(\eta - v) + (1-\lambda)(\mu_\infty^L - v)\|^2 = \|u^{L,o}(v)\|^2,$$
$$m^{F,o}(v) = \| -v + \alpha\mu_\infty^L + (1-\alpha)\mu_\infty^F\|^2 = \|u^{F,o}(v)\|^2.$$

Now by a direct computation for $\rho^{L,o}$ we get

$$\rho^{L,o} = \limsup_{T\to\infty} \frac{1}{T}\int_0^T \int_{\mathbb{R}} \left(m^{L,o}(v) + \|u^{L,o}\|^2\right) f_\infty^L(v) dv = 2E_{f_\infty^L}\|u^{L,o}\|^2$$
$$= 2E_{f_\infty^L}\| -v + \lambda\eta + (1-\lambda)\mu_\infty^L\|^2 = 2\pi_\infty^L,$$

and in a similar way $\rho^{F,o} = 2\pi_\infty^F$, where π_∞^L and π_∞^F are the solutions of the Lyapunov equations for the steady-state covariances corresponding to the systems (33)-(34). Therefore, by the algebraic Lyapunov equations we have $\pi_\infty^L = \pi_\infty^F = \varsigma^2/2$.

On the other hand, $f_\infty^L(v)$ and $f_\infty^F(v)$ are the solutions of the steady-state MF-FPK equations,

$$v \cdot \nabla_x f^L + \nabla_v \cdot (u^{L,o} f^L) - \frac{1}{2}\varsigma^2 \Delta f^L = 0,$$
$$v \cdot \nabla_x f^F + \nabla_v \cdot (u^{F,o} f^F) - \frac{1}{2}\varsigma^2 \Delta f^F = 0.$$

which have Gaussian distributions:

$$f_\infty^L(v) = \frac{1}{\sqrt{2\pi\pi_\infty^L}} \exp\{-\frac{\|v - (\lambda\eta + (1-\lambda)\mu_\infty^L)\|^2}{2\pi_\infty^L}\},$$
$$f_\infty^F(v) = \frac{1}{\sqrt{2\pi\pi_\infty^F}} \exp\{-\frac{\|v - (\alpha\mu_\infty^L + (1-\alpha)\mu_\infty^F)\|^2}{2\pi_\infty^F}\}.$$

Hence, our result in the linear coupling cost function case ($\beta = 0$) is consistent with the standard MF stochastic control for LQG dynamic game problems [18].

References

1. Reynolds, C.W.: Flocks, herds, and schools: A distributed behavioral model. Computer Graphics 21(4), 25–34 (1987)
2. Perea, L., Gómez, G., Elosegui, P.: Extension of the Cucker–Smale control law to space flight formations. Journal of Guidance, Control, and Dynamics 32(2), 297–325 (2009)
3. Vicsek, T., Czirók, A., Jacob, E.B., Cohen, I., Shochet, O.: Novel type of phase transition in a system of self-driven particles. Phys. Rev. Lett. 75(6), 1226–1229 (1995)
4. Cucker, F., Smale, S.: Emergent behavior in flocks. IEEE Transactions on Automatic Control 52(5), 852–862 (2007)
5. Carrillo, J.A., Fornasier, M., Toscani, G., Vecil, F.: Particle, kinetic, and hydrodynamic models of swarming (preprint 2009)
6. Topaz, C.M., Bertozzi, A.L.: Swarming patterns in a two-dimensional kinematic model for biological groups. SIAM Journal on Applied Mathematics 65(1), 152–174 (2004)
7. Ha, S.Y., Liu, J.G.: A simple proof of the Cucker-Smale flocking dynamics and mean-field limit. Communications in Mathematical Sciences 7(2) (2009)
8. Carrillo, J.A., Fornasier, M., Rosado, J., Toscani, G.: Asymptotic flocking dynamics for the kinetic Cucker–Smale model. SIAM Journal on Mathematical Analysis 42, 218 (2010)
9. Dyer, J.R.G., Johansson, A., Helbing, D., Couzin, I.D., Krause, J.: Leadership, consensus decision making and collective behaviour in humans. Philosophical Transactions of the Royal Society B: Biological Sciences 364(1518), 781–789 (2009)
10. Couzin, I.D., Krause, J., Franks, N.R., Levin, S.A.: Effective leadership and decision-making in animal groups on the move. Nature 433(7025), 513–516 (2005)
11. Simaan, M., Cruz, J.B.: On the Stackelberg strategy in nonzero-sum games. Journal of Optimization Theory and Applications 11(5), 533–555 (1973)
12. Gu, D., Wang, Z.: Leader–follower flocking: algorithms and experiments. IEEE Transactions on Control Systems Technology 17(5), 1211 (2009)
13. Shen, J.: Cucker-Smale flocking under hierarchical leadership. SIAM Journal on Applied Mathematics 68(3), 694–719 (2008)
14. Nagy, M., Ákos, Z., Biro, D., Vicsek, T.: Hierarchical group dynamics in pigeon flocks. Nature 464(7290), 890–893 (2010)
15. Caines, P.E.: Bode lecture: Mean field stochastic control. In: The 48th IEEE Conference on Decision and Control, Shanghai, China (December 2009), `http://www.ieeecss.org/CAB/conferences/cdc2009/shanghai2009.31.pdf`
16. Huang, M., Malhamé, R.P., Caines, P.E.: Individual and mass behaviour in large population stochastic wireless power control problems: centralized and Nash equilibrium solutions. In: 42th IEEE Conference on Decision and Control, Hawaii, pp. 98–103 (December 2003)
17. Huang, M., Caines, P.E., Malhamé, R.P.: The NCE (Mean Field) principle with locality dependent cost interactions. IEEE Transactions on Automatic Control (December 2010)
18. Huang, M., Caines, P.E., Malhamé, R.P.: Large-population cost-coupled LQG problems with nonuniform agents: Individual-mass behavior and decentralized ε-Nash equilibria. IEEE Transactions on Automatic Control 52(9), 1560–1571 (2007)

19. Li, T., Zhang, J.F.: Asymptotically optimal decentralized control for large population stochastic multiagent systems. IEEE Transactions on Automatic Control 53(7), 1643–1660 (2008)
20. Nourian, M., Caines, P.E., Malhamé, R.P., Huang, M.: Derivation of consensus algorithm dynamics from mean-field stochastic control NCE equations. In: Proc. of 1st IFAC Workshop on Estimation and Control of Networked Systems (NecSys 2009), Venice, Italy (September 2009), http://www.ifac-papersonline.net/Detailed/40520.html.
21. Nourian, M., Caines, P.E., Malhamé, R.P., Huang, M.: A solution to the consensus problem via stochastic mean field control. In: Proc. of 2st IFAC Workshop on Estimation and Control of Networked Systems (NecSys 2010), Annency, France (September 2010)
22. Nourian, M., Caines, P.E., Malhamé, R.P., Huang, M.: A game theoretic analysis of large population consensus problem via stochastic mean field control theory. IEEE Transactions on Automatic Control (submitted)
23. Huang, M., Malhamé, R.P., Caines, P.E.: Nash certainty equivalence in large population stochastic dynamic games: Connections with the physics of interacting particle systems. In: Proc. of the 45th IEEE Conference on Decision and Control, San Diego, USA, pp. 4921–4926 (December 2006)
24. Huang, M., Malhamé, R.P., Caines, P.E.: Large population stochastic dynamic games: closed-loop Mckean-Vlasov systems and the Nash certainty equivalence principle. Communications in Information and Systems 6(3), 221–252 (2006)
25. Yin, H., Mehta, P.G., Meyn, S.P., Shanbhag, U.V.: Synchronization of coupled oscillators is a game. In: American Control Conference (June 2010)
26. Nourian, M., Malhamé, R.P., Huang, M., Caines, P.E.: Mean field (NCE) formulation of estimation based leader-follower collective dynamics. International Journal of Robotics and Automation, Special Issue on New Advances in Nonlinear and Optimal Controls of Robotic and Autonomous Systems (to appear)
27. Nourian, M., Malhamé, R.P., Huang, M., Caines, P.E.: Optimality of adaption based Mean Field (NCE) control laws in Leader-Follower stochastic dynamic games. In: 49th IEEE Conference on Decision and Control, Atlanta, Georgia (December 2010)
28. Nourian, M., Caines, P.E., Malhamé, R.P., Huang, M.: Adaptive mean field control in leader-follower stochastic dynamic games. IEEE Transactions on Automatic Control (submitted)
29. Nourian, M., Caines, P.E., Malhamé, R.P.: Synthesis of Cucker-Smale type flocking via mean field stochastic control theory: Nash equilibria. In: Proc. the 48th Allerton Conference on Communication, Control and Computing, Monticello, Illinois (September 2010)
30. Nourian, M., Caines, P.E., Malhamé, R.P., Huang, M.: Synthesis of Cucker-Smale type flocking via mean field stochastic control theory. Technical Report, McGill Uni. (August 2010)

Dynamic Locomotion with a Wheeled-Legged Quadruped Robot

I. Sharf

Abstract. In this paper, we present an overview of the work carried out in the Mechatronic Locomotion Laboratory at McGill University on a quadruped robotic platform, PAW. This robot features four springy legs with rotary actuation at the hips and driven wheels mounted at the distal ends of the legs. The robot was designed to explore hybrid modes of locomotion, where it makes use of both wheels and legs to achieve novel behaviors. As well, the robot's simple construction allows PAW to exploit the dynamics of a mass-spring system to achieve gaits such as bounding, galloping and jumping. We begin by describing the basic design of the robot and its sensing capabilities. We then discuss several modes of locomotion that have been developed on the robot over the past five years. Specifically, results are presented for inclined turning and sprawled breaking achieved with the robot, as part of exploiting the leg capability in the rolling behaviors. This is followed by the presentation of the bounding gait implemented on the robot: the basic version and the intelligent version. The most recent addition to the robot's repertoire of behaviors is a dynamic jump. We will discuss the main stages of the jumping maneuver and present the results of the jump. The paper concludes with a summary and discussion of our future goals for PAW.

1 Introduction

Interest in legged robotic systems has increased significantly among roboticists since the ground-breaking work by Raibert in the 80's [1]. Much of the work on legged systems has been inspired by animal locomotion and also motivated by the possibilities offered by legged vehicles to traverse rough terrain and get over obstacles. At the same time, wheeled robotic systems still provide unparalleled efficiency for locomotion, especially on paved terrain.

Also pursued by researchers in the mobile robotics community is the development of *hybrid* systems, specifically, vehicles that incorporate both wheeled and legged capabilities. Examples of these are Walk'n Roll [2], the Roller-Walker [3],

I. Sharf
Centre for Intelligent Machines, McGill University

J. Angeles et al. (Eds.): Brain, Body and Machine, AISC 83, pp. 299–310.
springerlink.com

the Shrimp [4], the ANT [5] platforms, as well as the four-bar-based and the SDCSC-based design configurations in [6]. These robots combine aspects of traditional wheeled and legged systems into a single articulated suspension platform, thereby taking advantage of both locomotion modalities.

The platform employed in this work, PAW (Platform for Ambulating Wheels), is a quadruped robot and it was also designed to exploit the benefits of both legged and wheeled modes of locomotion (see Fig. 1). PAW features two-degree-of-freedom springy legs actuated with a single motor at the hip joint, thus exhibiting strong passive dynamics that is beneficial for dynamically stable legged gaits. In addition, PAW's legs are equipped with actuated wheels at the distal ends. Unlike many of the existing quadrupedal platforms, PAW is power and computationally autonomous: it has onboard batteries and computer with no umbilical connections to the outside world. PAW's versatility has been demonstrated in the development of purely legged behaviors, such as bound [7], pronk and gallop [8], as well as primarily wheeled behaviors augmented with legged capability, such as inclined turning [9]. More recently, a new biologically-inspired behavior has been developed for PAW---the jump [10]; it further expands the robot's locomotion capabilities and takes advantage of its dual (wheeled and legged) nature. In this paper, we summarize the present PAW's repertoire of legged and hybrid behaviors and discuss some directions for further research with this versatile and unique research platform.

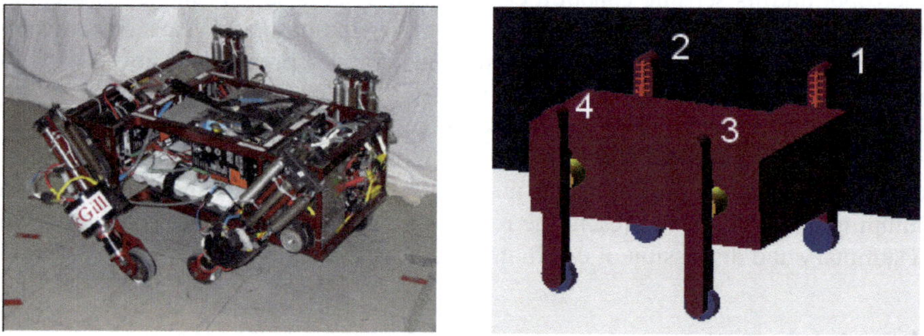

Fig. 1. PAW robot: physical prototype (left) and MSC ADAMS model (right)

2 PAW Design Overview

PAW has been designed as a platform for the study of both wheeled and dynamically stable legged modes of locomotion [9]. The legs of the robot have a unique compliant design incorporating a passive linear spring between the actuated hip and wheel joints.

The PAW robot has a T-shaped body which, with the four leg assemblies, adds up to a total weight of about 20 kg. The aluminum frame of the body is approximately 0.5 m long and 0.36 m wide at the widest location. The hip joints of the four legs are each driven by 90 Watt Maxon 118777 brushed DC motors with

73.5:1 gearheads. A toothed belt and pair of sprockets provide a further 32:24 reduction ratio. The legs are 0.212 m long and each is equipped with a pair of extension springs rated up to 3200 N/m. At the end of each leg is a 20 Watt Maxon 118751 brushed DC motor with a 4.8:1 Maxon 233147 planetary gearbox and a custom 3:1 ratio bevel gear pair. This end-leg assembly is connected to a 0.066 m diameter, hard rubber wheel [7]. Power and signal wires to the motors and sensors on each leg are passed through a hollow hip axle to prevent the cables from becoming entangled in the legs. However, the legs are not continuously recirculating, as is the case in robots such as RHex [9].

Other relevant electrical components on the robot include a PC/104 computer stack, four AMC 25A8 brushed DC motor amplifiers for driving the hip motors, a custom amplifier board containing Apex SA60 motor amplifiers and three NiMH battery packs. The PC/104 computer stack contains a Pentium-compatible processor board running the QNX 6.1 real-time operating system and control code, a PCMCIA board with wireless Ethernet card for teleoperation, a power supply board, a quadrature decoder board for obtaining motor angles and two I/O boards, one custom and the other off-the-shelf.

The PAW robot is controlled by relying on minimal sensing and, hence, it houses very few sensors. In addition to battery voltage and current sensors, the robot uses one quadrature encoder in each of its eight motors (2000 cpr effective resolution), one linear potentiometer with up to 0.10 m range in each of its four legs and a current sensor on each hip motor amplifier. The motor encoders are used to determine the angle of the eight motor shafts, while the linear potentiometers are used to measure leg compression. By measuring current consumption and battery voltage it is possible to determine power usage in various subsystems of the robot. In the case of the hip motor amplifiers, current measurement also provides a proportional estimate of motor torque applied to the hip. Finally, affixed to the body of PAW is a BAE SiIMU-01inertial measurement unit (IMU) which provides data on the attitude and acceleration of the vehicle.

3 Rolling Behaviors

The rolling behaviors described in this section were explored with the view to utilizing the leg actuation to increase the robot performance in standard wheeled behaviors. It is noted that the control of both hip and wheel motors for the rolling behaviors is accomplished with a PID control law which is responsible for either maintaining a desired position or a desired velocity at the joints. Transitions between one set of desired velocities and/or positions and another is resolved using cycloidal functions to provide smooth motion [11]. Because the robot is redundantly actuated when all legs are on the ground and there is no coordination between individual motor controllers during wheeled locomotion, it is not possible to have high gains on all the joints. Therefore, high gains are used on the hip actuators to ensure proper positioning of the legs with respect to the body and lower gains are used at the wheels, resulting in relatively compliant wheel motion.

3.1 Inclined Turning

3.1.1 Controller

A new method has been developed for achieving turns on the PAW robot [9] which takes advantage of its hybrid mobility. In particular, rather than applying differential wheel speeds on either side of the robot with the legs fixed, the legs are used to reposition the wheels to reduce shear forces on them and to produce tighter turns. Effectively, this means that while the legs on the outside of the turn are kept at right angle to the body, the legs on the inside of the turn are brought together, lowering the center of mass (COM) and leaning the robot into the turn. Figure 2 shows the views of the robot in a turn, and the important geometric parameters. The turn is implemented on the robot by prescribing the required inner hip joint angle (ϕ) and individual wheel speeds to produce the desired turning radius, COM speed and height (*H*). Detailed derivation of the control set points can be found in [8].

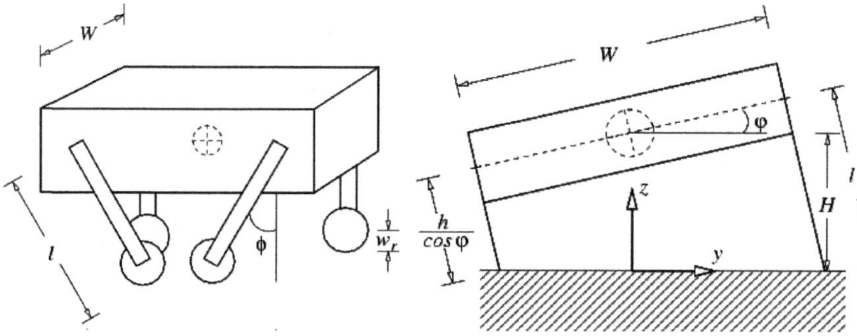

Fig. 2. Simplified views of PAW in inclined turning configuration and relevant variables.

3.1.2 Results

The results of six experiments conducted on the turning behavior for different settings of desired COM radii, COM speed and two settings of leg angles (29.6° and 62.0°) are summarized in Table 1. Note that Columns 3 and 4 of Table 1 represent percentages of the values found in Column 5, the desired forward speed. For all experiments, desired radii and speed of the COM were matched to within approximately 10%. It should be noted that increasing the roll-over stability via the proposed turning maneuver is not critical at the speeds and turn radii demonstrated on PAW to date; however, this will become more important at higher speeds for this or a scaled-up version of the vehicle.

Table 1. Experimental results for inclined turning

Exp. #	Leg angle [deg]	Outer Wheel Spds (%)	Inner Wheel Spds (%)	COM Speed Des'd (m/s)	COM Speed Ach'd (m/s)	Turn Radius Des'd (m)	Turn Radius Ach'd (m)
1	29.6	154, 141	56, 69	0.50	0.43	0.50	0.50
2	62.0	160, 147	58, 71	1.25	1.17	0.50	0.55
3	29.6	135, 127	70, 79	0.50	0.47	0.75	0.83
4	62.0	139, 131	72, 81	1.25	1.20	0.75	0.85
5	29.6	126, 120	78, 84	0.50	0.47	1.00	1.13
6	62.0	129, 123	79, 86	1.25	1.19	1.00	1.13

3.2 Sprawled Braking

3.2.1 Controller

An important aspect of PAW's maneuverability is the stopping distance during breaking. While driving forwards or backwards, it is important to apply braking action in order to prevent the robot from slipping and from pitching over. Pitching motion can result from braking too suddenly or by having the legs oriented vertically or in a tucked configuration [9]. To improve the pitching stability during the breaking maneuver, we experimented with the robot placing its legs in a sprawled posture, at about ±11.5° with respect to the vertical reference. When a brake command is issued, the motors are used to dissipate the kinetic energy of the robot through the use of low-gain PID controllers, which also prevents wheel slip.

3.2.2 Results

Experimental trials were performed with the robot driving at 1.5 m/s and with the legs tucked in or sprawled out at 11.5° relative to the vertical. As well, alternatively high and low control gains were used for the wheel motors to demonstrate the effect of controller gains on the breaking distance and wheel slipping. Ten trials were conducted for each of the four configurations (legs tucked and high gains, legs tucked and low gains, legs sprawled and high gains, legs sprawled and low gains) and the average braking distances were measured. In the first set of experiments, the robot repeatedly tipped over due to high gain wheel control and a tucked-in leg posture. In the second set of trials, the wheel control gains were lowered and the robot did not tip over, but minor pitching was observed in the video footage. The braking distance was measured at 0.24 m. A sprawled posture and high wheel gains used in the third set of experiments resulted in wheel slip, but a relatively short braking distance of 0.15 m. In the fourth set of experiments with lowered wheel gains and a sprawled posture, stable braking was achieved with little slip or pitching but with an increased braking distance of 0.23 m.

4 Legged Behaviors: Bounding Gait

Differently from the rolling behaviors discussed in the previous section, our work on legged behaviors for PAW has focused on *dynamically stable* gaits. These gaits are sustained in a robot like PAW through exchange between kinetic and potential energies, with additional energy injected during stance phases by the motors. The bound gait in particular, is a two-beat dynamic gait in which there is a large degree of symmetry in both control and overall motion along the spinal or *sagittal* plane [7]. A schematic of the sagittal plane representation of the PAW robot in a bound is found in Fig. 3. Following the diagram from left to right, starting from the flight phase, the gait involves front-leg touchdown, followed by rear-leg touchdown resulting in double-stance phase; then, front-leg liftoff and rear-leg liftoff, this last step launching the robot into flight again. In our research on the bounding gait with PAW, we investigated the effects of touchdown and takeoff leg angles on the performance of the robot, under conditions where the wheels are passively or actively locked. As well, an intelligent controller was implemented to allow PAW to track a desired forward velocity by automatically adjusting touchdown and takeoff leg angles.

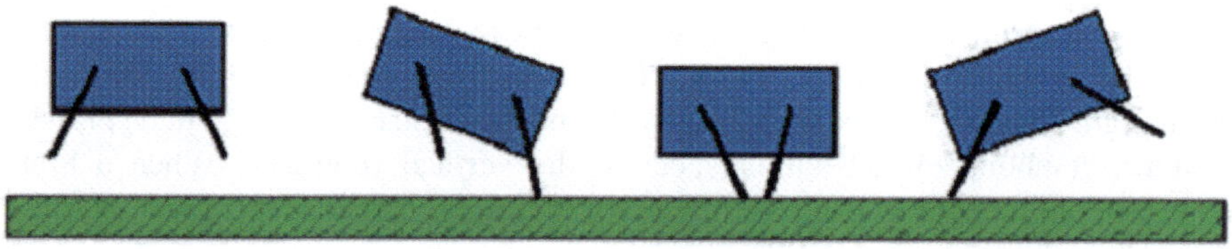

Fig. 3. Schematic of main phases of bounding gait

4.1 Basic Bounding Controller

4.1.1 Finite-State Machine

To implement the basic bounding gait, the hip angles of PAW's rear and front leg pairs are controlled by two separate and independent finite-state machines (one for each leg pair) [7]. Referring to Fig. 4, in the flight phase, the leg is actuated to a desired touchdown angle using a position-based PD controller. During the stance retraction phase, a constant desired torque is commanded in order to drive the motors to saturation. When the takeoff angle is reached, the state switches to stance brake where a position PD controller engages, holding the leg at the takeoff angle. Note that if the takeoff angle is not reached prior to takeoff being detected, the stance brake phase is bypassed. The switching between these modes is determined by the bounding state machine. The leg length potentiometers are responsible for detecting the transitions to the flight and stance retraction states, while the hip motor encoders are used for transition to the stance brake.

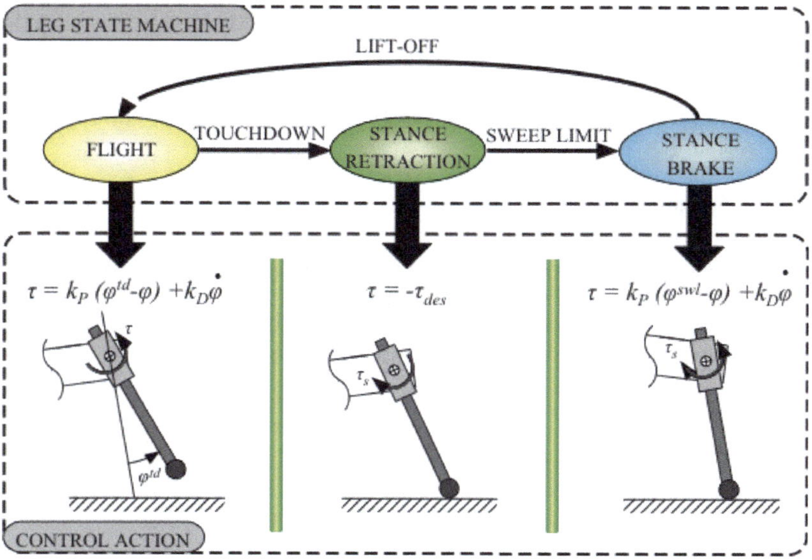

Fig. 4. Finite-state machine and control laws for main phases of PAW's bounding gait

4.1.2 Results

We present the results of the bounding gait experiments with PAW for the case when the wheels are mechanically blocked (passive locking). Performance of PAW is slightly worse when the wheels are actively locked by commanding them to zero motion during the stance phase [7]. Table 2 displays the results from three experiments for a particular set of touchdown angles (rear and front) and three sets of liftoff angles. The corresponding speeds of the robot vary from 0.87 to 1.18 m/s, demonstrating that one can affect the speed of the robot by changing the lift-off angles. Repeatability of the experiments is excellent, although, it deteriorates slightly for actively locked wheel experiments [7].

Table 2. Experimental results: PAW bounding with mechanically blocked wheels

Exp.	Front-leg touchdown, liftoff angles (deg)	Rear-leg touchdown, liftoff angles (deg)	COM speed (m/s)	Repeatability (successful/total trials)
1	(−20°,4°)	(−22°,12°)	0.87	10/10
2	(−20°,6°)	(−22°,14°)	0.99	10/10
3	(−20°,10°)	(−22°,18°)	1.18	10/11

4.2 Intelligent Bounding Controller

The bounding controller described in the previous section uses specific hip angle combinations for the leg touchdown and liftoff instants to achieve a stable bounding gait. These hip angle values are predetermined through a trial-and-error process which is undesirable in practice. Using intelligent control allows the robot to modify these fixed values in order to adapt to the desired velocity set-point and possible changes in the terrain, thereby increasing the robot autonomy and versatility.

4.2.1 Controller

The intelligent controller implemented on PAW to achieve velocity tracking is a Levenberg-Marquardt (LM) learning algorithm that modifies the controller developed by Raibert [1]. The control action is defined by the following equation for the foot placement x_f of the robot, for the specified desired velocity v_d:

$$x_f = \alpha_1 v + \alpha_2 (v - v_d) + \alpha_3 \tag{1}$$

The desired leg angles can then be computed directly from the foot placement locations and the leg lengths [12]. The LM intelligent algorithm solves the least squares problem to tune gains α_1 and α_2; this is done at the top of flight during each gait cycle, for front- and rear-leg touchdown and liftoff combinations. The α_3 term aids in compensating for the loss in energy; this term remains constant for small velocity changes [12].

Implementation of the controller on the PAW robot requires the measurement of the velocity of the robot, v. As noted previously, PAW is equipped with an inertial measurement unit which in addition to measuring the attitude of the robot, also outputs its acceleration. To obtain an optimal estimate of forward velocity, we use an Extended Kalman Filter which combines the IMU data with the velocity estimate obtained from the measured leg angles and leg lengths, through the kinematics relations for the robot. A detailed description of the velocity estimator and controller implementation can be found in [13].

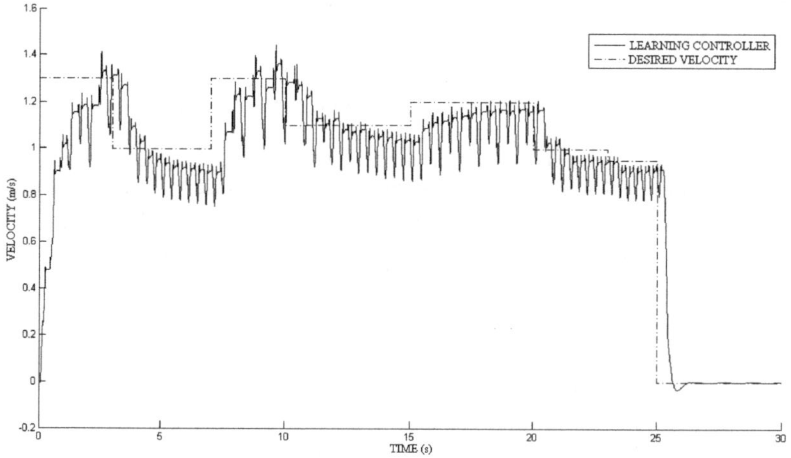

Fig. 5. Simulation results showing tracking velocity variations with intelligent controller

4.2.2 Results

The intelligent controller was first tested in simulation with the MSC Adams model of PAW (see Fig. 1). Figure 5 displays the simulated velocity results, showing fairly good tracking of the variable velocity set points. Implementation on the physical robot showed successful performance of the intelligent controller for desired velocities in the range 0.9-1.3 m/s, with tracking to within 0.1 m/s after convergence of the controller is achieved (see Fig. 6 for results at $v_d = 1.2$ m/s).

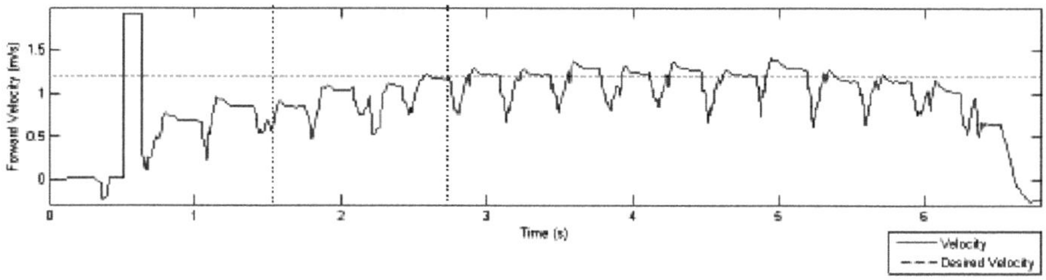

Fig. 6. Experimental results showing tracking desired velocity of 1.2 m/s. The dashed vertical lines signify the beginning of intelligent controller and the time of convergence.

5 Dynamic Jump

The dynamic jump described here represents the latest addition to PAW's repertoire of behaviors and indeed, to the author's knowledge, the first implementation of the jump on a quadrupedal robot. Previous investigations on jumping quadrupeds did not progress beyond simulation or simplified experimental stages as, for example, in Wong and Orin [14]. Most relevant to our work is the recently published work of Krasny and Orin [15], who investigate in simulation a running jump of quadruped robot with articulated knees from a galloping gait.

5.1 Jumping Maneuver

The jumping maneuver presented here was arrived at by considering the specific structure and design of PAW: PAW's legs lack the knee joint and PAW cannot actively control the leg thrusts or lengths. Moreover, the proposed jump maneuver relies on PAW having actuated wheels at the ends of its legs.

The solution is to use PAW's wheels to generate a "run-up" phase, so that the robot has a significant amount of forward momentum before the leg rotation even begins. By positioning the rear legs properly and applying sudden braking to the rear wheels, a large fraction of the body's kinetic energy is transferred to the rear leg springs, causing compression. The rear legs are then rapidly rotated into position for takeoff, redirecting the stored spring energy and launching PAW into the air. Thus, the proposed jumping sequence is separated into five principal phases:

Phase a. forward momentum gain,
Phase b. body positioning,
Phase c. rear leg compression,
Phase d. rear leg rotation, and
Phase e. flight.

The above phases of the maneuver are implemented by parameterizing the hip desired angles and wheel velocities, as well as the time during the jumping sequence. The resulting parameterizations based on the piecewise linear functions of time are illustrated in Fig. 7, where the phases and the relevant parameters are indicated. Out of these, eight parameters were identified as important to the jumping performance. Their values were initially selected by manual tuning, but ultimately optimized by employing a genetic algorithm. In both cases, we employed the MSC Adams/Simulink model of PAW to determine whether the jump performance, defined as the obstacle height cleared by the robot, is acceptable [10].

5.2 Jump Implementation and Results

Experiments were conducted with PAW in the laboratory setting at Defence R&D Canada – Suffield, where PAW was operated on a concrete floor and the jumping behaviors were tested with an elevated wooden platform, 0.105 m high. For all trials, the jump maneuver was initiated at a fixed distance of 1.1 m from the obstacle; in a real-life setting, distance to obstacle could be provided by vision or laser-range sensing on the robot. Figure 8 shows a photo sequence of PAW jumping onto the platform and a stable landing produced with the optimal parameter values, thus demonstrating the feasibility of the jump on the physical robot. The highest jump observed with PAW is 0.156 m from the ground, which represents 74% of the leg length.

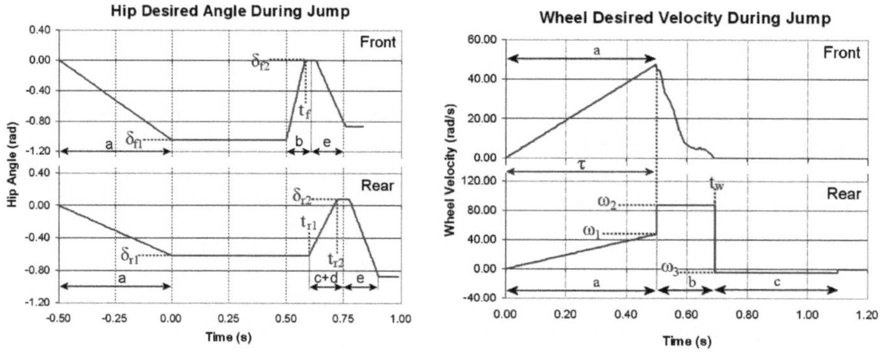

Fig. 7. Desired hip angles and wheel rates during jump

Fig. 8. Photo sequence of PAW jumping onto a platform

6 Conclusions and Future Work

We have summarized our work over the past five years on the development of mobility for the hybrid robotic platform PAW. This research demonstrates some of the possibilities offered by wheel-legged robots. Our future work with PAW will involve a systematic study, optimization and experimental evaluation of maneuverability of hybrid vehicles and further development of hybrid dynamic gaits.

Acknowledgements. The author would like to thank former graduate students Michele Faragalli, Adam Harmat and James Smith for their dedication and perseverance with PAW. Funding support from NSERC, PRECARN and Defence R&D Canada – Suffield is also gratefully acknowledged.

References

[1] Raibert, M.H.: Legged Robots That Balance. MIT Press, Cambridge (1986)

[2] Adachi, H., Koyachi, N., Arai, T., Shimiza, A., Nogami, Y.: Mechanism and control of a leg-wheel hybrid mobile robot. In: Proc. IEEE/RSJ Int. Conf. on Intelligent Robots and Systems, Kyongju, South Korea, vol. 3, pp. 1792–1797 (1999)

[3] Endo, G., Hirose, S.: Study on roller-walker (system integration and basic experiments). In: Proc. IEEE Int. Conf. on Robotics and Automation, Detroit, MI, vol. 3, pp. 2032–2037 (1999)

[4] Estier, T., et al.: An innovative space rover with extended climbing abilities. In: Proc. Space and Robotics 2000, Albuquerque, NM, pp. 333–339 (2000)

[5] Brosinsky, C., Hanna, D., Penzes, S.: Articulated navigation testbed (ANT): an example of intrinsic mobility. In: Proc. of the SPIE 14th Annual Int. symposium on Aerospace/Defence Sensing, Simulation and Controls, Orlando, FL, vol. 4024, pp. 86–95 (2000)

[6] Jun, S.K., White, G.D., Krovi, V.N.: Kinetostatic design considerations for an articulated leg-wheel locomotion subsystem. ASME Journal of Dynamic Systems Measurement and Control 128, 112–121 (2005)

[7] Smith, J.A., et al.: Bounding with active wheels and liftoff angle velocity adjustment. International Journal of Robotics Research (2010)

[8] Smith, J.: Galloping, bounding and wheeled-leg modes of locomotion on underactuated quadrupedal robots, PhD Thesis, McGill University (2006a)

[9] Smith, J.A., Sharf, I., Trentini, M.: PAW: a hybrid wheeled-leg robot. In: Proc. 2006 IEEE International Conference on Robotics and Automation, Orlando, FL, USA, pp. 4043–4048 (May 2006b)

[10] Harmat, A.: Jumping behavior for a wheeled quadruped robot: analysis and experiments. MEng. Thesis, McGill University (2008)
[11] Angeles, J.: Fundamentals of Robotic Mechanical Systems - Theory, Methods, and Algorithms, 2nd edn. Springer, Heidelberg (2003)
[12] Faragalli, M., Sharf, I., Trentini, M.: Velocity control of a hybrid quadruped bounding robot. In: Proc. IEEE/RSJ International Conference on Intelligent Robots and Systems, Nice, France, September 22-26, pp. 1–6 (2008)
[13] Faragalli, M.: Intelligent velocity control of a bounding quadruped robot. MEng Thesis, McGill University (2009)
[14] Wong, H.C., Orin, D.E.: Dynamic control of a quadruped standing jump. In: Proc. IEEE International Conference on Robotics and Automation (Cat. No.93CH3247-4), Atlanta, GA, USA, vol. 3, pp. 346–351 (1993)
[15] Krasny, D.P., Orin, D.E.: Evolution of dynamic maneuvers in a 3D galloping quadruped robot. In: Proc. 2006 Conference on International Robotics and Automation, Orlando, FL, USA, pp. 1084–1089 (2006)

Underactuated Cable-Driven Robots: Machine, Control and Suspended Bodies

Clément Gosselin, Simon Lefrançois, and Nathaniel Zoso

Abstract. This paper introduces a novel family of robots that consist of cable-suspended bodies whose motion is not fully constrained. The robots are underactuated and exhibit a pendulum-like behavior. Based on the dynamic model, a technique is proposed to allow the planning of point-to-point trajectories with zero-velocity landing by making the robot swing itself similarly to children on playground swings. A three-degree-of-freedom planar robot is studied as an example and a prototype of the robot and its controller are presented. Experimental results clearly demonstrate the effectiveness of the proposed control technique. Underactuated cable-suspended robots have the potential to lead to low-cost solutions in applications that require the performance of point-to-point trajectories in a large workspace.

1 Introduction

Cable-driven robots have been proposed in the literature for applications requiring large workspaces (see for instance [1, 2] for pioneer designs). Indeed, cables can be wound on reels, thereby providing large motion ranges. Moreover, replacing rigid links with cables greatly reduces the weight and required actuation power. However, cable-driven robots are generally based on parallel architectures comprising more (or at least as many) actuators than degrees of freedom since cables can only pull and

Clément Gosselin
Dép. de génie mécanique, Université Laval, Québec, Qc, Canada
e-mail: gosselin@gmc.ulaval.ca

Simon Lefrançois
Dép. de génie mécanique, Univ. Laval, Québec, Qc, Canada
e-mail: simon.lefrancois.1@ulaval.ca

Nathaniel Zoso
Dép. de génie mécanique, Univ. Laval, Québec, Qc, Canada
e-mail: nathaniel.zoso.1@ulaval.ca

J. Angeles et al. (Eds.): Brain, Body and Machine, AISC 83, pp. 311–323.
springerlink.com

not push. Implementing such mechanisms requires installing actuators at different locations in space. A support structure as well as some calibration [3] are thereby needed.

On the other hand, underactuated mechanisms are systems with fewer actuators than degrees of freedom whose control has attracted significant attention. The motion of the free (unactuated) joints is generally related to that of the other joints by complex dynamics, which makes the control problem challenging [4]. Moreover, due to underactuation, only a subset of the kinematically possible global trajectories are achievable. Nevertheless, such mechanisms are well-suited for point-to-point operations since the latter involve mainly getting from a point to another regardless of the path.

In this paper, it is proposed to combine the advantages of underactuation (few actuators, design simplicity) with those of cable transmissions (agility, large workspace, low mass) by introducing a family of underactuated cable-suspended robots. The general principle is to suspend one or a few connected rigid bodies by one or several cables whose length is actively controlled. Using an appropriate trajectory planning strategy for the actuated joints, the passive swinging motion of the suspended bodies can be exploited to reach predefined target poses. Hence, the robot is capable of performing point-to-point trajectories, i.e., trajectories in which only the initial and final configurations are of interest. Potential applications of such mechanisms include dock loading, construction, field robotics, domotics, surveillance systems, automated greenhouses as well as entertainment and human-robot interactions. In the context of the latter, it can be argued that since underactuated cable-driven robots are lightweight and minimally constrained, they cannot lead to potential hazardous clamping incidents [5].

A first two-degree-of-freedom (dof) cable-driven pendulum-like robot, referred to as the Winch-Bot, was introduced in [6]. The authors used off-line trajectory planning to control the position of the robot. However, the proposed technique requires specific initial conditions, which in turn require either moving the payload or the attachment point of the cable.

In this paper, the general concept of underactuated cable-suspended robot is introduced through two planar three-dof cable-suspended robots having two actuators each and fixed attachment points. The paper is structured as follows: first, the architecture of the two proposed robots is described and the equations of motion are provided. Then, the actuated joint trajectory design for swing-up as well as an optimization technique used to control the behavior of the unactuated dof are proposed. A prototype of one of the robots and its real-time controller are also presented with experimental results for point-to-point trajectories. Finally, conclusions are drawn and future work is briefly discussed.

2 Robot Architectures

The general concept of underactuated cable-suspended robots consists in suspending a rigid body or a set of articulated bodies using a number of cables smaller than that needed to fully constrain the motion of the moving bodies. As a general rule, the cables are mounted on actuated spools which allow the control of the cable extensions. Other actuators can also be included at the joints of the suspended articulated bodies. This is illustrated schematically in Fig. 1. For instance, a rigid body in the plane can be suspended by two cables, knowing that three cables would be required to fully constrain its pose. Pendulum-type mechanisms are thereby obtained and their swinging motion can be controlled in order to generate point-to-point trajectories. These robots can achieve point-to-point motion (from one prescribed pose to another) with zero-velocity landing by swinging themselves as children do on playground swings. They can use the two techniques pointed out by Case [7],[8] for effective swinging, namely, *i*) leg-stretching as in seated swinging (when articulated bodies are suspended) and *ii*) center of mass motion as in standing swinging, using cable extension. Moreover, the control technique proposed here does not require specific initial conditions.

In this paper, two specific architectures are studied. They are now briefly described.

2.1 Planar Three-Degree-of-Freedom Underactuated Serial Cable-Suspended Robot

A planar three-dof serial cable-suspended robot is illustrated schematically in Fig. 2. It consists of two articulated bodies suspended on a single cable. The end-effector has a length L, a mass m_2 and a moment of inertia I_2 about its center of mass. The joint between the articulated bodies is driven by a motor mounted on the body

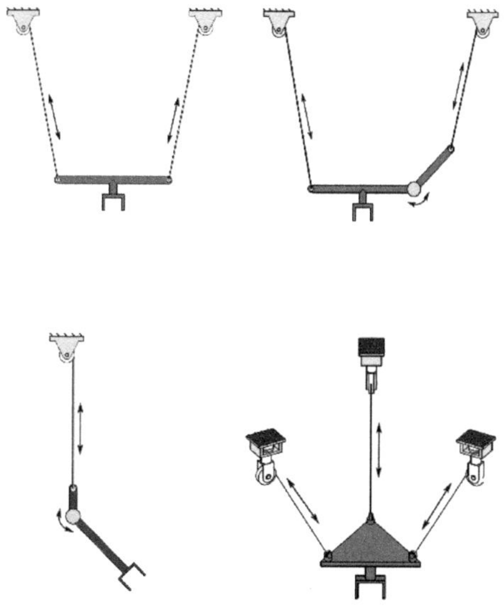

Fig. 1. Conceptual representation of underactuated cable-suspended robots.

directly attached to the cable. The latter has a mass m_1 and an inertia I_1. In practice, the last section of the cable is replaced with a rod which is rigidly attached to this body, thereby ensuring that it is always aligned with the cable. The center of mass of the end-effector is located a distance d_2 from the actuated joint while the center of mass of the proximal body is located a distance d_1 from the same joint. The angle of the cable with respect to a vertical axis is noted θ_1 while the angle between the cable and the end-effector is noted θ_2 and the angle between the vertical axis and the end-effector is noted ϕ. Angle θ_2 is associated with the actuated joint between the two suspended bodies while angle θ_1 is an unactuated coordinate. The proximal suspended body is attached to a cable of length ρ passing through a pulley and wound on a reel actuated by a second fixed motor. Therefore, the length of the cable can be controlled using the latter. Globally, the mechanism can be thought of as a planar serial RPR robot whose first R joint is not actuated. The robot has three dofs and only two actuators.

2.2 *Planar Three-Degree-of-Freedom Underactuated Parallel Cable-Suspended Robot*

A planar three-dof parallel cable-suspended robot is illustrated schematically in Fig. 3. It consists of a single body suspended on two cables. The suspended body has a length b, a mass m, a moment of inertia I with respect to its center of mass and

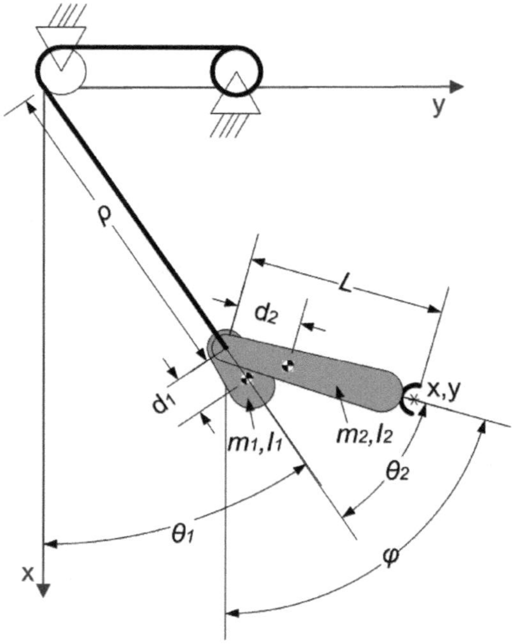

Fig. 2. Architecture of the planar three-dof underactuated serial cable-suspended robot.

is assumed to be inertially symmetric with respect to the cable attachment points. Its center of mass is located a distance d from the line passing through the cable attachment points. The joints on the suspended body as well as the fixed revolute joints are not actuated. The distance between the fixed revolute joints is noted a. The angles of the cables with respect to a vertical axis — associated with an unactuated motion — are noted θ_1 and θ_2, respectively. The two cables have controllable lengths noted ρ_1 and ρ_2. Similar to the serial robot presented above, the cables pass through a pulley and are wound on actuated reels controlling the length of the cables. Globally, the mechanism can be thought of as a planar parallel 2-RPR robot whose P joints are actuated. The robot has three dofs and only two actuators. The main difference between the serial robot represented in Fig. 2 and this parallel robot is that in the latter, both actuators are fixed to the ground and no actuator needs to be mounted on the suspended body. Also, since the end-effector is suspended on two cables, its swinging motion is more complex.

3 Equations of Motion

In this section, the kinematics and dynamics equations governing the motion of the underactuated cable-driven robots described above are derived.

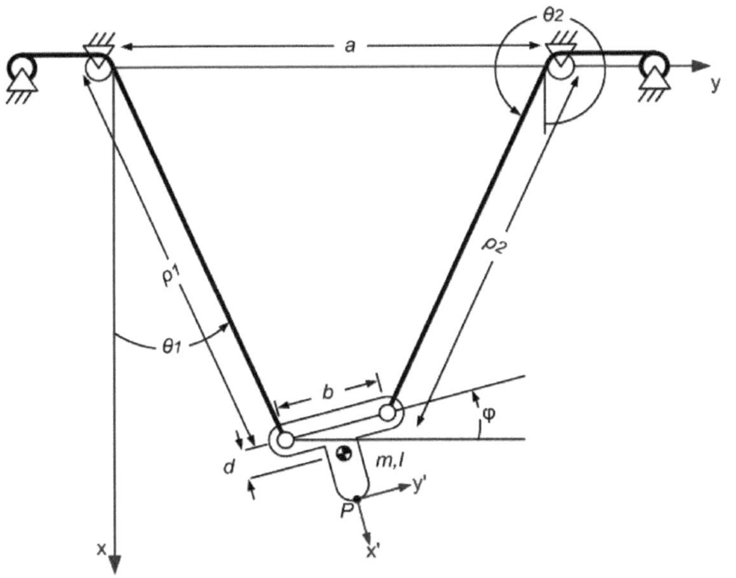

Fig. 3. Architecture of the planar three-dof underactuated parallel cable-suspended robot.

3.1 Planar Serial Cable-Suspended Robot

The direct kinematics of the robot shown in Fig. 2 can be written as:

$$\begin{aligned} x &= \rho\cos\theta_1 + L\cos(\theta_1+\theta_2) \\ y &= \rho\sin\theta_1 + L\sin(\theta_1+\theta_2) \\ \phi &= \theta_1+\theta_2 \end{aligned} \tag{1}$$

where x and y are the coordinates of the reference point on the end-effector. The solution to the inverse kinematics is readily obtained from the above system of equations.

In the derivation of the dynamic model, it is assumed that the cable is massless. This implies that the tension T in the cable is always sufficient to avoid sagging, i.e., the cable always remains straight. Moreover, dissipative effects are neglected. Using Lagrangian dynamics, the equations of motion are readily obtained and can be written as:

$$\begin{aligned} &\left(I_1+I_2+m_1(\rho+d_1)^2+m_2(\rho+d_2\cos\theta_2)^2\right)\ddot{\theta}_1 \\ &-m_2 d\sin\theta_2\ddot{\rho}+\left(I_2+m_2(d_2^2+\rho d_2\cos\theta_2)\right)\ddot{\theta}_2 \\ &+2\left(m_1(\rho+d_1)+m_2(\rho+d_2\cos\theta_2)\right)\dot{\rho}\dot{\theta}_1 - m_2\rho\dot{\theta}_2\sin\theta_2\left(2\dot{\theta}_1+\dot{\theta}_2\right) \\ &+m_1 g(\rho+d_1)\sin\theta_1+m_2 g(\rho\sin\theta_1+d_2\sin(\theta_1+\theta_2)) = 0 \end{aligned} \tag{2}$$

$$\begin{aligned} &m_2 d_2\sin\theta_2\left(\ddot{\theta}_1+\ddot{\theta}_2\right)-(m_1+m_2)\ddot{\rho}+(m_1(\rho+d_1)+m_2\rho)\dot{\theta}_1^2 \\ &+m_2 d_2\cos\theta_2\left(\dot{\theta}_1+\dot{\theta}_2\right)^2+(m_1+m_2)g\cos\theta_1 = T \end{aligned} \tag{3}$$

$$\begin{aligned} &\left(I_2+m_2\left(d_2^2+\rho d_2\cos\theta_2\right)\right)\ddot{\theta}_1 - m_2 d_2\sin\theta_2\ddot{\rho}+\left(m_2 d^2+I_2\right)\ddot{\theta}_2 \\ &+2m_2 d_2\dot{\rho}\dot{\theta}_1\cos\theta_2 m_2 d_2\rho\dot{\theta}_1^2\sin\theta_2+m_2 d_2 g\sin(\theta_1+\theta_2) = \tau \end{aligned} \tag{4}$$

where m_i, I_i and d_i are respectively the mass, inertia and position of the center of mass of body i, T is the tension in the cable, τ is the torque at the actuator mounted on the end-effector and g is the gravitational acceleration.

Equation (2) is the differential equation corresponding to the unactuated (free) motion of the system. Indeed, no control input appears in this equation since the fixed revolute joint is not actuated.

3.2 Planar Parallel Cable-Suspended Robot

For the robot shown in Fig. 3, the controlled inputs are the extensions of the two cables. Because of the parallel structure of the robot, the direct kinematic problem is not simple. However, the inverse kinematics, for a given pose of the platform, is readily obtained as:

$$\rho_i^2 = (\mathbf{p} + \mathbf{Q}\mathbf{s}_i^o - \mathbf{r}_i)^T (\mathbf{p} + \mathbf{Q}\mathbf{s}_i^o - \mathbf{r}_i), \quad i = 1,2 \tag{5}$$

where $\mathbf{p}$ is the position vector of the prescribed end-effector point in the fixed reference frame, $\mathbf{r}_i$ is the position vector of the fixed attachment point of the ith cable in the fixed reference frame and $\mathbf{s}_i^o$ is the position vector of the ith attachment point on the platform in the moving reference frame. Also, $\mathbf{Q}$ is the rotation matrix from the fixed frame to the moving frame and can be written as

$$\mathbf{Q} = \begin{bmatrix} \cos\phi & -\sin\phi \\ \sin\phi & \cos\phi \end{bmatrix} \tag{6}$$

where ϕ is the angle between the x axis of the fixed frame and the x axis of the moving frame.

In practice, encoders are mounted on the fixed pivots and the angles that the cables are making with the vertical, noted θ_1 and θ_2, are known. Similar to the case of the serial robot, it is assumed that the cable is massless and that dissipative effects can be neglected. Because of the closed-loop architecture of the robot, it is more convenient to use the Newton-Euler formulation to derive the dynamic model. Based on a free-body diagram of the end-effector, the motion equations are readily obtained as:

$$T_1\mathbf{e}_1 + T_2\mathbf{e}_2 + mg\mathbf{e} = m\ddot{\mathbf{p}}_c \tag{7}$$
$$T_1\mathbf{r}_{c1}^T\mathbf{E}\mathbf{e}_1 + T_2\mathbf{r}_{c2}^T\mathbf{E}\mathbf{e}_2 = I\ddot{\phi} \tag{8}$$

where eq.(7) arises from the sum of the forces on the end-effector while eq.(8) arises from the sum of the moments about the center of mass of the end-effector. In the latter equations, T_i denotes the tension in cable i, $\mathbf{e}_i$ is a unit vector in the direction of cable i, $\mathbf{e}$ is a unit vector in the direction of gravity, m is the mass of the end-effector, $\ddot{\mathbf{p}}_c$ is the acceleration vector of the center of mass of the end-effector, $\mathbf{r}_{ci}$ is the vector connecting the center of mass of the end-effector to the ith cable attachment point, I is the moment of inertia of the end-effector with respect to its center of mass and $\mathbf{E}$ is the matrix defined as:

$$\mathbf{E} = \begin{bmatrix} 0 & -1 \\ 1 & 0 \end{bmatrix}. \tag{9}$$

Equation (7) is readily solved for T_1 and T_2 since it constitutes a set of two linear equations in these variables, which leads to:

$$T_i = \mathbf{u}_i^T\mathbf{R}^{-1}\mathbf{w}, \quad i = 1,2 \tag{10}$$

where

$$\mathbf{u}_1 = \begin{bmatrix} 1 \\ 0 \end{bmatrix}, \mathbf{u}_2 = \begin{bmatrix} 0 \\ 1 \end{bmatrix}, \mathbf{R} = \begin{bmatrix} \mathbf{e}_1 & \mathbf{e}_2 \end{bmatrix}, \text{ and } \mathbf{w} = m\ddot{\mathbf{p}}_c - mg\mathbf{e} \tag{11}$$

Equations (10) can be substituted into eq.(8) to obtain a differential equation for ϕ. The latter represents the dynamics of the unactuated degree of freedom of the robot, similar to eq.(2) for the serial robot. However, the expression obtained for the parallel robot is much more complex due to the dependencies arising from the closed kinematic loop. Therefore, it is more convenient to keep some of the dependent variables in this equation (i.e. angles θ_1 and θ_2) and to handle the dependencies separately, using the loop closure equations. The detailed equations are not given here because of space limitations.

The remainder of this paper deals with the serial cable-suspended robot, taken as an example. The proposed approach can also be applied to the parallel robot.

4 Trajectory Planning

For the serial robot, based on the dynamic model formulated above and assuming known initial conditions, the cable angle $\theta_1(t)$ can be predicted for given trajectories of the actuated joints $[\rho(t), \theta_2(t)]^T$ by integrating eq.(2) numerically. However, for the predictions to be accurate, the prescribed actuated joint trajectories must also satisfy the following conditions: *i*) the cable is maintained in tension (T is positive), *ii*) the actuator torque limits and the joint limits are not exceeded, *iii*) the joint trajectories and their derivatives are smooth to avoid discontinuities and *iv*) the prediction time is small enough to ensure that the model errors remain small. From that foregoing condition, it becomes obvious that planning entire trajectories from zero to large cable angles is impossible. On the other hand, since the motion of the robot is governed by pendulum-like dynamics, planning trajectories over a half-period of oscillation is appropriate since the prediction horizon is relatively small and the dynamics remain similar between two zero-velocity states $\left(\dot{\theta}_1 = 0\right)$.

Trajectory Planning of the Free Parameters

Defining a goal $[x_g, y_g, \phi_g]$ to be reached at zero final velocity, the corresponding joint coordinates $[\theta_{1g}, \rho_g, \theta_{2g}]$ are obtained from the solution of the inverse kinematics. Since actuated joint goals are easily reachable, the objective is to find actuated joint trajectories that will produce the desired cable angle in the goal configuration (at zero velocity) or, at least, within half a period, a cable angle that is closer to the goal. These trajectories must also satisfy initial conditions to maintain continuity and final conditions for objective fulfillment. Since the dynamics of the robot are governed by nonlinear equations, it is clear that optimal trajectories cannot be obtained within acceptable computation time.

Thus, a technique similar to the one presented by Tortopidis and Papadopoulos [9] for space manipulators is used. Indeed, $n+p$ parameter trajectories are used satisfying the n boundary conditions of each actuated joint and leaving p free parameters to be tuned for controlling the behavior of the unactuated degree of freedom. Here, there are $n=6$ boundary conditions (initial/final positions, velocities and accelerations) and, for simplicity reasons, $(p=)$ 1 free parameter was left for each actuated joint.

Since zero velocity is desired for smooth landing, the extrema of θ_1 will be considered as starting/ending trajectory points. This leads to a cosine-like function for $\theta_1(t)$ over a half-period. Moreover, since large workspaces are desired, large values of θ_1 must be reachable and the robot must be able to efficiently excite itself to achieve such motions. Special attention must then be given to trajectory design, which is now addressed independently for each actuated joint.

Cable Extension

Neglecting the rotation of the end-effector, the system is similar to a variable-length (Lorentz) pendulum studied for swing-up in [10],[11] or to the famous Santiago de Compostela censer studied in [12]. It was suggested by Burns [13], and proven optimal for instantaneous variation of length by Piccoli and Kulkarni [14], that lengthening the cable when the angular velocity of the cable is minimum and shortening it when it is maximum amplifies the energy of the system.

Hence, given the mathematical form of θ_1 prescribed above, a sine-like function was chosen for the cable extension. It was determined by numerical simulation [15] that exciting the system at twice its natural frequency with a $\pm\pi/2$ phase from cable-angle position is best-suited to increase (or decrease) the cable angle, which is consistent with Burns' results. Since the cable angle describes a cosine-like function between two zero-velocity states ($\dot{\theta}_1 = 0$), the following function was chosen for cable extension:

$$\rho(t) = A_1 \sin(2\omega t) + B_1 \sin(3\omega t) + C_1 \sin(4\omega t) + \rho_i + \frac{(\rho_f - \rho_i)\,\omega t}{\pi} \tag{12}$$

for $0 \leq t \leq \pi/\omega$, which matches the initial and final conditions.

In eq.(12), indices i and f refer to initial/final conditions, t is the time, ω is the system natural frequency and A_1 is a free parameter which is optimized at each half period in order to control the behavior of θ_1. Coefficients B_1 and C_1 are obtained by prescribing the time derivative of eq. (12) at times $t = 0$ and $t = \pi/\omega$ to be equal to the initial/final velocities $\dot{\rho}_i$ and $\dot{\rho}_f$. In practice, the natural frequency is simply estimated by calculating the time between two consecutive zero velocity points $(\dot{\theta}_1 = 0)$. For stability reasons, the initial and final acceleration is prescribed to be zero from eq.(12). Initial conditions are measured from encoders, while final conditions are defined as the goal configuration or the maximum step taken towards this goal if the current configuration is too far from the objective. Hence, a target cable length is set at first and maintained until the target cable angle θ_{1g} is reached. The final velocity is chosen to match basic sine conditions in order to limit accelerations and is set to zero for smooth landing when the target is reached.

End-Effector Swing

On the other hand, considering a fixed-length cable, the problem is similar to a double pendulum with a long first link. The swing-up problem of a double-pendulum with a passive first joint was widely studied in the literature in the context of the Acrobot [16],[17],[18] and Brachiation Robots [19]. Spong [16] suggests that, in order to increase the system energy, the motion of the lower link must be in-phase with that of the upper link. Thus, a sine-like function was chosen for the end-effector swing since the upper link motion is of that form. It was determined by numerical simulation [15] that exciting the system at its natural frequency is best-suited to increase (or decrease) cable angle as pointed out by Spong. A phase of $-3\pi/4$ or $\pi/4$ from the cable angle is also preferable. However, in order to synchronize the goal-reaching with the cable extension and to impose a zero final acceleration, a sine function with a phase of $\pm\pi/2$ was chosen since it generates almost the same amplitude of excitation. Therefore, the following function was chosen for the end-effector swing:

$$\theta_2(t) = A_2 \sin(\omega t) + B_2 \sin(2\omega t) + C_2 \sin(3\omega t) + \theta_{2i} + \frac{(\theta_{2f} - \theta_{2i})\,\omega t}{\pi} \qquad (13)$$

for $0 \leq t \leq \pi/\omega$. The rest of the procedure is similar to what was presented above for the cable extension.

5 Implementation

A prototype of the underactuated robot was built as shown in Fig. 4. The robot was designed using a parallelogram shape in order to constrain the robot to planar motion. DC motors with encoders are used for the end-effector actuated joint and the cables. The two cables are wound on a single reel in order to ensure that they have the same length. The cable angle θ_1 is measured using an encoder mounted on the robot-supporting pulleys.

A controller was implemented on a real-time QNX computer with a servo-rate of 500 Hz. Close-loop PIDs were used to control the actuated joint positions and a 12-steps fixed-step fourth order Runge-Kutta [20] was used to integrate the equations of motion. Four iterations ($M = 4$) of the Nelder-Mead algorithm were executed per step and a total of $N = 6$ steps (0.012s) were used to define the free parameters (A_1, A_2).

In practice, it was difficult to make predictions for small cable angles because the dynamic behavior is slightly different under these conditions [21]. Indeed, when θ_1 is small, the system is more subject to cable bending and vibrations. Therefore, pre-planned excitation trajectories with fixed amplitude and frequency were used for end-effector swing in order to initiate motion.

A sequence of end-effector Cartesian coordinates $[x, y, \phi]^T$ simulating point-to-point trajectories were used to evaluate performances. Points were chosen to include various amplitudes of θ_1 as well as various cable lengths.

Using the controller structure and the trajectory planning described above, actuated joint trajectories and cable angle predictions were computed in real-time for theses objectives. Figure 5 shows joint trajectories for point-to-point motion. It can be observed from the graph that the goal configurations are precisely reached.

Cable angle predictions for the underactuated robot were generally sufficiently accurate for excellent controllability. The goal-reaching precision was also very good since joint motions were well synchronized with the frequency estimation for pick-and-place actions. The predictions were slightly less accurate when decreasing

Fig. 4. Prototype of the underactuated cable-driven robot

the system energy since the tension in the cable is then reduced. This causes off-axis rotations which cannot be predicted by the model. Nevertheless, the experimental results were sufficiently convincing to confirm the effectiveness of the method introduced above.

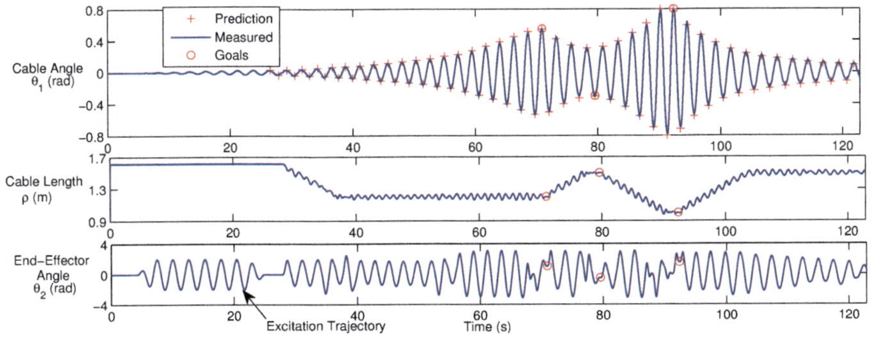

Fig. 5. Joints trajectories for point-to-point motion

6 Conclusions

A new family of cable-suspended underactuated robots was proposed in this paper. The proposed robots combine the advantages of cable-driven systems (light weight, large workspace) with those of underactuation (minimal set of actuators). More specifically, two three-dof planar underactuated cable-suspended robots were described and the corresponding dynamic models were developed. One of the robots was then used as an example and studied in more detail. The actuated joint trajectory design for swing-up was studied and an optimization technique was proposed to control the free joint motion. Although not discussed above because of lack of space, constraints on actuator torque, joint limits and positive tension in the cable were also included in the formulation. Finally, a prototype of the robot and its real-time controller were introduced. It was shown through experiments that the strategy developed was successful in reaching targets and that precision was sufficient for point-to-point trajectories. The implementation of the algorithms described above on the planar parallel underactuated cable-suspended mechanism is currently underway. Future work includes the implementation of $x-y$ path-tracking strategies and the development of other prototypes. The extension of this work to three-dimensional 6-dof cable-suspended robots is also planned.

Acknowledgement. The authors would like to acknowledge the financial support of the Natural Sciences and Engineering Research Council of Canada (NSERC) and of the Canada Research Chair Program.

References

1. Cone, L.L.: Skycam: An aerial robotic camera system. Byte, 122–132 (1985)
2. Albus, J., Bostelman, R., Dagalakis, N.: The NIST ROBOCRANE. J. Res. Natl. Inst. Stand. Technol. 97, 373 (1992)
3. Bosscher, P.R., Tummino, M.: A concept for rapidly deployable cable robot search and rescue systems. In: Proceedings of the ASME IDETC/CIE, Long Beach CA, USA, pp. 1–10 (2005)
4. Fantoni, I., Lozano, R.: Non-linear control for underactuated mechanical systems. Springer, Heidelberg (2001)
5. Haddadin, S., Albu-Schaffer, A., Frommberger, M., Hirzinger, G.: The role of the robot mass and velocity in physical human-robot interaction-Part II: Constrained blunt impacts. In: Proc. of the IEEE ICRA, Pasadena CA, USA, 1339–1345 (2008)
6. Cunningham, D., Asada, H.: The Winch-Bot: A Cable-Suspended, Under-Actuated Robot Utilizing Parametric Self-Excitation. In: Proc. of the IEEE ICRA, Kobe, Japan, pp. 1844–1850 (2009)
7. Case, W.: The pumping of a swing from the standing position. American Journal of Physics 64, 215 (1996)
8. Case, W., Swanson, M.: The pumping of a swing from the seated position. American Journal of Physics 58, 463 (1990)
9. Tortopidis, I., Papadopoulos, E.: On point-to-point motion planning for underactuated space manipulator systems. Robotics and Autonomous Systems 55(2), 122–131 (2007)
10. Curry, S.: How children swing. American Journal of Physics 44, 924 (1976)

11. Pinsky, M., Zevin, A.: Oscillations of a pendulum with a periodically varying length and a model of swing. International Journal of Non Linear Mechanics 34(1), 105–110 (1999)
12. Sanmartin Losada, J.: La physique de l'encensoir. Pour la science 155, 96–104 (1990)
13. Burns, J.: More on pumping a swing. American Journal of Physics 38, 920 (1970)
14. Piccoli, B., Kulkarni, J.: Pumping a swing by standing and squatting: do children pump time optimally? IEEE Control Systems Magazine 25(4), 48–56 (2005)
15. Lefrançois, S., Gosselin, C.: Point-to-point motion control of a pendulum-like 3-dof underactuated cable-driven robot. In: Proc. of the IEEE ICRA, Anchorage (2010)
16. Spong, M.: The swing up control problem for the acrobot. IEEE Control Systems Magazine 15(1), 49–55 (1995)
17. Brown, S., Passino, K.: Intelligent control for an acrobot. Journal of Intelligent and Robotic Systems 18(3), 209–248 (1997)
18. Boone, G.: Minimum-time control of the acrobot. In: In Proc. of the IEEE ICRA, Albuquerque NM, USA, pp. 111–111 (1997)
19. Nakanishi, J., Fukuda, T., Koditschek, D.: A brachiating robot controller. IEEE Transactions on Robotics and Automation 16(2), 109–123 (2000)
20. Scheid, F.: Schaum's outline of theory and problems of numerical analysis. Schaum's Outline Series (1989)
21. Gore, B.: Starting a swing from rest. American Journal of Physics 39, 347 (1971)

Computing the Rigid-Body Acceleration Field from Nine Accelerometer Measurements

Philippe Cardou

Abstract. Among other applications, accelerometer arrays have been used in crashworthiness studies to measure the acceleration field of the head of a dummy subjected to an impact. In previous analyzes, the centripetal component of the rigid-body acceleration was estimated *linearly* from point-acceleration measurements, that is, by considering the quadratic products of the angular-velocity components as independent variables. Although this assumption greatly simplifies the estimation process, it has two drawbacks: (i) it raises the minimum number of accelerometers from nine to 12, and, when more than the minimum number of accelerometers are available, (ii) it ignores some of the constraints between the kinematic parameters, which would otherwise help in filtering the data. In this paper, we solve the *nonlinear* problem of estimating the rigid-body acceleration field from point-acceleration measurements. To this end, we partition the associated system of equations into two subsystems, one linear, the other nonlinear. The nonlinear subsystem of three equations in three unknowns represents three quadrics in 3D space, whose intersection contains the rigid-body angular velocity. This intersection is computed using a readily-available technique, which yields eight closed-form solutions to the problem. A criterion for the selection of the proper solution is given. The proposed nonlinear method should prove useful when the number of accelerometers is limited, or to improve the robustness of an array of 12 or more accelerometers by taking into account the constraints between the quadratic terms of the angular velocity.

1 Introduction

An accelerometer array may be regarded as a type of inertial measurement unit (IMU), since it consists of a set of accelerometers that are "strapped" onto a rigid body. From the accelerometer measurements, one may infer the complete

Philippe Cardou
Laval University, Adrien-Pouliot Building, room #1504, Quebec City, Quebec, Canada, G1V 0A6
e-mail: pcardou@gmc.ulaval.ca

J. Angeles et al. (Eds.): Brain, Body and Machine, AISC 83, pp. 325–339.
springerlink.com

rigid-body acceleration field or some of its components, and, thence, kinematic parameters of the rigid-body motion. This measurement technique seems to have been originally proposed by Corey (1962), who was quickly followed by Grammatikos (1965) and Schuler (1965). Accelerometer arrays were first implemented by Mertz (1967) and Padgaonkar et al (1975) to estimate the angular acceleration and the angular velocity of rigid bodies undergoing planar and spatial motions, respectively.

Accelerometer arrays are now used in several applications, most notably in biomechanics. For instance, these devices have been used extensively in crashworthiness studies (Yoganandan et al, 2006; McIntosh et al, 2007), sports (Walilko et al, 2005), rehabilitation exercises (Baselli et al, 2001), and general human-movement analysis (van den Bogert et al, 1996). Applications outside the field of biomechanics have recently been reported. As examples, we may cite human-machine interfaces (Di Puccio and Forte, 2004), robotics (Kröger et al, 2008; Lin et al, 2006), and projectile guidance (Pickel, 2005).

For the most part, the theory of accelerometer arrays has been limited to the analysis of specific geometries (Grammatikos, 1965; Schuler, 1965). Exceptions to this are the works of Genin et al (1997), Zappa et al (2001), Parsa (2003), Williams and Fyfe (2004), Cappa et al (2005), and Cardou (2008). Williams and Fyfe (2004) analyzed planar accelerometer arrays. Other authors considered the spatial case and obtained the n input-output equations of a generic n-accelerometer array[1], namely, the set of equations relating the n measured point-accelerations to the nine parameters of the rigid-body acceleration field. These equations are nonlinear in three of the nine acceleration-field parameters, the components of the rigid-body angular velocity, which appear generally as six second-degree terms in each input-output equation. Rather than solving directly the n input-output equations, previous authors chose to simply consider the six second-degree terms as six independent variables, turning the nonlinear problem into a linear one, but with three additional unknowns. The obvious drawback of this approach is that the minimum number of accelerometers is then 12 instead of nine. Moreover, in the case where $n \geq 12$, the linear approach does not use all the available information to filter the data, since it does not take into account the relations between the six nonlinear terms.

In this paper, we solve the nonlinear input-output equations of any array of $n \geq 9$ accelerometers. We show that a closed-form solution exists for any such array, provided it is nonsingular, thereby answering concerns raised by some authors (Genin et al, 1997) regarding the iterative nature of a nonlinear solution.

2 Input-Output Equations of a Generic Accelerometer Array

We start from the general case where n accelerometers are attached at n points $\{P_i\}_{i=1}^n$ of a rigid body moving in space, as shown in Fig. 1. The sensitive directions of these accelerometers are represented by unit vectors $\mathbf{e}_i$, $i = 1, \ldots, n$.

[1] Notice that Parsa (2003) and Zappa et al (2001) did not consider the completely general case, since they assumed that all accelerometers were triaxial.

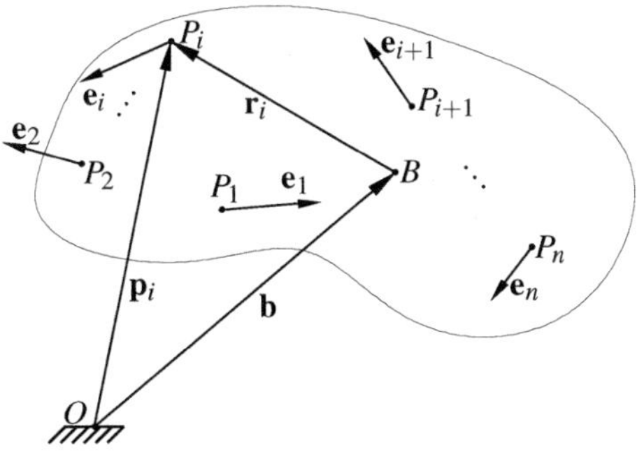

Fig. 1. A rigid body equipped with n accelerometers moving in space

Reference point O is fixed to the ground, while reference point B pertains to the moving rigid body. Vectors $\mathbf{p}_i$, $\mathbf{b}$, and $\mathbf{r}_i$ point from O to P_i, from O to B, and from B to P_i, respectively. The problem is to estimate the rigid-body acceleration field from the accelerometer measurements. To this end, we compute the acceleration of P_i as

$$\ddot{\mathbf{p}}_i = \ddot{\mathbf{b}} + \dot{\omega} \times \mathbf{r}_i + \omega \times \omega \times \mathbf{r}_i = \ddot{\mathbf{b}} + \mathbf{W}\mathbf{r}_i, \tag{1}$$

where $\mathbf{W} \equiv \dot{\Omega} + \Omega^2$ is the angular acceleration matrix Angeles (1999), $\Omega \equiv \mathrm{CPM}(\omega)$, ω is the rigid-body angular-velocity vector, and $\mathrm{CPM}(\cdot)$ denotes the cross-product matrix[2] of its argument.

It is convenient to factor out the unknowns, which are the acceleration-field parameters $\ddot{\mathbf{b}}$ and $\mathbf{W}$. Let us regroup the entries of $\mathbf{W}$ into the arrays $\dot{\omega}$ and $\xi \equiv [\omega_1^2 \;\; \omega_2^2 \;\; \omega_3^2 \;\; \omega_2\omega_3 \;\; \omega_3\omega_1 \;\; \omega_1\omega_2]^T$, which correspond, respectively, to the skew-symmetric and symmetric components of $\mathbf{W}$. Upon substituting in eq. (1), we obtain

$$\ddot{\mathbf{p}}_i = \ddot{\mathbf{b}} + \mathbf{R}_i^T \dot{\omega} + \Sigma_i^T \xi, \tag{2}$$

where

$$\mathbf{R}_i \equiv \mathrm{CPM}(\mathbf{r}_i), \quad \Sigma_i \equiv \begin{bmatrix} 0 & -r_{i1} & -r_{i1} & 0 & r_{i3} & r_{i2} \\ -r_{i2} & 0 & -r_{i2} & r_{i3} & 0 & r_{i1} \\ -r_{i3} & -r_{i3} & 0 & r_{i2} & r_{i1} & 0 \end{bmatrix}^T,$$

and r_{i1}, r_{i2}, and r_{i3} are the first, second, and third component of $\mathbf{r}_i$, respectively.

In the following analysis, we assume that an accelerometer is sensitive only to the component of its acceleration that is parallel to its sensitive direction. In industry, the sensitivity of an accelerometer to acceleration components orthogonal to its sensitive direction is called the *cross-axis sensitivity*, and it amounts at most to a few percentage points of the accelerometer sensitivity to acceleration components parallel to its sensitive direction. Hence, upon neglecting these cross-axis effects, the readout a_i of the ith accelerometer is modeled as

[2] CPM($\mathbf{a}$) is defined as $\partial(\mathbf{a} \times \mathbf{x})/\partial\mathbf{x}$, for any $\mathbf{a}, \mathbf{x} \in \mathbb{R}^3$.

$$a_i = \mathbf{e}_i^T \ddot{\mathbf{p}}_i. \tag{3}$$

Upon substituting eq. (2) into eq. (3), we obtain

$$a_i = \mathbf{e}_i^T \ddot{\mathbf{b}} + \mathbf{e}_i^T \mathbf{R}_i^T \dot{\omega} + \mathbf{e}_i^T \Sigma_i^T \xi, \tag{4}$$

We stack eq. (4), $i = 1, \ldots, n$, into a linear system of equations, which yields the same system as that obtained by Cappa et al (2005), namely,

$$\mathbf{a} = \mathbf{E}^T \ddot{\mathbf{b}} + \mathbf{F}^T \mathbf{R}^T \dot{\omega} + \mathbf{F}^T \Sigma^T \xi, \tag{5}$$

where $\mathbf{a} \equiv [a_1 \quad a_2 \quad \cdots \quad a_n]^T$, $\mathbf{E} \equiv [\mathbf{e}_1 \quad \mathbf{e}_2 \quad \cdots \quad \mathbf{e}_n]$, $\mathbf{R} \equiv [\mathbf{R}_1 \quad \mathbf{R}_2 \quad \cdots \quad \mathbf{R}_n]$, $\Sigma \equiv [\Sigma_1 \quad \Sigma_2 \quad \cdots \quad \Sigma_n]$, and

$$\mathbf{F} \equiv \begin{bmatrix} \mathbf{e}_1 & \mathbf{0}_3 & \cdots & \mathbf{0}_3 \\ \mathbf{0}_3 & \mathbf{e}_2 & \cdots & \mathbf{0}_3 \\ \vdots & \vdots & \ddots & \vdots \\ \mathbf{0}_3 & \mathbf{0}_3 & \cdots & \mathbf{e}_n \end{bmatrix}.$$

Equation (5) gives a linear relation between the rigid-body acceleration field parameters and the accelerometer measurements. Therefore, it may be referred to as the accelerometer-array input-output equation. For the sake of clarity, let us rewrite eq. (5) as

$$\mathbf{a} = \mathbf{A}_P \ddot{\mathbf{b}} + \mathbf{A}_T \dot{\omega} + \mathbf{A}_R \xi, \tag{6}$$

where $\mathbf{A}_P \equiv \mathbf{E}^T$, $\mathbf{A}_T \equiv \mathbf{F}^T \mathbf{R}^T$, and $\mathbf{A}_R \equiv \mathbf{F}^T \Sigma^T$ may be termed the accelerometer-array point, tangential and radial matrices, respectively. Upon defining the accelerometer-array matrix $\mathbf{A} \equiv [\mathbf{A}_P \quad \mathbf{A}_T \quad \mathbf{A}_R]$, we may state that the linear system admits an exact or a least-squares solution if and only if $\text{rank}(\mathbf{A}) = 12$. This condition implies that $n \geq 12$ accelerometers are necessary to *linearly* determine the full rigid-body acceleration field, but not sufficient.[3]

The rigid-body acceleration field, however, is completely described by nine independent parameters, which is less than the number of unknowns appearing (linearly) in eq. (6). Hence, it may be possible to identify the full acceleration field from nine accelerometer measurements only, or, for accelerometer arrays with $n > 9$, to use the additional information provided by the nonlinear relations to further filter the data.

3 Solution for a Nonlinearly Determined Accelerometer Array

Solutions to such nonlinear relations for the magnitudes of the angular-velocity components were obtained (Grammatikos, 1965; Schuler, 1965; Zappa et al, 2001)

[3] The reader is referred to Cardou and Angeles (2009) for an analysis of fully and partially linearly determined accelerometer arrays.

for special geometries. What we propose here is a method that can be applied to *any* accelerometer array for which $\text{rank}(\mathbf{A}) \geq 9$.

3.1 Decoupling the System of Equations

Under the latter condition, one may compute a QR factorization of $\mathbf{A}$ using, for example, Gram-Schmidt orthogonalization, which gives

$$\mathbf{A} = \mathbf{QS} = \begin{bmatrix}\mathbf{Q}_1 \ \mathbf{Q}_2 \ \mathbf{Q}_3\end{bmatrix} \begin{bmatrix} \mathbf{S}_{1,1} & \mathbf{S}_{1,2} \\ \mathbf{O}_{3\times 6} & \mathbf{S}_{2,2} \\ \mathbf{O}_{(n-9)\times 6} & \mathbf{S}_{3,2} \end{bmatrix}, \tag{7}$$

where $\mathbf{Q} \equiv [\mathbf{Q}_1 \quad \mathbf{Q}_2 \quad \mathbf{Q}_3] \in \mathbb{R}^{n\times n}$ is orthogonal; $\mathbf{Q}_1 \in \mathbb{R}^{n\times 6}$, $\mathbf{Q}_2 \in \mathbb{R}^{n\times 3}$, and $\mathbf{Q}_3 \in \mathbb{R}^{n\times(n-9)}$; $\mathbf{S} \in \mathbb{R}^{n\times 12}$, $\mathbf{S}_{1,1} \in \mathbb{R}^{6\times 6}$, and $\mathbf{S}_{2,2} \in \mathbb{R}^{3\times 6}$ are upper-triangular; $\mathbf{S}_{1,2} \in \mathbb{R}^{6\times 6}$; $\mathbf{O}_{r\times s}$ is the $r \times s$ zero matrix; and $\mathbf{S}_{3,2} \in \mathbb{R}^{(n-9)\times 6}$ contains $\text{rank}(\mathbf{A}) - 9$ non-zero rows. Hence, when $\text{rank}(\mathbf{A}) = 9$, we have $\mathbf{S}_{3,2} = \mathbf{O}_{(n-9)\times 6}$, and the nonlinear problem is determined. When $\text{rank}(\mathbf{A}) > 9$, the nonlinear system is overdetermined. In this case, we propose to solve the first nine equations as a determined nonlinear system, leaving the remaining $\text{rank}(\mathbf{A}) - 9$ equations for a later use, to filter the obtained estimates.

Upon defining $\mathbf{x}_{PT} \equiv [\ddot{\mathbf{b}}^T \quad \dot{\omega}^T]^T$ and premultiplying both sides of eq. (7) with $\mathbf{Q}^T$, we obtain the partitioned determined system of equations

$$\mathbf{S}_{1,1}\mathbf{x}_{PT} + \mathbf{S}_{1,2}\xi = \mathbf{Q}_1^T\mathbf{a}, \tag{8}$$

$$\mathbf{S}_{2,2}\xi = \mathbf{Q}_2^T\mathbf{a}. \tag{9}$$

Equations (8) and (9) allow for a solution in cascade, starting with eq. (9), which represents a nonlinear system of three equations into the three unknowns ω, and ending with eq. (8), a linear system of six equations into the six unknowns $\ddot{\mathbf{b}}$ and $\dot{\omega}$. We follow this order, and address the nonlinear problem first.

3.2 Solution to the Nonlinear Subsystem of Equations

Since the three polynomials into three unknowns on the left-hand side of eq. (9) are of degree two, solving this system of equations amounts to finding the intersections of three quadrics in three-dimensional Euclidean space. This can be done numerically using the Newton-Raphson method, for example. However, since, in general, one wants to compute the angular velocity in real time, it is preferable to transform this system of multivariate polynomials into one univariate polynomial bearing the same roots. The computational burden associated with the solution of this resultant is generally lighter than that of the system of multivariate polynomials. Moreover, as will be shown, the resultant to the system of eq. (9) admits a closed-form solution. This is important for real-time implementation, since the solutions can then be computed in a predetermined number of operations.

The problem of finding the resultant of three quadric equations into three unknowns has already been addressed by other researchers (Morgan and Sarraga, 1982; Roth, 1993). In the following analysis, we use the method proposed by Roth (1993), as it results in a coefficient matrix of minimum dimensions.

Let us first rewrite the three quadrics of eq. (9) in the more familiar form

$$
\begin{aligned}
s_{1,1}\omega_1^2 + s_{1,2}\omega_2^2 + s_{1,3}\omega_3^2 + s_{1,4}\omega_2\omega_3 + s_{1,5}\omega_3\omega_1 + s_{1,6}\omega_1\omega_2 - \mathbf{q}_7^T\mathbf{a} &= 0,\\
s_{2,2}\omega_2^2 + s_{2,3}\omega_3^2 + s_{2,4}\omega_2\omega_3 + s_{2,5}\omega_3\omega_1 + s_{2,6}\omega_1\omega_2 - \mathbf{q}_8^T\mathbf{a} &= 0,\\
s_{3,3}\omega_3^2 + s_{3,4}\omega_2\omega_3 + s_{3,5}\omega_3\omega_1 + s_{3,6}\omega_1\omega_2 - \mathbf{q}_9^T\mathbf{a} &= 0,
\end{aligned}
\tag{10}
$$

where ω_1, ω_2, and ω_3 are the unknowns; $s_{i,j}$ is the (i,j) entry of $\mathbf{S}_{2,2}$; and $\mathbf{Q} \equiv [\mathbf{q}_1 \quad \mathbf{q}_2 \quad \cdots \quad \mathbf{q}_n]$. Notice that these equations are somewhat more specific than those of general quadrics, since they are free of linear terms.

3.2.1 Solving for ω_3

The objective is to obtain a resultant in ω_3 only. Therefore, we convert all other variables into homogeneous coordinates, that is, we substitute $\omega_1 = \Omega_1/W$ and $\omega_2 = \Omega_2/W$ into eq. (10), to obain

$$
\begin{aligned}
s_{1,1}\Omega_1^2 + s_{1,2}\Omega_2^2 + s_{1,6}\Omega_1\Omega_2 + s_{1,5}\omega_3\Omega_1 W + s_{1,4}\omega_3\Omega_2 W + \left(s_{1,3}\omega_3^2 - \mathbf{q}_7^T\mathbf{a}\right)W^2 &= 0,\\
s_{2,2}\Omega_2^2 + s_{2,6}\Omega_1\Omega_2 + s_{2,5}\omega_3\Omega_1 W + s_{2,4}\omega_3\Omega_2 W + \left(s_{2,3}\omega_3^2 - \mathbf{q}_8^T\mathbf{a}\right)W^2 &= 0,\\
s_{3,6}\Omega_1\Omega_2 + s_{3,5}\omega_3\Omega_1 W + s_{3,4}\omega_3\Omega_2 W + \left(s_{3,3}\omega_3^2 - \mathbf{q}_9^T\mathbf{a}\right)W^2 &= 0.
\end{aligned}
\tag{11}
$$

Three additional independent equations are required for the associated coefficient matrix to be of full column rank. To generate these additional equations, Roth (1993) proposed to compute the gradient of the determinant of the Jacobian matrix of eqs. (11) with respect to the homogeneous coordinates Ω_1, Ω_2, and W. As shown by Salmon (1964), the polynomials computed from this method share the same roots as the original ones.

Hence, we first compute the Jacobian matrix

$$
\mathbf{J} = \begin{bmatrix}
2s_{1,1}\Omega_1 + s_{1,6}\Omega_2 + s_{1,5}\omega_3 W & 2s_{1,2}\Omega_2 + s_{1,6}\Omega_1 + s_{1,4}\omega_3 W & s_{1,5}\omega_3\Omega_1 + s_{1,4}\omega_3\Omega_2 + 2\left(s_{1,3}\omega_3^2 - \mathbf{q}_7^T\mathbf{a}\right)W\\
s_{2,6}\Omega_2 + s_{2,5}\omega_3 W & 2s_{2,2}\Omega_2 + s_{2,6}\Omega_1 + s_{2,4}\omega_3 W & s_{2,5}\omega_3\Omega_1 + s_{2,4}\omega_3\Omega_2 + 2\left(s_{2,3}\omega_3^2 - \mathbf{q}_8^T\mathbf{a}\right)W\\
s_{3,6}\Omega_2 + s_{3,5}\omega_3 W & s_{3,6}\Omega_1 + s_{3,4}\omega_3 W & s_{3,5}\omega_3\Omega_1 + s_{3,4}\omega_3\Omega_2 + 2\left(s_{3,3}\omega_3^2 - \mathbf{q}_9^T\mathbf{a}\right)W
\end{bmatrix}.
\tag{12}
$$

We then compute the determinant of the Jacobian,

$$\begin{aligned}\det(\mathbf{J}) =& u_1(\omega_3)\Omega_1^3 + u_2(\omega_3)\Omega_1^2\Omega_2 + u_3(\omega_3^2,\mathbf{a})\Omega_1^2 W + u_4(\omega_3)\Omega_1\Omega_2^2 \\ &+ u_5(\omega_3^2,\mathbf{a})\Omega_1\Omega_2 W + u_6(\omega_3^3,\mathbf{a})\Omega_1 W^2 + u_7(\omega_3)\Omega_2^3 \\ &+ u_8(\omega_3^2,\mathbf{a})\Omega_2^2 W + u_9(\omega_3^3,\mathbf{a})\Omega_2 W^2 + u_{10}(\omega_3^4,\mathbf{a})W^3,\end{aligned} \tag{13}$$

where $u_i(\omega_3^{d_i})$ is a degree-d_i polynomial in ω_3 whose complete expression is too long to be reproduced in this paper. The gradient of this determinant with respect to the homogeneous coordinates has to be zero, which yields the three new equations

$$\begin{aligned}\frac{\partial \det(\mathbf{J})}{\partial \Omega_1} = 0 =& 3u_1(\omega_3)\Omega_1^2 + 2u_2(\omega_3)\Omega_1\Omega_2 + 2u_3(\omega_3^2,\mathbf{a})\Omega_1 W \\ &+ u_4(\omega_3)\Omega_2^2 + u_5(\omega_3^2,\mathbf{a})\Omega_2 W + u_6(\omega_3^3,\mathbf{a})W^2,\end{aligned} \tag{14}$$

$$\begin{aligned}\frac{\partial \det(\mathbf{J})}{\partial \Omega_2} = 0 =& u_2(\omega_3)\Omega_1^2 + 2u_4(\omega_3)\Omega_1\Omega_2 + u_5(\omega_3^2,\mathbf{a})\Omega_1 W \\ &+ 3u_7(\omega_3)\Omega_2^2 + 2u_8(\omega_3^2,\mathbf{a})\Omega_2 W + u_9(\omega_3^3,\mathbf{a})W^2,\end{aligned} \tag{15}$$

$$\begin{aligned}\frac{\partial \det(\mathbf{J})}{\partial W} = 0 =& u_3(\omega_3^2,\mathbf{a})\Omega_1^2 + u_5(\omega_3^2,\mathbf{a})\Omega_1\Omega_2 + 2u_6(\omega_3^3,\mathbf{a})\Omega_1 W \\ &+ u_8(\omega_3^2,\mathbf{a})\Omega_2^2 + 2u_9(\omega_3^3,\mathbf{a})\Omega_2 W + 3u_{10}(\omega_3^4,\mathbf{a})W^2.\end{aligned} \tag{16}$$

We cast eqs. (11) and eqs. (14–16) in matrix form, which yields

$$\mathbf{C}\mathbf{w} = \mathbf{0}_6, \tag{17}$$

where

$$\mathbf{C} \equiv \begin{bmatrix} s_{1,1} & s_{1,6} & s_{1,5}\omega_3 & s_{1,2} & s_{1,4}\omega_3 & s_{1,3}\omega_3{}^2 - \mathbf{q}_7^T\mathbf{a} \\ 0 & s_{2,6} & s_{2,5}\omega_3 & s_{2,2} & s_{2,4}\omega_3 & s_{2,3}\omega_3{}^2 - \mathbf{q}_8^T\mathbf{a} \\ 0 & s_{3,6} & s_{3,5}\omega_3 & 0 & s_{3,4}\omega_3 & s_{3,3}\omega_3{}^2 - \mathbf{q}_9^T\mathbf{a} \\ 3u_1(\omega_3) & 2u_2(\omega_3) & 2u_3(\omega_3^2,\mathbf{a}) & u_4(\omega_3) & u_5(\omega_3^2,\mathbf{a}) & u_6(\omega_3^3,\mathbf{a}) \\ u_2(\omega_3) & 2u_4(\omega_3) & u_5(\omega_3^2,\mathbf{a}) & 3u_7(\omega_3) & 2u_8(\omega_3^2,\mathbf{a}) & u_9(\omega_3^3,\mathbf{a}) \\ u_3(\omega_3^2,\mathbf{a}) & u_5(\omega_3^2,\mathbf{a}) & 2u_6(\omega_3^3,\mathbf{a}) & u_8(\omega_3^2,\mathbf{a}) & 2u_9(\omega_3^3,\mathbf{a}) & 3u_{10}(\omega_3^4,\mathbf{a}) \end{bmatrix}$$

is the coefficient matrix sought, and $\mathbf{w} \equiv [\Omega_1^2 \quad \Omega_1\Omega_2 \quad \Omega_1 W \quad \Omega_2^2 \quad \Omega_2 W \quad W^2]^T$. Finally, since $\mathbf{C}$ must be singular, we set its determinant to zero. This yields a resultant polynomial of the form

$$\det(\mathbf{C}) = v_4\omega_3^8 + v_3\omega_3^6 + v_2\omega_3^4 + v_1\omega_3^2 + v_0 = 0, \tag{18}$$

where all the coefficients of odd powers of ω_3 are null. This interesting fact stems from the symmetry of eq. (6) with respect to the rigid-body angular velocity. Indeed, this symmetry requires that the opposite $-\omega$ of any root ω of eq. (6) be a root as well, since both result in the same array ξ, or, from a mechanical standpoint, in the same centripetal component of the rigid-body acceleration field.

We exploit the evenness of the exponents in eq. (18), and perform the change of variable

$$\zeta_3 \equiv \omega_3^2, \tag{19}$$

which yields

$$\nu_4\zeta_3^4 + \nu_3\zeta_3^3 + \nu_2\zeta_2^2 + \nu_1\zeta_3 + \nu_0 = 0. \tag{20}$$

This reduction of the degree of the resultant is important, since the ensuing quartic allows for a closed-form solution (Cardano, 2007). Without rewriting the four solutions to a quartic equation, we may write symbolically

$$\zeta_{3,i} = \phi_i(\nu_0, \nu_1, \nu_2, \nu_3, \nu_4), \tag{21}$$

$i = 1, \ldots, 4$. Finally, the eight roots of the resultant of eq. (18) are

$$\omega_{3,i} = \sqrt{\zeta_{3,i}} \text{ and } \omega_{3,i+4} = -\sqrt{\zeta_{3,i}},\ i = 1, \ldots, 4. \tag{22}$$

3.2.2 Solving for ω_1 and ω_2

Having computed all possible solutions in ω_3, we turn our attention to the remaining unknowns of eqs. (10), which still appear in eq. (17) under the form of homogeneous coordinates. We perform the substitutions $\Omega_1 = W\omega_1$ and $\Omega_2 = W\omega_2$ into $\mathbf{w}$, which yields

$$\mathbf{w} = W^2\mathbf{w}', \tag{23}$$

where $\mathbf{w} \equiv [\omega_1^2 \quad \omega_1\omega_2 \quad \omega_1 \quad \omega_2^2 \quad \omega_2 \quad 1]^T$. Since we are not interested in infinite solutions, we may safely assume that $W \neq 0$, and eq. (17) takes the form

$$\mathbf{C}\mathbf{w}' = \mathbf{0}_6. \tag{24}$$

We partition the matrices and vectors as $\mathbf{C} \equiv [\mathbf{C}_1 \quad \mathbf{c}_2]$, $\mathbf{C}_1 \in \mathbb{R}^{6\times5}$, $\mathbf{c}_2 \in \mathbb{R}^6$, and $\mathbf{w}' \equiv [\xi_{1,2}^T \quad 1]^T$. Equation (24) can then be rewritten in the usual form

$$\mathbf{C}_1(\omega_3, \mathbf{a})\xi_{1,2} = -\mathbf{c}_2(\omega_3, \mathbf{a}), \tag{25}$$

where $\mathbf{C}_1$ and $\mathbf{c}_2$ depend on both ω_3 and $\mathbf{a}$. For a single exact solution to exist, the condition $\text{rank}(\mathbf{C}_1(\omega_3, \mathbf{a})) = 5$ must be fulfilled. Only in special cases will this condition not be satisfied, that is, when one or more roots of the polynomials of eq. (18) are the same, given a set of measurements $\mathbf{a}$. Although this was never experienced in simulation, we cannot overlook this possible singularity in the algorithm.

To detect a rank-deficiency of $\mathbf{C}_1$, we perform a QR factorization of $\mathbf{C}_1$ online using Householder reflections with column pivoting. If it is unique, the solution to eq. (25) is then obtained through a backsubstitution. Notice that this solution is not an approximate least-squares solution, but an exact one, since $\mathbf{c}_2$ is bound to lie in the range of $\mathbf{C}_1$.

This generally yields eight solutions $\xi_{1,2,i}$, $i = 1, \ldots, 8$, where ω_1 and ω_2 directly appear as the third and fifth components. Hence, we obtain

$$\omega_i \equiv [\omega_{1,i} \quad \omega_{2,i} \quad \omega_{3,i}]^T, \quad i = 1, \ldots, 8, \tag{26}$$

in which we have four symmetric pairs, i.e., $\omega_{3,i} = -\omega_{3,i+4}$, $i = 1, \ldots, 4$.

If, for one or more of the solutions listed in eq. (22), we have $\text{rank}(\mathbf{C}_1(\omega_{3,i},\mathbf{a})) < 5$, then the arrays $\xi_{1,2,i}$ corresponding to these solutions cannot be identified uniquely. As a result, one cannot extract the associated angular-velocity vectors ω_i.

3.3 Solution to the Linear System of Equations

From the eight solutions obtained in eq. (26), we compute four distinct arrays

$$\xi_i \equiv \left[\omega_{1,2i}^2 \ \omega_{2,2i}^2 \ \omega_{3,2i}^2 \ \omega_{2,2i}\omega_{3,2i} \ \omega_{3,2i}\omega_{1,2i} \ \omega_{1,2i}\omega_{2,2i}\right]^T, \tag{27}$$

$i = 1,\ldots,4$. These arrays are substituted back in eq. (8), which we rewrite as

$$\mathbf{S}_{1,1}\mathbf{x}_{PT,i} = \mathbf{Q}_1^T\mathbf{a} - \mathbf{S}_{1,2}\xi_i, \tag{28}$$

$i = 1,\ldots,4$. The linear system of eqs. (28) being already in upper-triangle form, it is easily solved through backsubstitution, which yields $\dot{\omega}_i$ and $\ddot{\mathbf{b}}_i$, $i = 1,\ldots,4$.

3.4 Selection of the Proper Estimate

We rely on the time-integration of the angular acceleration from the previous angular-velocity estimate to choose among the eight possible values ω_i, $i = 1,\ldots,8$. First, four *a priori* angular-velocity estimates are computed as

$$\omega_{TA,i,k} = \omega_{k-1} + \frac{t_k - t_{k-1}}{2}\left(\dot{\omega}_{k-1} + \dot{\omega}_{i,k}\right), \tag{29}$$

$i = 1,\ldots,4$, where k refers to the acquisition performed at time t_k, and the subscript "TA" stands for tangential acceleration. We then select the estimate ω_i corresponding to the smallest of the errors $\|\omega_i - \omega_{TA,i}\|_2$, $\|\omega_{4+i} - \omega_{TA,i}\|_2$, $i = 1,\ldots,4$.

3.5 Summary of the Algorithm

Because of the generality of the approach proposed in this paper, the associated calculations are more elaborate than those proposed by other researchers (Padgaonkar et al, 1975; Genin et al, 1997). For this reason, we include a summary of the offline and online calculations.

3.5.1 Offline Calculations

1. From the geometry $\{\mathbf{r}_i, \mathbf{e}_i\}_{i=1}^n$, compute the accelerometer-array input-output equations:
   $\mathbf{A} \xleftarrow{\text{eq. (6)}} \mathbf{E}, \mathbf{R}, \Sigma \xleftarrow{\text{eq. (5)}} \{\mathbf{r}_i, \mathbf{e}_i\}_{i=1}^n$
2. Compute a QR factorization of the accelerometer-array matrix $\mathbf{A}$:
   $\mathbf{Q}_1, \mathbf{S}_{1,1}, \mathbf{S}_{1,2}, \mathbf{Q}_2, \mathbf{S}_{2,2}, \mathbf{S}_{3,2} \xleftarrow{\text{eq. (7)}} \texttt{GramSchmidt}(\mathbf{A})$

3. Following the procedure detailed in section 3.2, calculate symbolic expressions for the five coefficients defining the octic of eq. (18), which are functions of the accelerometer measurements **a**:
   $\{v_i(\mathbf{a})\}_{i=0}^4 \xleftarrow{\text{eq. (18)}} \mathbf{C} \xleftarrow{\text{eqs. (13–17)}} \mathbf{J} \xleftarrow{\text{eqs. (9–12)}} \mathbf{Q}_2, \mathbf{S}_{2,2}, \mathbf{a}$

3.5.2 Online Calculations

The following caculations need to be performed online, often in real time, as they depend on the accelerometer measurements. We start from the accelerometer measurements $\mathbf{a}_k$ taken at time t_k.

1. $\{v_{i,k}\}_{i=0}^4 \xleftarrow{\text{eq. (18)}} \mathbf{a}_k$
2. $\{\omega_{3,i,k}\}_{i=1}^8 \xleftarrow{\text{eq. (22)}} \{\zeta_{3,i,k}\}_{i=1}^4 \xleftarrow{\text{eq. (21)}} \texttt{FerrariCardano}(\{v_{i,k}\}_{i=0}^4)$
3. $\{\omega_{i,k}\}_{i=1}^8 \leftarrow \{\xi_{1,2,i,k}\}_{i=1}^8 \xleftarrow{\text{eq. (24)}} \{\omega_{3,i,k}\}_{i=1}^8$
4. $\{\dot{\omega}_{i,k}, \ddot{\mathbf{b}}_{i,k}\}_{i=1}^4 \xleftarrow{\text{eq. (28)}} \{\xi_{i,k}\}_{i=1}^4 \xleftarrow{\text{eq. (27)}} \{\omega_{i,k}\}_{i=1}^8$
5. $\{\omega_{TA,i,k}\}_{i=1}^4 \xleftarrow{\text{eq. (29)}} \omega_{k-1}, \dot{\omega}_{k-1}, \{\dot{\omega}_{i,k}\}_{i=1}^4$
6. Choose i corresponding to the minimum of $\|\omega_i - \omega_{TA,i}\|_2$, $\|\omega_{4+i} - \omega_{TA,i}\|_2$, $i = 1,\ldots,4$.

In general, the number of flops associated with this algorithm comes mainly from steps 1 and 3. Since the symbolic expressions for v_i, $i = 0,\ldots,4$, vary importantly from one accelerometer array to the other, it is impossible to associate one computation cost to all accelerometer arrays.

4 Example: A Nonlinearly Determined Nine-Accelerometer Array

In this example, we estimate the acceleration field of a brick that rotates freely in space by solving the nonlinear input-output equations of the nine-accelerometer array proposed by Huijser and Vreeburg (1993). The geometry of this accelerometer array and the brick on which it is mounted are depicted in Fig. 2, and its dimensions are given in Table 1, where we choose $\rho = 0.1$ m.

Because it is composed of nine accelerometers, this array does not allow for the linear estimation of all the rigid-body acceleration-field parameters. Moreover, it may be shown that it is impossible to estimate linearly *any* of the parameters ω, $\dot{\omega}$, and $\ddot{\mathbf{b}}$, making a nonlinear solution inevitable.

4.1 Offline Computations

The point, tangential, and radial matrices are computed from their definitions in eqs. (6) and (5), which yields

Table 1. Geometric parameters of the nine-accelerometer array Huijser and Vreeburg (1993)

i	$\mathbf{e}_i$	$\mathbf{r}_i$
1	$[0\ \ 1\ \ 0]^T$	$[\rho\ \ 0\ \ 0]^T$
2	$[0\ \ 0\ \ 1]^T$	$[\rho\ \ 0\ \ 0]^T$
3	$[1\ \ 0\ \ 0]^T$	$[0\ \ \rho\ \ 0]^T$
4	$[0\ \ 0\ \ 1]^T$	$[0\ \ \rho\ \ 0]^T$
5	$[0\ \ 1\ \ 0]^T$	$[0\ \ 0\ \ \rho]^T$
6	$[1\ \ 0\ \ 0]^T$	$[0\ \ 0\ \ \rho]^T$
7	$[1/\sqrt{2}\ \ 0\ \ 1/\sqrt{2}]^T$	$[\rho/2\ \ \rho/2\ \ 0]^T$
8	$[0\ \ 1/\sqrt{2}\ \ 1/\sqrt{2}]^T$	$[0\ \ \rho/2\ \ \rho/2]^T$
9	$[1/\sqrt{2}\ \ 1/\sqrt{2}\ \ 0]^T$	$[\rho/2\ \ \rho/2\ \ 0]^T$

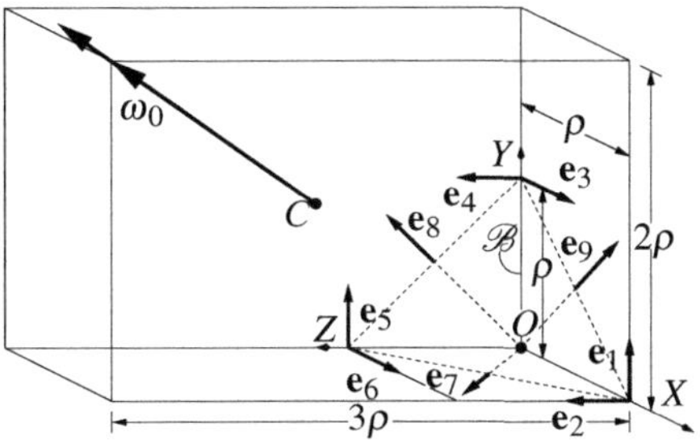

Fig. 2. A nine-accelerometer array Huijser and Vreeburg (1993)

$$\mathbf{A}_P = \begin{bmatrix} 0 & 0 & 1 & 0 & 0 & 1 & \frac{1}{\sqrt{2}} & 0 & \frac{1}{\sqrt{2}} \\ 1 & 0 & 0 & 0 & 1 & 0 & 0 & \frac{1}{\sqrt{2}} & \frac{1}{\sqrt{2}} \\ 0 & 1 & 0 & 1 & 0 & 0 & \frac{1}{\sqrt{2}} & \frac{1}{\sqrt{2}} & 0 \end{bmatrix}^T, \quad \mathbf{A}_T = \rho \begin{bmatrix} 0 & 0 & 0 & 1 & -1 & 0 & 0 & 0 & 0 \\ 0 & -1 & 0 & 0 & 0 & 1 & 0 & 0 & 0 \\ 1 & 0 & -1 & 0 & 0 & 0 & 0 & 0 & 0 \end{bmatrix}^T, \tag{30}$$

$$\mathbf{A}_R = \rho \begin{bmatrix} 0 & 0 & 0 & 0 & 0 & 0 & -\frac{1}{2} & -1 & -\frac{1}{2} \\ 0 & 0 & 0 & 0 & 0 & 0 & -1 & -\frac{1}{2} & -\frac{1}{2} \\ 0 & 0 & 0 & 0 & 0 & 0 & -\frac{1}{2} & -\frac{1}{2} & -1 \\ 0 & 0 & 0 & 1 & 1 & 0 & 0 & 1 & 0 \\ 0 & 1 & 0 & 0 & 0 & 1 & 1 & 0 & 0 \\ 1 & 0 & 1 & 0 & 0 & 0 & 0 & 0 & 1 \end{bmatrix}^T, \tag{31}$$

and $\mathbf{A} = [\mathbf{A}_P \quad \mathbf{A}_T \quad \mathbf{A}_R]$. A QR factorization of $\mathbf{A}$ is then computed by means of the Gram-Schmidt procedure, which yields the same form as the one obtained in eq. (7), except that matrices $\mathbf{Q}_3$ and $\mathbf{S}_{3,1}$ are absent. Thence, matrix $\mathbf{J}$ is computed from eqs. (9–12), which yields $\mathbf{J} = [\mathbf{j}_1 \quad \mathbf{j}_2 \quad \mathbf{j}_3]$, where

$$\mathbf{j}_1 = -\frac{\sqrt{3}}{60}\begin{bmatrix} \left(-6\Omega_1 - (\sqrt{2}-1)\Omega_2 - (\sqrt{2}-1)\omega_3 W\right) \\ \left(-(\sqrt{2}-1)\Omega_1 - 5\Omega_2 - 2(\sqrt{2}-1)\omega_3 W\right) \\ \left(-(\sqrt{2}-1)\omega_3\Omega_1 - 2(\sqrt{2}-1)\omega_3\Omega_2 \cdots \right. \\ +10\sqrt{2}\left(a_1 + a_2 + a_3 + 2a_4 + 2a_5 + a_6 \cdots \right. \\ \left. \left. -\sqrt{2}a_7 - 2\sqrt{2}a_8 - \sqrt{2}a_9\right) W - 5\omega_3{}^2 W\right) \end{bmatrix},$$

$$\mathbf{j}_2 = -\frac{\sqrt{33}}{660}\begin{bmatrix} \left(-\sqrt{2}\Omega_2 + \Omega_2 - 7(\sqrt{2}-1)\omega_3 W\right) \\ \left(-(\sqrt{2}-1)\Omega_1 - 11\Omega_2 + 4(\sqrt{2}-1)\omega_3 W\right) \\ \left(-7(\sqrt{2}-1)\omega_3\Omega_1 + 4(\sqrt{2}-1)\omega_3\Omega_2 \cdots \right. \\ +10\sqrt{2}\left(a_1 + 7a_2 + a_3 - 4a_4 - 4a_5 + 7a_6 \cdots \right. \\ \left. \left. -7\sqrt{2}a_7 + 4\sqrt{2}a_8 - \sqrt{2}a_9\right) W - 5\omega_3^2 W\right) \end{bmatrix},$$

$$\mathbf{j}_3 = -\frac{\sqrt{11}}{220}\begin{bmatrix} \left(-3(2-\sqrt{2})\Omega_2 + (2-\sqrt{2})\omega_3 W\right) \\ \left(-3(2-\sqrt{2})\Omega_1 + (2-\sqrt{2})\omega_3 W\right) \\ \left(-(2-\sqrt{2})\omega_3\Omega_1 + (2-\sqrt{2})\omega_3\Omega_2 \cdots \right. \\ +20\left(3a_1 - a_2 + 3a_3 - a_4 - a_5 - a_6 \cdots \right. \\ \left. \left. +\sqrt{2}a_7 + \sqrt{2}a_8 - 3\sqrt{2}a_9\right) W - 4\sqrt{2}\omega_3{}^2 W\right) \end{bmatrix}.$$

Matrix **C** and the coefficients $\{v_i\}_{i=0}^4$ of the resulting polynomial are computed in turn from eqs. (13–18), but are not reported here because of space constraints. The symbolic expressions of coefficients $\{v_i\}_{i=0}^4$ form a total approximately 500 terms.

4.2 Online Computations

Assume the set of acceleration measurements $\mathbf{a} = \pi^2 \cdot [16/175\ 27/175\ 4/175\ 108/455\ 48/455\ 3/175\ -6/35\ -3/70\ -19/70]^T$ m/s^2, from which we are to compute the acceleration field parameters ω, $\dot{\omega}$ and $\ddot{\mathbf{b}}$.

The resulting polynomial takes the form

$$0 = \left(177856476\sqrt{2} - 251468735\right)\zeta_3{}^4 - 902776\left(419\sqrt{2} - 592\right)\pi^2\zeta_3{}^3 \cdots$$
$$- 10584\left(17282\sqrt{2} - 24585\right)\pi^4\zeta_3{}^2 - 18144\pi^6\left(-548 + 461\sqrt{2}\right)\zeta_3 \cdots$$
$$- 11664\pi^8\left(-13841 + 9744\sqrt{2}\right), \qquad (32)$$

and its roots are readily computed from the Ferrari-Cardano formula (Cardano, 2007). From the relation $\zeta_3 = \omega_3^2$, we obtain the eight values of ω_3 that are listed in Table 2.

We successively substitute these solutions into the symbolical expression of matrix **C**. We then solve eq. (25) for $\xi_{1,2}$, which yields, for $\omega_3 = \omega_{3,1}$, $\xi_{1,2} =$

Table 2. The numerical values of $\omega_{3,i}$, $i = 1,\ldots,8$

i	1	2	3	4	5	6	7	8
$\omega_{3,i}$ (rad/s)	2.5476	5.0378	5.0563	5.2394	-2.5476	-5.0378	-5.0563	-5.2394

$[19.4084 \quad 19.9869 \quad -4.4055 \quad 20.5827 \quad -4.5368]^T$. Recall, from the definition of $\xi_{1,2}$ in eqs. (23) and (24), that ω_1 and ω_2 are the third and fifth components of $\xi_{1,2}$. Hence, we obtain the set of eight angular velocities that are listed in Table 3.

Table 3. The computed values of ω_i, $i = 1,\ldots,8$, in rad/s

ω_1	ω_2	ω_3	ω_4	ω_5	ω_6	ω_7	ω_8
-4.4055	1.6793	1.5247	3.7418	4.4055	-1.6793	-1.5247	-3.7418
-4.5368	3.3585	3.4270	0.1900	4.5368	-3.3585	-3.4270	-0.1900
2.5476	5.0378	5.0563	5.2394	-2.5476	-5.0378	-5.0563	-5.2394

Finally, we form ξ_i, $i = 1,\ldots,8$, from ξ_{12} and ω_3, and we compute $\mathbf{x}_{PT,i}$ by solving eq. (28) through a backsubstitution. This yields the values of $\ddot{\mathbf{b}}_i$ and $\dot{\omega}_i$ that are listed in Table 4.

Table 4. The computed values of $\dot{\omega}_i$ in rad/s^2 and $\ddot{\mathbf{b}}_i$ in m/s^2, $i = 1,\ldots,4$

$\ddot{\mathbf{b}}_1$	$\ddot{\mathbf{b}}_2$	$\ddot{\mathbf{b}}_3$	$\ddot{\mathbf{b}}_4$	$\dot{\omega}_1$	$\dot{\omega}_2$	$\dot{\omega}_3$	$\dot{\omega}_4$
-2.3141	0.0000	0.1574	-2.2140	-27.5229	6.5074	6.1716	22.5814
-0.5553	-0.0000	-0.0745	3.1998	36.0568	-6.7677	-7.5911	4.2272
6.2508	0.0000	-0.0073	-0.0150	-5.4103	3.3839	4.5431	-23.6850

Each of the eight solutions displayed in Tables 3 and 4 satisfies eq. (6), the original input-output equation of the accelerometer array. In other words, there are eight distinct choices of $\ddot{\mathbf{b}}$, $\dot{\omega}$ and ω that produce the same acceleration measurements. Among these solutions, we can choose the right solution by computing another estimate of ω from the tangential acceleration, as proposed in section 3.4.

5 Conclusions

In summary, this paper presented a method for the nonlinear estimation of nine independent parameters representing the rigid-body acceleration field from any nonsingular n-accelerometer array, with $n \geq 9$. This problem was shown to admit up to eight solutions, which, quite unexpectedly, could be expressed in closed forms.

We see two main advantages of using the closed form solutions over iterative methods for computing the acceleration-field parameters ω, $\dot{\omega}$ and $\ddot{\mathbf{b}}$. (i) The number of operations required to solve this problem can be known in advance, which is critical for real-time implementations. (ii) We are guaranteed to find all the solutions, which is not the case with Newton methods, for instance.

References

Angeles, J.: The angular-acceleration tensor of rigid-body kinematics and its properties. Archive of Applied Mechanics 69(3), 204–214 (1999)

Baselli, G., Legnani, G., Franco, P., Brognoli, F., Marras, A., Quaranta, F., Zappa, B.: Assessment of inertial and gravitational inputs to the vestibular system. Journal of Biomechanics 34, 821–826 (2001)

van den Bogert, A.J., Read, L., Nigg, B.M.: A method for inverse dynamic analysis using accelerometry. Journal of Biomechanics 29(7), 949–954 (1996)

Cappa, P., Masia, L., Patanè, F.: Numerical validation of linear accelerometer systems for the measurement of head kinematics. ASME Journal of Biomechanical Engineering 127(6), 919–928 (2005)

Cardano, G.: The Rules of Algebra: Ars Magna. Dover Publications, New York (2007)

Cardou, P.: Design of multiaxial accelerometers with simplicial architectures for rigid-body pose-and-twist estimation. PhD thesis, McGill University, Montreal, QC, Canada (2008)

Cardou, P., Angeles, J.: Linear estimation of the rigid-body acceleration field from point-acceleration measurements. ASME Journal of Dynamic Systems, Measurement and Control 131(4) (2009) 041,013–1–041,013–10

Corey, V.B.: Measuring angular acceleration with linear accelerometers. Control Engineering 9(3), 79–80 (1962)

Di Puccio, F., Forte, P.: Identification of the 3d vibratory motion of a rigid body by accelerometer measurements. Shock and Vibration 11(3-4), 281–293 (2004)

Genin, J., Hong, J., Xu, W.: Accelerometer placement for angular velocity determination. ASME Journal of Dynamic Systems, Measurements and Control 119, 474–477 (1997)

Grammatikos, A.: Gimballess inertial systems for space navigation. PhD thesis, University of Pennsylvania, PA (1965)

Huijser, R.H., Vreeburg, J.P.B.: Assembly of linear accelerometers mounted on a rigid body remote from the center of mass measuring three dimensional movements (1993)

Kröger, T., Kubus, D., Wahl, F.M.: 12d force and acceleration sensing: A helpful experience report on sensor characteristics. In: Proceedings of the IEEE International Conference on Robotics and Automation, Pasadena, CA, USA, pp. 3455–3462 (2008)

Lin, P.C., Komsuoḡlu, H., Koditschek, D.: Sensor data fusion for body state estimation in a hexapod robot with dynamical gaits. IEEE Transactions on Robotics 22(5), 932–943 (2006)

McIntosh, A.S., Kallieris, D., Frechede, B.: Neck injury tolerance under inertial loads in side impacts. Accident Analysis and Prevention 39(2), 326–333 (2007)

Mertz, H.J.: Kinematics and kinetics of whiplash. PhD thesis, Wayne State University, Detroit, MI (1967)

Morgan, A.P., Sarraga, R.F.: A method for computing three surface intersection points in gmsolid (1982)

Padgaonkar, A.J., Krieger, K.W., King, A.I.: Measurement of angular acceleration of a rigid body using linear accelerometers. ASME Journal of Applied Mechanics, 552–556 (1975)

Parsa, K.: Dynamics, state estimation, and control of manipulators with rigid and flexible subsystems. PhD thesis, McGill University, Montreal, Canada (2003)

Pickel, W.C.: Estimation of postlaunch angular motion for kinetic energy projectiles. Journal of Guidance, Control, and Dynamics 28(4), 604–610 (2005)

Roth, B.: Computational Kinematics. chap Computations in Kinematics, pp. 3–14. Kluwer Academic Publishers, Dordrecht (1993)

Salmon, G.: Higher Algebra, 5th edn. Chelsea Publishing Co., New York (1964)

Schuler, A.R.: Design and analysis of analytic platform inertial navigation systems. PhD thesis, University of Pennsylvania, PA (1965)

Walilko, T.J., Viano, D.C., Bir, C.A.: Biomechanics of the head for olympic boxer punches to the face. British Journal of Sports Medicine 39(10), 710–719 (2005)

Williams, T.R., Fyfe, K.R.: Planar accelerometer configurations. ASME Journal of Applied Mechanics 71, 10–14 (2004)

Yoganandan, N., Zhang, J., Pintar, F.A., Liu, Y.K.: Lightweight low-profile nine-accelerometer package to obtain head angular accelerations in short-duration impacts. Journal of Biomechanics 39(7), 1347–1354 (2006)

Zappa, B., Legnani, G., van den Bogert, A.J., Adamini, R.: On the number and placement of accelerometers for angular velocity and acceleration determination. ASME Journal of Dynamic Systems, Measurement and Control 123(3), 552–554 (2001)

Singularity Analysis of a Six-Dof Parallel Manipulator Using Grassmann-Cayley Algebra and Gröbner Bases

Stéphane Caro, Guillaume Moroz, Thibault Gayral,
Damien Chablat, and Chao Chen

Abstract. The subject of this paper deals with the singularity analysis of a six-dof three-legged parallel manipulator for force-feedback interface. To this end, a geometric condition for the manipulator singularities is obtained by means of Grassmann-Cayley algebra; the parallel singularities of the manipulator are computed using Jacobian and Gröbner basis. As a result, the algebraic relations of the singularities satisfied by the orientation variables are reported. Finally, the parallel singularities of the manipulator are plotted in its orientation workspace.

1 Introduction

The challenge in Virtual Reality (VR) today is to develop "transparent" haptic interface. VR aims at creating high-fidelity reproductions of a real environment, where a user perceives the illusion of being in this environment and interacts with it to perform specific tasks in an effective and comfortable way [1]. Haptics is critical in VR applications, where the user actively manipulates the simulated world, such as in surgical simulators [2]. There are two main classes of haptic devices: admittance devices and impedance devices [3]. Admittance devices sense the force applied by the operator and then control the operator's position. They are often more expensive because of the cost of the required force sensors, and more complex because of the weight of the force sensors. Impedance devices, also termed force feedback devices, are simpler, cheaper and more common. The fidelity in VR requires light and stiff interfaces. In this vein, Chen *et al.* developed a new six-dof three-legged parallel

Stéphane Caro · Guillaume Moroz · Thibault Gayral · Damien Chablat
Institut de Recherche en Communications et Cybernétique de Nantes, France
e-mail: {stephane.caro,guillaume.moroz}@irccyn.ec-nantes.fr, damien.chablat@irccyn.ec-nantes.fr

Thibault Gayral · Chao Chen
Department of Mechanical and Aerospace Engineering, Monash University, Australia
e-mail: thibault.gayral@ens-cachan.org, chao.chen@monash.edu

J. Angeles et al. (Eds.): Brain, Body and Machine, AISC 83, pp. 341–352.
springerlink.com

manipulator for force-feedback interface [4]. The advantage of the proposed device is that all motors are mounted on the ground. This feature significantly reduces the overall mass and inertia of the moving parts, which will eventually yield a more transparent haptic device.

Here, we focus on the singularity analysis of the foregoing manipulator by using Grassmann-Cayley algebra (GCA) and Gröbner bases (GB). Singularities can be found using either numeric, symbolic or geometric methods. For Gough-Stewart parallel manipulators, the rows of the backward Jacobian matrix are the Plücker coordinate vectors of six finite lines that are the lines of action of six actuation forces applied by the actuators to the moving platform. The parallel singularities of such manipulators that Merlet [5] analyzed using Grassmann line geometry occur when those lines become linearly dependent. Ben-Horin and Shoham analyzed the parallel singularities of six-dof parallel manipulators using GCA [6, 7]. Kanaan et al. [8] enlarged the application of GCA to lower-mobility manipulators, in which the rows of the backward Jacobian are either finite lines (zero pitch wrenches) or infinite lines (infinite pitch wrenches). These wrenches, also known as governing lines, are actuation and constraint wrenches applied to the moving platform.

This paper is organized as follows. Section 2 describes the manipulator under study. Sections 3 and 4 are devoted to its singularity analysis with Grassmann-Cayley algebra and Gröbner bases, respectively. Finally, the parallel singularities of the manipulator are plotted in its orientation workspace.

2 Manipulator Description

Figure 1 illustrates the parallel manipulator under study, which is a simplified kinematic version of the manipulator proposed in [4]. It is composed of an equilateral moving platform connected to the base by means of three identical legs. Each leg is composed of three orthogonal prismatic joints and one spherical joint, the first two prismatic joints being actuated. P stands for a prismatic joint whereas S stands for a spherical joint. An underline letter denotes an actuated joint. As a consequence, the manipulator is named 3-<u>PP</u>PS manipulator and provides six-degree-of-freedom motions, i.e., three translations and three rotations.

2.1 *Parameterization*

Let C_1, C_2 and C_3 be the corners of the moving platform (MP) of side length r. Let $\mathscr{F}_p$ (C_p, X_p, Y_p, Z_p) be the frame attached to the moving platform, its origin C_p being the centroid of the MP. Y_p is parallel to line (C_2C_3) and Z_p is normal to the MP. Accordingly,

$$\mathbf{c}_{1p} = \begin{bmatrix} 2r\sqrt{3}/6 \\ 0 \\ 0 \end{bmatrix}, \mathbf{c}_{2p} = \begin{bmatrix} -r\sqrt{3}/6 \\ r/2 \\ 0 \end{bmatrix}, \mathbf{c}_{3p} = \begin{bmatrix} -r\sqrt{3}/6 \\ -r/2 \\ 0 \end{bmatrix} \tag{1}$$

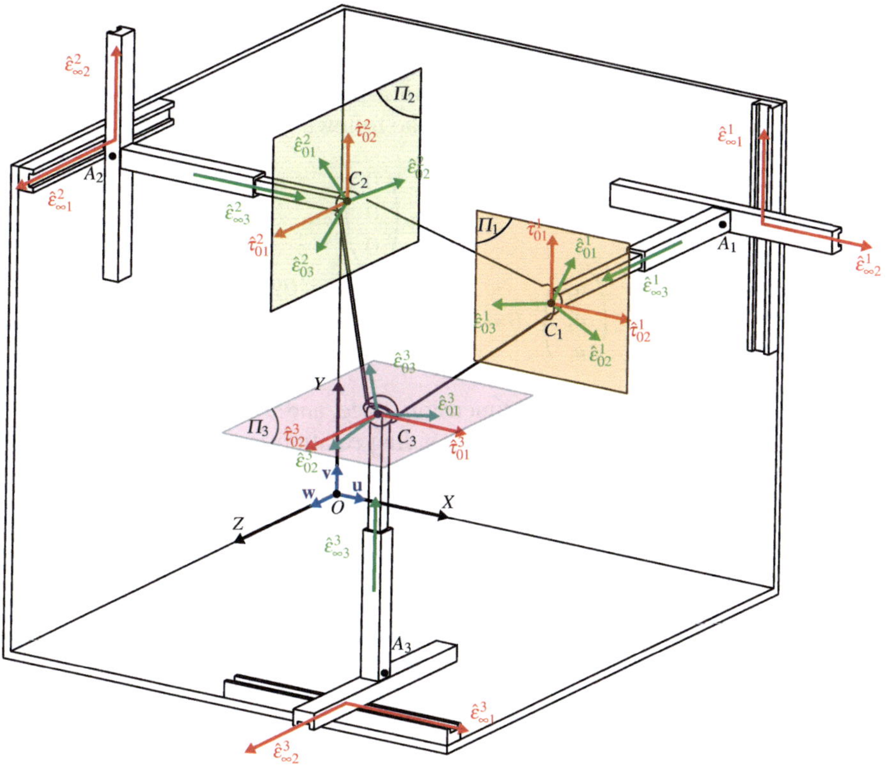

Fig. 1. The 3-PPPS manipulator

are the Cartesian coordinate vectors of points C_1, C_2 and C_3 expressed in $\mathscr{F}_p$. Likewise, let $\mathscr{F}_b$ (O,X,Y,Z) be the frame attached to the base and

$$\mathbf{a}_{1b} = \begin{bmatrix} x_1 \\ y_1 \\ 0 \end{bmatrix}, \mathbf{a}_{2b} = \begin{bmatrix} 0 \\ y_2 \\ z_2 \end{bmatrix}, \mathbf{a}_{3b} = \begin{bmatrix} x_3 \\ 0 \\ z_3 \end{bmatrix} \tag{2}$$

be the Cartesian coordinate vectors of points A_1, A_2 and A_3.

2.2 Orientation Space

The orientation space can be fully represented with the variables (Q_2, Q_3, Q_4), a subset of the quaternions coordinates. Indeed, the quaternions represent the rotations of the platform with a rotation axis $\mathbf{u}$ and an angle θ. The relation between the quaternions and the axis and angle representation can be found in [9]:

$$Q_1 = \cos(\theta/2),\ Q_2 = u_x \sin(\theta/2),\ Q_3 = u_y \sin(\theta/2),\ Q_4 = u_z \sin(\theta/2) \tag{3}$$

where $u_x^2 + u_y^2 + u_z^2 = 1$ and $0 \leq \theta \leq \pi$.

Thus, each rotation can be mapped onto a point of the unit ball in the space defined by the variables (Q_2, Q_3, Q_4) with the following bijection:

$$\Phi : \mathscr{S} \times]0, \pi] \rightarrow \mathscr{B} \backslash \{ \begin{pmatrix} 0 \\ 0 \\ 0 \end{pmatrix} \}$$

$$\begin{pmatrix} u_x \\ u_y \\ u_z \end{pmatrix}, \theta \mapsto \begin{pmatrix} Q_2 := u_x \sin(\theta/2) \\ Q_3 := u_y \sin(\theta/2) \\ Q_4 := u_z \sin(\theta/2) \end{pmatrix} \tag{4}$$

where $\mathscr{S}$ is the unit sphere in a 3-dimension space, and $\mathscr{S}$ is the closed unit ball.

When θ is equal to zero, the corresponding rotation matrix is the identity that does not depend on the rotation axis $\mathbf{u}$. It also maps to the center of $\mathscr{B}$ in the quaternions representation.

2.3 *Geometric Model*

Let $\mathbf{c}_b = \begin{bmatrix} c_x & c_y & c_z \end{bmatrix}^T$ be the Cartesian coordinate vector of point C, the centroid of the MP, expressed in $\mathscr{F}_b$. The following equations characterize the geometric model of the 3-P̲P̲PS manipulator:

$$c_x - 1/3\sqrt{3}{Q_1}^2 - 1/3\sqrt{3}{Q_2}^2 + 1/6\sqrt{3} - Q_2Q_3 + Q_1Q_4 - x_1 = 0 \tag{5a}$$
$$c_y - 1/3\sqrt{3}Q_2Q_3 - 1/3\sqrt{3}Q_1Q_4 - {Q_1}^2 - {Q_3}^2 + 1/2 - y_1 = 0 \tag{5b}$$
$$c_y - 1/3\sqrt{3}Q_2Q_3 - 1/3\sqrt{3}Q_1Q_4 + {Q_1}^2 + {Q_3}^2 - 1/2 - y_2 = 0 \tag{5c}$$
$$c_z - 1/3\sqrt{3}Q_2Q_4 + 1/3\sqrt{3}Q_1Q_3 + Q_3Q_4 + Q_1Q_2 - z_2 = 0 \tag{5d}$$
$$c_x + 2/3\sqrt{3}{Q_1}^2 + 2/3\sqrt{3}{Q_2}^2 - 1/3\sqrt{3} - x_3 = 0 \tag{5e}$$
$$c_z + 2/3\sqrt{3}Q_2Q_4 - 2/3\sqrt{3}Q_1Q_3 - z_3 = 0 \tag{5f}$$
$${Q_1}^2 + {Q_2}^2 + {Q_3}^2 + {Q_4}^2 - 1 = 0 \tag{5g}$$

3 Singularity Analysis with Grassmann-Cayley Algebra

3.1 *Grassmann-Cayley Algebra*

The Grassmann-Cayley Algebra (GCA), also known as exterior algebra, was developed by H. Grassmann as a calculus for linear varieties operating on *extensors* with the *join* and *meet* operators. The latter are associated with the *span* and *intersection* of vector spaces of extensors. Extensors are symbolically denoted by Plücker coordinates of lines and characterized by their *step*. In the four-dimensional vector space V associated with the three-dimensional projective space P_3, extensors of step 1, 2 and 3 represent points, lines and planes, respectively. They are also associated with subspaces of V, of dimension 1, 2 and 3, respectively. Points are represented with their homogeneous coordinates, while lines and planes are represented with their Plücker coordinates. The notion of extensor makes it possible to work at the symbolic level and therefore, to produce coordinate-free algebraic expressions for the geometric singularity conditions of spatial parallel manipulators (PMs). For further details on GCA, the reader is referred to [6, 10, 11, 12, 13].

3.2 *Twist System of the 3-PPPS Manipulator*

A unit screw is given by $\hat{\$} = [\mathbf{s}, (\mathbf{s_0} \times \mathbf{s} + \lambda \mathbf{s})]^T$ where $\mathbf{s}$ is a unit vector along the screw axis, $\mathbf{s_0}$ is the position vector of a point on the screw axis, with respect to a reference frame and λ is the *pitch* of the screw. A screw of intensify ρ is written as: $\$ = \rho\, \hat{\$}$. A zero pitch screw $\hat{\$}_0 = (\mathbf{s}, \mathbf{r} \times \mathbf{s})^T$ $(\lambda = 0)$ corresponds to the Plücker coordinate vector of a finite line in P_3. An infinite pitch screw $\hat{\$}_\infty = (\mathbf{0}, \mathbf{s})^T$ $(\lambda \rightarrow \infty)$ corresponds to the Plücker coordinate vector of an infinite line in the projective space P_3.

A *twist* is a screw representing the instantaneous motion of a rigid body, a *wrench* is a screw representing a system of forces and moments acting on a rigid body. Let ε denote a twist and τ denote a wrench.

(a) An infinite pitch twist ε_∞ represents a pure translation;
(b) A zero pitch twist ε_0 represents a pure rotation;
(c) A pure force constrains the translation along its line of action and is represented by a zero pitch wrench τ_0;
(d) A pure moment constrains the rotation about its direction and is represented by an infinite pitch wrench τ_∞.

Each leg of the 3-PPPS manipulator provides three independent translations and three independent rotations, represented with six independent twists $\hat{\varepsilon}^i_{\infty 1}$, $\hat{\varepsilon}^i_{\infty 2}$, $\hat{\varepsilon}^i_{\infty 3}$, $\hat{\varepsilon}^i_{01}$, $\hat{\varepsilon}^i_{02}$, $\hat{\varepsilon}^i_{03}$, $i = 1, 2, 3$, that span its twist sytem T^i. As shown in Fig. 1, the independent twists can be defined as:

$$\hat{\varepsilon}^i_{01} = \begin{bmatrix} \mathbf{u} \\ \mathbf{c}_i \times \mathbf{u} \end{bmatrix}, \hat{\varepsilon}^i_{02} = \begin{bmatrix} \mathbf{v} \\ \mathbf{c}_i \times \mathbf{v} \end{bmatrix}, \hat{\varepsilon}^i_{03} = \begin{bmatrix} \mathbf{w} \\ \mathbf{c}_i \times \mathbf{w} \end{bmatrix} \tag{6}$$

and

$$\hat{\varepsilon}^1_{\infty 2} = \hat{\varepsilon}^2_{\infty 3} = \hat{\varepsilon}^3_{\infty 1} = [\mathbf{0}, \mathbf{u}]^T \tag{7a}$$
$$\hat{\varepsilon}^1_{\infty 1} = \hat{\varepsilon}^2_{\infty 2} = \hat{\varepsilon}^3_{\infty 3} = [\mathbf{0}, \mathbf{v}]^T \tag{7b}$$
$$\hat{\varepsilon}^1_{\infty 3} = \hat{\varepsilon}^2_{\infty 1} = \hat{\varepsilon}^3_{\infty 2} = [\mathbf{0}, \mathbf{w}]^T \tag{7c}$$

where $\mathbf{c}_1$, $\mathbf{c}_2$ and $\mathbf{c}_3$ are the Cartesian coordinate vectors of points C_1, C_2 and C_3, respectively; $\mathbf{u}$, $\mathbf{v}$ and $\mathbf{w}$ are the unit vectors along the X, Y and Z axes of $\mathscr{F}_b$.

Accordingly, the twist system T of the 3-PPPS Manipulator is the intersection of the twist systems associated with its three legs, i.e., a 6-system.

$$\mathrm{T} = \bigcap_{i=1}^{3} \mathrm{T}^i \tag{8}$$

3.3 Wrench System of the 3-PPPS Manipulator

The actuated joints of the 3-PPPS manipulator are the first two prismatic joints of each leg. The actuation wrench $\hat{\tau}_{01}^i$ corresponding to the first prismatic joint of the ith leg is reciprocal to all the twists in T^i, but to $\hat{\varepsilon}_{\infty 1}^i$. Likewise, the actuation wrench $\hat{\tau}_{02}^i$ corresponding to the second prismatic joint of the ith leg is reciprocal to all the twists in T^i, but to $\hat{\varepsilon}_{\infty 2}^i$. As a result,

$$\hat{\tau}_{01}^1 = \begin{bmatrix} \mathbf{v} \\ \mathbf{c}_1 \times \mathbf{v} \end{bmatrix}, \hat{\tau}_{01}^2 = \begin{bmatrix} \mathbf{w} \\ \mathbf{c}_2 \times \mathbf{w} \end{bmatrix}, \hat{\tau}_{01}^3 = \begin{bmatrix} \mathbf{u} \\ \mathbf{c}_3 \times \mathbf{u} \end{bmatrix} \tag{9}$$

and

$$\hat{\tau}_{02}^1 = \begin{bmatrix} \mathbf{u} \\ \mathbf{c}_1 \times \mathbf{u} \end{bmatrix}, \hat{\tau}_{02}^2 = \begin{bmatrix} \mathbf{v} \\ \mathbf{c}_2 \times \mathbf{v} \end{bmatrix}, \hat{\tau}_{02}^3 = \begin{bmatrix} \mathbf{w} \\ \mathbf{c}_3 \times \mathbf{w} \end{bmatrix} \tag{10}$$

In a non-singular configuration, the six actuation wrenches $\hat{\tau}_{01}^1$, $\hat{\tau}_{02}^1$, $\hat{\tau}_{01}^2$, $\hat{\tau}_{02}^2$, $\hat{\tau}_{01}^3$ and $\hat{\tau}_{02}^3$ span the actuation wrench system of the 3-PPPS manipulator. As the 3-PPPS manipulator does not have any constraint wrench, its global wrench system amounts to its actuation wrench, namely,

$$\mathrm{W}_{3-\underline{PP}PS} = span(\hat{\tau}_{01}^1, \hat{\tau}_{02}^1, \hat{\tau}_{01}^2, \hat{\tau}_{02}^2, \hat{\tau}_{01}^3, \hat{\tau}_{02}^3) \tag{11}$$

The legs of the 3-PPPS manipulator apply six actuation forces to its moving-platform. Its global wrench system is a six-system. A parallel singularity occurs when the wrenches in the six-system become linearly dependent and span a k-system with $k < 6$.

3.4 Wrench Diagram of the 3-PPPS Manipulator in P_3

The six actuation forces $\hat{\tau}_{01}^1$, $\hat{\tau}_{02}^1$, $\hat{\tau}_{01}^2$, $\hat{\tau}_{02}^2$, $\hat{\tau}_{01}^3$ and $\hat{\tau}_{02}^3$ form a basis of the global wrench system $\mathrm{W}_{3-\underline{PP}PS}$. Those wrenches are represented by six finite lines in P_3. To obtain the six extensors of the 3-PPPS superbracket, we have to select twelve projective points on the six projective lines, i.e., two points on each line. The extensor of a finite line can be represented by either two distinct finite points or one finite point and one infinite point.

Let c_1, c_2, c_3 be the intersection points of $\hat{\tau}_{01}^1$ and $\hat{\tau}_{02}^1$, $\hat{\tau}_{01}^2$ and $\hat{\tau}_{02}^2$, $\hat{\tau}_{01}^3$ and $\hat{\tau}_{02}^3$, respectively. From Fig. 1, $\hat{\tau}_{02}^1$ and $\hat{\tau}_{01}^3$ are parallel and intersect at the infinite plane Π_∞ at point $\underline{\mathrm{x}} = (\mathbf{x}, 0)^T$, which corresponds to the X direction. Likewise, $\hat{\tau}_{01}^1$ and $\hat{\tau}_{02}^2$ are parallel and intersect at the infinite plane Π_∞ at point $\underline{\mathrm{y}} = (\mathbf{y}, 0)^T$, which corresponds to the Y direction. Similarly, $\hat{\tau}_{01}^2$ and $\hat{\tau}_{02}^3$ are parallel and intersect at the infinite plane Π_∞ at point $\underline{z} = (\mathbf{z}, 0)^T$, which corresponds to the Z direction.

Therefore, the six points have been selected to express the 3-PPPS superbracket and $\hat{\tau}_{01}^1 \equiv c_1\underline{y}$, $\hat{\tau}_{02}^1 \equiv c_1\underline{x}$, $\hat{\tau}_{01}^2 \equiv c_2\underline{z}$, $\hat{\tau}_{02}^2 \equiv c_2\underline{y}$, $\hat{\tau}_{01}^3 \equiv c_3\underline{x}$ and $\hat{\tau}_{02}^3 \equiv c_3\underline{z}$. The wrench diagram of the 3-PPPS manipulator is shown in Fig. 2.

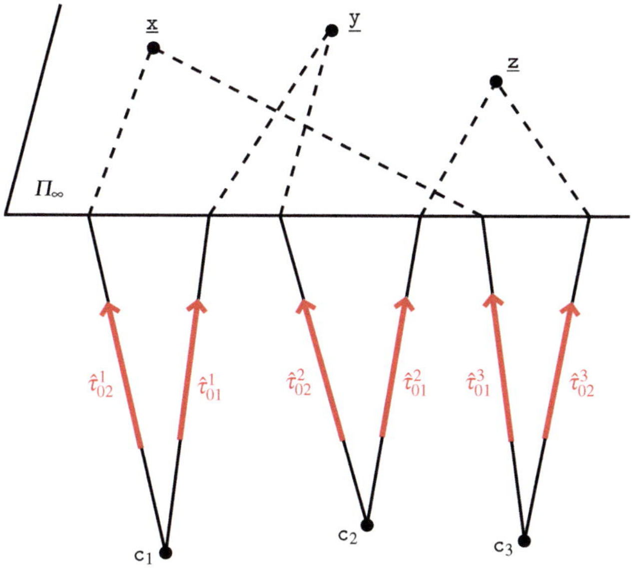

Fig. 2. Wrench diagram in P_3 of the 3-PPPS manipulator

3.5 *Superbracket of the 3-PPPS Manipulator*

The rows of the backward Jacobian matrix of a parallel manipulator are the Plücker coordinates of six lines in P_3. The superjoin of these six vectors in P_5 corresponds to the determinant of their six Plücker coordinate vectors up to a scalar multiple, which is the superbracket in GCA $\Lambda(V^{(2)})$ [12]. Thus, a singularity occurs when these six Plücker coordinate vectors are dependent, which is equivalent to a superbracket equal to zero.

The expression of the 3-PPPS superbracket is $[c_1\underline{x}\, c_1\underline{y}\, c_2\underline{y}\, c_2\underline{z}\, c_3\underline{x}\, c_3\underline{z}]$, which corresponds to six points selected in the robot wrench diagram. This expression can be developed into a linear combination of 24 bracket monomials [6, 14], each one being the product of three brackets of four projective points. The 3-PPPS superbracket was simplified by means of the user interface developed in [15]:

$$[c_1\underline{x}\, c_1\underline{y}\, c_2\underline{y}\, c_2\underline{z}\, c_3\underline{x}\, c_3\underline{z}] = [c_3\,\underline{x}\,\underline{z}\,\underline{y}]\Big([c_3\, c_2\,\underline{z}\, c_1][c_2\,\underline{x}\, c_1\,\underline{y}] - [c_3\,\underline{x}\,\underline{z}\, c_2][c_3\, c_2\,\underline{y}\, c_1]\Big) \tag{12}$$

3.6 Geometric Condition for the 3-PPPS Manipulator Singularities

Let Π_1 be the plane passing through point C_1 and normal to vector **w**. Let Π_2 be the plane passing through point C_2 and normal to vector **u**. Let Π_3 be the plane passing through point C_3 and normal to vector **v**. Let Π_4 be the plane passing through points C_1, C_2 and C_3. From Eq. (12) and the user interface developed in [15], it turns out that the 3-PPPS manipulator reaches a singular configuration if and only if planes Π_1, Π_2, Π_3 and Π_4 intersect at least at one point.

4 Singularity Analysis with Gröbner Bases

In this section, we focus on the computation of the parallel singularities of the 3-PPPS manipulator using the Jacobian and Gröbner bases. We derive the algebraic relations of the singularities satisfied by the orientation variables.

4.1 Jacobian Formulation

The formula we use to define the parallel singularities is the determinant of a Jacobian matrix. This criterion was introduced in [16], where parallel singularities were referred to singularities of the first type. Equations (5a)-(g) depend on six joint variables $\mathbf{T} = (x_1, y_1, y_2, z_2, x_3, z_3)$, six pose variables $(c_x, c_y, c_z, Q_2, Q_3, Q_4)$ and one passive variable (Q_1). We denote by $\mathbf{X}$ the union of the pose and the passive variables. Let $\mathbf{A}$ be the Jacobian matrix of these seven equations with respect to $\mathbf{X}$, i.e.,

$$\mathbf{A} = \left(\frac{\partial F_i}{\partial \mathbf{X}}\right) =$$

$$\begin{bmatrix}
1\;0\;0 & -\frac{2}{3}\sqrt{3}Q_2 - Q_3 & -Q_2 & Q_1 & -\frac{2}{3}\sqrt{3}Q_1 + Q_4 \\
0\;1\;0 & -\frac{1}{3}\sqrt{3}Q_3 & -\frac{1}{3}\sqrt{3}Q_2 - 2Q_3 & -\frac{1}{3}\sqrt{3}Q_1 & -\frac{1}{3}\sqrt{3}Q_4 - 2Q_1 \\
0\;1\;0 & -\frac{1}{3}\sqrt{3}Q_3 & -\frac{1}{3}\sqrt{3}Q_2 + 2Q_3 & -\frac{1}{3}\sqrt{3}Q_1 & -\frac{1}{3}\sqrt{3}Q_4 + 2Q_1 \\
0\;0\;1 & -\frac{1}{3}\sqrt{3}Q_4 + Q_1 & \frac{1}{3}\sqrt{3}Q_1 + Q_4 & -\frac{1}{3}\sqrt{3}Q_2 + Q_3 & \frac{1}{3}\sqrt{3}Q_3 + Q_2 \\
1\;0\;0 & \frac{4}{3}\sqrt{3}Q_2 & 0 & 0 & \frac{4}{3}\sqrt{3}Q_1 \\
0\;0\;1 & \frac{2}{3}\sqrt{3}Q_4 & -\frac{2}{3}\sqrt{3}Q_1 & \frac{2}{3}\sqrt{3}Q_2 & -\frac{2}{3}\sqrt{3}Q_3 \\
0\;0\;0 & 2Q_2 & 2Q_3 & 2Q_4 & 2Q_1
\end{bmatrix} \tag{13}$$

Moreover, let **B** be the Jacobian matrix of Eqs. (5a)-(g) with respect to the joint variables. It appears that **B** is the negative identity matrix. Denoting by $F(\mathbf{X},\mathbf{T})$ the vector of seven polynomials on the left-hand side of Eqs. (5a)-(g), we have:

$$F(\mathbf{X},\mathbf{T}) = 0 \tag{14}$$

Differentiating Eq. (14) with respect to time we obtain:

$$\mathbf{A}\dot{\mathbf{X}} - \dot{\mathbf{T}} = 0 \tag{15}$$

In particular, as in [16], we can infer from Eq. (15) that parallel singularities occur when the determinant of **A** vanishes:

$$\begin{aligned}\det(\mathbf{A}) = & -8\,Q_4\sqrt{3}Q_3{}^3 + 48\,Q_2{}^2Q_3Q_1 - 48\,Q_2Q_3{}^2Q_4 \\ & -24\,\sqrt{3}Q_2{}^2Q_3Q_4 + 48\,Q_4Q_2Q_1{}^2 + 24\,\sqrt{3}Q_2Q_1Q_4{}^2 \\ & +24\,Q_3\sqrt{3}Q_1{}^2Q_4 - 48\,Q_3Q_1Q_4{}^2 - 24\,Q_2Q_1\sqrt{3}Q_3{}^2 \\ & +8\,\sqrt{3}Q_4{}^3Q_3 - 8\,\sqrt{3}Q_2{}^3Q_1 + 8\,Q_2\sqrt{3}Q_1{}^3 = 0\end{aligned} \tag{16}$$

Besides, it turns out that the 3-PPPS manipulator does not have any serial singularity as matrix **B** is always invertible.

4.2 Singularities in the Workspace

The singularities of this mechanism can be represented in terms of the pose variables. To this end, we need to eliminate the joint and passive variables from Eq. (16) and Eqs. (5a)-(g). This can be achieved with methods based on Gröbner basis theory.

4.2.1 Gröbner Basis for Elimination

Let $\mathscr{P}$ be a set of polynomials in the variables $(x_1, y_1, y_2, z_2, x_3, z_3)$ and $(c_x, c_y, c_z, Q_2, Q_3, Q_4)$. Moreover, let $\mathscr{V}$ be the set of common roots of the polynomials in $\mathscr{P}$, and let $\mathscr{W}$ be the projection of $\mathscr{V}$ on the workspace. It might not be possible to represent $\mathscr{W}$ by polynomial equations. Let $\overline{\mathscr{W}}$ be the smallest set defined by polynomial equations that contains $\mathscr{W}$. Our goal is to compute the polynomial equations defining $\overline{\mathscr{W}}$.

These polynomial equations are computed with Gröbner-basis theory. A Gröbner basis of a polynomial system is a polynomial system equivalent to the first one, and satisfying some additional specific properties. The Gröbner basis of a system depends on an ordering on the monomials. In our case, if we choose an elimination ordering eliminating **X**, then the Gröbner basis of $\mathscr{P}$ will contain exactly the polynomials defining $\overline{\mathscr{W}}$. This theory is not the subject of this article and shall not be further detailed here. An introduction on elimination with Gröbner basis is available in [17, Chapter 3].

4.2.2 Equations of the Parallel Singularities in the Workspace

We can now use the elimination of the previous paragraph for our problem to obtain the polynomial equations defining implicitly the parallel singularities in the workspace. Let us consider the polynomial set:

$$\begin{cases} F(\mathbf{X},\mathbf{T}) = [0,0,0,0,0,0,0]^T \\ \det(\mathbf{A}) = 0 \end{cases} \tag{17}$$

We compute a Gröbner basis of system (17) with respect to elimination ordering eliminating **X**. This computation yields directly the relation satisfied by the parallel singularities in the orientation workspace, namely,

$$\begin{aligned}
&- Q_2{}^2 + 9\,Q_2{}^2Q_4{}^2 + 5\,Q_2{}^4 + 9\,Q_3{}^2Q_2{}^2 \\
&+ 16\,Q_3{}^6Q_2{}^2 + 44\,Q_3{}^4Q_2{}^4 + 32\,Q_3{}^2Q_2{}^6 - 24\,Q_3{}^4Q_2{}^2 \\
&- 40\,Q_3{}^2Q_2{}^4 + 16\,Q_4{}^6Q_3{}^2 + 28\,Q_4{}^4Q_3{}^4 + 16\,Q_4{}^2Q_3{}^6 \\
&+ 16\,Q_4{}^6Q_2{}^2 + 48\,Q_4{}^4Q_3{}^2Q_2{}^2 + 72\,Q_4{}^2Q_3{}^4Q_2{}^2 + 20\,Q_4{}^4Q_2{}^4 \\
&+ 64\,Q_4{}^2Q_3{}^2Q_2{}^4 + 8\,Q_4{}^2Q_2{}^6 - 24\,Q_4{}^4Q_3{}^2 - 24\,Q_4{}^2Q_3{}^4 \\
&- 48\,Q_4{}^2Q_3{}^2Q_2{}^2 - 16\,Q_4{}^2Q_2{}^4 + 9\,Q_4{}^2Q_3{}^2 + 16\,Q_4{}^2\sqrt{3}Q_3Q_2 \\
&+ 40\sqrt{3}Q_4{}^4Q_3{}^3Q_2 + 48\sqrt{3}Q_4{}^2Q_3{}^5Q_2 + 40\sqrt{3}Q_4{}^4Q_3Q_2{}^3 + 80\sqrt{3}Q_4{}^2Q_3{}^3Q_2{}^3 \\
&+ 32\sqrt{3}Q_4{}^2Q_3Q_2{}^5 - 16\sqrt{3}Q_4{}^4Q_3Q_2 - 60\sqrt{3}Q_4{}^2Q_3{}^3Q_2 - 52\sqrt{3}Q_4{}^2Q_3Q_2{}^3 \\
&- 8\,Q_2{}^6 + 4\,Q_2{}^8 - 4\sqrt{3}Q_2{}^3Q_3 - 16\sqrt{3}Q_3{}^5Q_2{}^3 \\
&- 24\sqrt{3}Q_3{}^3Q_2{}^5 - 8\sqrt{3}Q_3Q_2{}^7 + 20\sqrt{3}Q_3{}^3Q_2{}^3 + 12\sqrt{3}Q_3Q_2{}^5 \\
&- 24\,Q_4{}^4Q_2{}^2 = 0
\end{aligned} \tag{18}$$

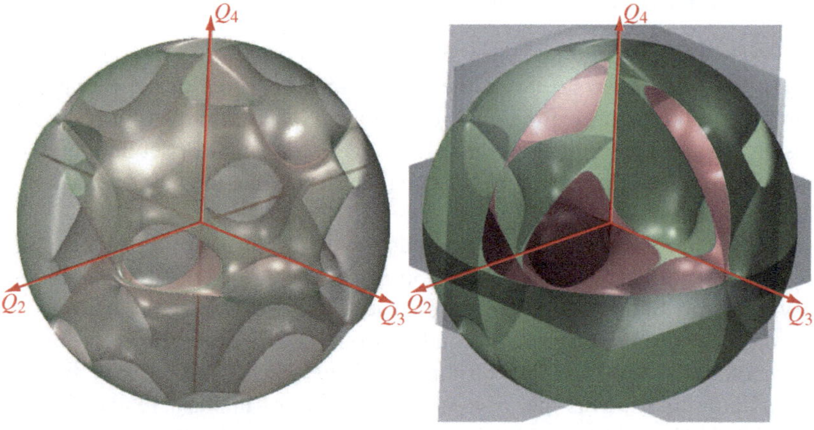

Fig. 3. The parallel singularities of the 3-PPPS manipulator in the orientation workspace: given a point M, vector $\overrightarrow{OM}$ defines the orientation axis and its Euclidean norm $\|\overrightarrow{OM}\|$ is the sine of the half-angle of rotation

We can notice that these equations depend only on the orientation variables (Q_2, Q_3, Q_4). This means that the parallel singularities do not depend on the position of the centroid of the moving platform. As a matter of fact, the parallel singularities of the 3-PPPS manipulator can be represented in its orientation workspace only, the latter being characterized with variables (Q_2, Q_3, Q_4) as shown in Fig. 3.

Finally, note that Eq. (18) was also obtained from the geometric condition for the 3-PPPS manipulator singularities given in Sec. 3.6.

5 Conclusions

The subject of this paper was the singularity analysis of a six-dof three-legged parallel mechanism to be used as a force-feedback interface. A geometric condition for the manipulator singularities was obtained by means of Grassmann-Cayley algebra. The parallel singularities of the manipulator were computed using the Jacobian and Gröbner bases. Consequently, the algebraic relations of the singularities satisfied by the orientation variables were derived. Finally, the parallel singularities of the manipulator were plotted in its orientation workspace.

Acknowledgements. This work has been supported by the "Agence Nationale de la Recherche (ANR SIROPA)".

References

1. Alonso, M.G., Vexo, F., Thalmann, D.: Stepping into Virtual Reality. Springer, London (2008)
2. Burdea, G.: Force and Touch Feedback for Virtual Reality. John Wiley and Sons, New York (1996)
3. Siciliano, B., Khatib, O.: Handbook of Robotics. Springer, Heidelberg (2008)
4. Chen, C., Jackson, D., Heyne, W.: A new 6-dof 3-legged parallel mechanism for force-feedback interface. In: Proceeding of the 2010 IEEE/ASME International Conference on Mechatronic and Embedded Systems and Applications, Qingdao (July 2010)
5. Merlet, J.P.: Singular Configurations of Parallel Manipulators and Grassmann Geometry. The International Journal of Robotics Research 8, 45–56 (1989)
6. Ben-Horin, P., Shoham, P.: Singularity Condition of Six-Degree-Of-Freedom Three-Legged Parallel Robots Based on Grassmann-Cayley Algebra. IEEE Transactions on Robotics 22, 577–590 (2006)
7. Ben-Horin, P., Shoham, P.: Application of Grassmann-Cayley Algebra to Geometrical Interpretation of Parallel Robot Singularities. International Journal of Robotics Research 1, 127–141 (2009)
8. Kanaan, D., Wenger, P., Caro, S., Chablat, D.: Singularity Analysis of Lower-Mobility Parallel Manipulators Using Grassmann-Cayley Algebra. IEEE Transactions on Robotics 25, 995–1004 (2009)
9. Khalil, W., Dombre, E.: Modeling, identification and control of robots. Kogan Page Science (2004)
10. White, N.L.: The Bracket Ring of a Combinatorial Geometry I. Transactions of the American Mathematical Society 202, 79–95 (1975)

11. White, N.L.: The bracket of 2-extensors. Congressus Numerantium 40, 419–428 (1983)
12. White, N.L.: Grassmann-Cayley Algebra and Robotics Applications. In: Handbook of Geometric Computing, Part VIII, pp. 629–656 (2005)
13. Amine, S., Kanaan, D., Caro, S., Wenger, P.: Singularity Analysis of Lower-Mobility Parallel Robots with an Articulated Nacelle. In: On Advances in Robot Kinematics, pp. 273–282. Springer, Heidelberg (2010)
14. McMillan, T., White, N.L.: The Dotted Straightening Algorithm. Journal of Symbolic Computation 11, 471–482 (1991)
15. Ben-Horin, P., Shoham, M., Caro, S., Chablat, D., Wenger, P.: User Interface for the Singularity Analysis of Parallel Robots Based on GrassmannCayley Algebra. In: On Advances in Robot Kinematics, pp. 49–58. Springer, Heidelberg (2008)
16. Gosselin, C., Angeles, J.: Singularity analysis of closed-loop kinematic chains. IEEE Journal of Robotics and Automation 6(3), 281–290 (1990)
17. Cox, D., Little, J., O'Shea, D.: Ideals, Varieties, and Algorithms. Undergraduate Texts in Mathematics. Springer, Heidelberg (1992)

GPSR Compliance

The European Union's (EU) General Product Safety Regulation (GPSR) is a set of rules that requires consumer products to be safe and our obligations to ensure this.

If you have any concerns about our products, you can contact us on ProductSafety@springernature.com

In case Publisher is established outside the EU, the EU authorized representative is:

Springer Nature Customer Service Center GmbH
Europaplatz 3
69115 Heidelberg, Germany

Batch number: 10755023

Printed by Printforce, the Netherlands